PEARSON EDEXCEL INTERNATIONAL GCSE (9–1)

BIOLOGY

Student Book

Philip Bradfield
Steve Potter

Published by Pearson Education Limited, 80 Strand, London, WC2R 0RL.

www.pearsonglobalschools.com

Copies of official specifications for all Edexcel qualifications may be found on the website: https://qualifications.pearson.com

Text © Pearson Education Limited 2017
Edited by Daniel Gill and Anna Wallis
Designed by Cobalt id
Typeset by Tech-Set Ltd, Gateshead, UK
Original illustrations © Pearson Education Limited 2017
Illustrated by Tech-Set Ltd, Gateshead, UK
Cover design by Pearson Education Limited
Cover photo/illustration © EQ

The rights of Philip Bradfield and Steve Potter to be identified as authors of this work have been asserted by them in accordance with the Copyright, Designs and Patents Act 1988.

First published May 2017

20 19 18
10 9 8 7 6 5 4 3

British Library Cataloguing in Publication Data
A catalogue record for this book is available from the British Library

ISBN 978 0 435185 08 4

Printed in Italy by Grafica Veneta S.p.A.

Endorsement Statement
In order to ensure that this resource offers high-quality support for the associated Pearson qualification, it has been through a review process by the awarding body. This process confirms that this resource fully covers the teaching and learning content of the specification or part of a specification at which it is aimed. It also confirms that it demonstrates an appropriate balance between the development of subject skills, knowledge and understanding, in addition to preparation for assessment.

Endorsement does not cover any guidance on assessment activities or processes (e.g. practice questions or advice on how to answer assessment questions), included in the resource nor does it prescribe any particular approach to the teaching or delivery of a related course.

While the publishers have made every attempt to ensure that advice on the qualification and its assessment is accurate, the official specification and associated assessment guidance materials are the only authoritative source of information and should always be referred to for definitive guidance.

Pearson examiners have not contributed to any sections in this resource relevant to examination papers for which they have responsibility.

Examiners will not use endorsed resources as a source of material for any assessment set by Pearson. Endorsement of a resource does not mean that the resource is required to achieve this Pearson qualification, nor does it mean that it is the only suitable material available to support the qualification, and any resource lists produced by the awarding body shall include this and other appropriate resources.

Picture Credits
The authors and publisher would like to thank the following individuals and organisations for permission to reproduce photographs:

(Key: b-bottom; c-centre; l-left; r-right; t-top)

123RF.com: 123rf.com 187l, Eric Isselee 271cl; **Alamy Stock Photo:** Ashley Cooper 191cr, Blend-Memento 226, blickwinkel 3 (bcl), 3 (bcr), 3bl, 3br, 28 (a), 271c, Bill Brooks 191c, David Colbran 257, Dorling Kindersley ltd 283tl, FineArt 262, Genevieve Vallee 168c, Hayley Evans 25 (b), IanDagnall Computing 26 (b), Jan Wlodarczyk 187r, JLImages 27tl, Juniors Bildarchiv GmbH 271tr, Loop Images Ltd 272br, Nathan Allred 25 (c), National Geographic Creative 77br, Naturfoto-Online 264, Nigel Cattlin 30l, 270, 272tr, Nigel Dickinson 215c, Pictorial Press Ltd 249, Rachel Husband 25 (a), Rodney_X 168cr, Sergey Nivens 280, Trevor Smith 102, Zoonar GmbH 186; **Fotolia.com:** bit24 52, Jess8 271br, Dr_Kateryna 55t, Kateryna_Kon 234 (a), nengredeye 147, Pavla Zakova 271tc; **Getty Images:** BlackJack3D 38, qbanczyk 100cr, Muditha Madushan / EyeEm 2, Thomas Barwick 134; **Science Photo Library Ltd:** A. Barrington Brown / Gonville Caius College 228c, Adam Hart-Davis 169, Adrian Thomas 256, Alan L. Detrick 208, Andrew Lambert Photography 175cr, Animated Healthcare Ltd 118l, Biodisc, Biomedical Imaging Unit, Biophoto Associates 56, 57, 78 (b), 138, CNRI 265, David Parker 207, DENNIS KUNKEL MICROSCOPY 80, Dorling Kindersley / UIG 27cl, Dr Brad Mogan / Visuals Unlimited 294, Dr Jeremy Burgess 158, 199, DR GOPAL MURTI 6 (a), Frans Lanting / Mint Images 215b, Gastrolab 65 (b), GIPhotoStock 58, JC Revy, ISM 155tl, John Durham 6 (b), Keith R Porter 5, Lee D. Simon 30c, Leonard Lessin 85 (a), Martin Shields 168, MARTYN F. CHILLMAID 58, 189, MAURO FERMARIELLO 54b, 28 (c), Microscope 125b, MONTY RAKUSEN 211, National Library of Medicine 261, Natural History Museum, London 263 (a), 263 (b), Omikron 118r, POWER AND SYRED 26bc, Robert Brook 218, Rosenfeld Images Ltd 272cr, Saturn Stills 101, Science Photo Library 40tl, 41tl, 46r, 47, 79 (b), 137, Science Pictures Ltd 242, Science Source 228cr, Sinclair Stammers 28 (b), Southampton General Hospital 22, STEPHEN AUSMUS / US DEPARTMENT OF AGRICULTURE 26 (a), Steve Gschmeissner 155cr, 159bl, 241, 85 (b), Susumu Nishinaga 155c, 161tl, Trevor Clifford Photography 152, W.A Ritchie / Roslin Institute / Eurelios 274; **VISUALS UNLIMITED** 160br; **Shutterstock.com:** 279photo Studio 206, Africa Studio 54t, arka38 283 (b), Digieva 283tr, IANG HONGYAN 295, Juan Gaertner 21, Kateryna Kon 29tl, ktsdesign 29bl, Mauricio Graiki 100cl, Reinhard Tiburzy 285, Resul Muslu 283 (a), Robyn Mackenzie 135, sciencepics 86, Sozaijiten 26 (c).

Cover images: *Front:* **Alamy Stock Photo:** Cultura RM / Alamy Stock Photo tl
Inside front cover: **Shutterstock.com:** Dmitry Lobanov

All other images © Pearson Education

Disclaimer: **neither Pearson, Edexcel nor the authors take responsibility for the safety of any activity**. Before doing any practical activity you are legally required to carry out your own risk assessment. In particular, any local rules issued by your employer must be obeyed, regardless of what is recommended in this resource. Where students are required to write their own risk assessments they must always be checked by the teacher and revised, as necessary, to cover any issues the students may have overlooked. The teacher should always have the final control as to how the practical is conducted.

CONTENTS

UNIT 1

ORGANISMS AND LIFE PROCESSES

UNIT 2

ANIMAL PHYSIOLOGY

UNIT 3

PLANT PHYSIOLOGY

UNIT 4

ECOLOGY AND THE ENVIRONMENT

UNIT 5

VARIATION AND SELECTION

UNIT 6

MICROORGANISMS AND GENETIC MODIFICATION

ABOUT THIS BOOK

This book is written for students following the Pearson Edexcel International GCSE (9–1) Biology specification and the Edexcel International GCSE (9–1) Science Double Award specification. You will need to study all of the content in this book for your Biology examinations. However, you will only need to study some of it if you are taking the Double Award specification. The book clearly indicates which content is in the Biology examinations and not in the Double Award specification. To complete the Double Award course you will also need to study the Physics and Chemistry parts of the course.

In each unit of this book, there are concise explanations and worked examples, plus numerous exercises that will help you build up confidence. The book also describes the methods for carrying out all of the required practicals.

The language throughout this textbook is graded for speakers of English as an additional language (EAL), with advanced Biology-specific terminology highlighted and defined in the glossary at the back of the book. A list of command words, also at the back of the book, will help you to learn the language you will need in your examination.

You will also find that questions in this book have Progression icons and Skills tags. The Progression icons refer to Pearson's Progression scale. This scale – from 1 to 12 – tells you what level you have reached in your learning and will help you to see what you need to do to progress to the next level. Furthermore, Edexcel have developed a Skills grid showing the skills you will practise throughout your time on the course. The skills in the grid have been matched to questions in this book to help you see which skills you are developing. Both skills and Progression icons are not repeated where they are same in consecutive questions. You can find Pearson's Progression scale at www.pearsonglobalschools.com/igscienceprogression along with guidelines on how to use it.

Learning Objectives show what you will learn in each lesson.

Looking Ahead tells you what you would learn if you continued your study of Biology to a higher level, such as International A Level.

Biology only features show the content that is on the Biology specification only and not the Double Award specification. All other content in this book applies to Double Award students.

Key Point boxes summarise the essentials.

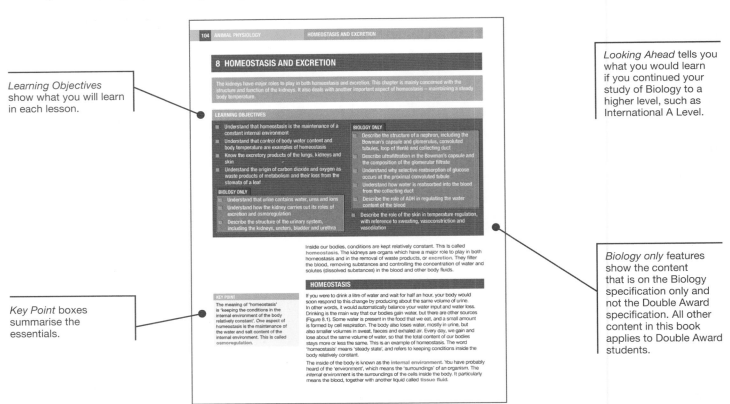

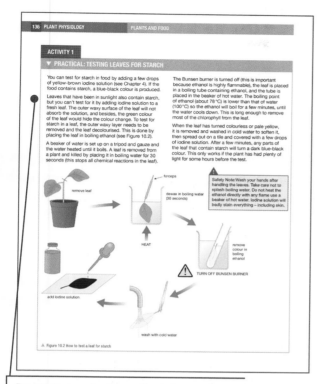

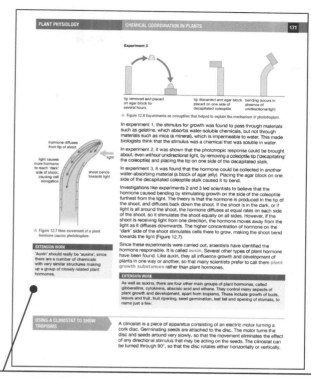

Practicals describe the methods for carrying out all of the practicals you will need to know for your examination.

Extension work boxes include content that is not on the specification and which you do not have to learn for your examination. However, the content will help to extend your understanding of the topic.

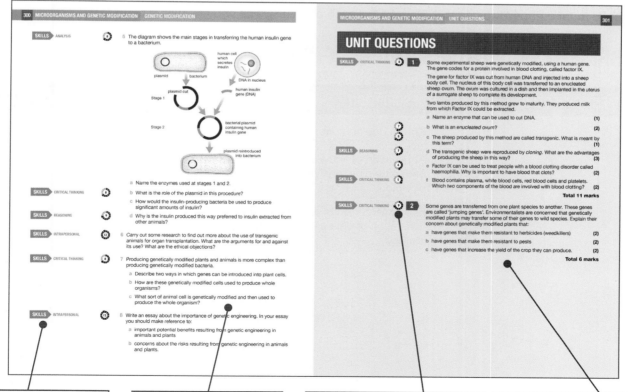

Skills tags tell you which skills you are practising in each question.

Chapter Questions test your knowledge of the topic in that chapter.

Progression icons show the level of difficulty according to the Pearson International GCSE Science Progression Scale.

Unit Questions test your knowledge of the whole unit and provide quick, effective feedback on your progress.

ASSESSMENT OVERVIEW

The following tables give an overview of the assessment for this course.

We recommend that you study this information closely to help ensure that you are fully prepared for this course and know exactly what to expect in the assessment.

PAPER 1	SPECIFICATION	PERCENTAGE	MARK	TIME	AVAILABILITY
Written examination paper Paper code 4BI1/1B and 4SD0/1B Externally set and assessed by Edexcel	Biology Science Double Award	61.1%	110	2 hours	January and June examination series First assessment June 2019
PAPER 2	**SPECIFICATION**	**PERCENTAGE**	**MARK**	**TIME**	**AVAILABILITY**
Written examination paper Paper code 4BI1/2B Externally set and assessed by Edexcel	Biology	38.9%	70	1 hour 15 mins	January and June examination series First assessment June 2019

If you are studying Biology then you will take both Papers 1 and 2. If you are studying Science Double Award then you will only need to take Paper 1 (along with Paper 1 for each of the Physics and Chemistry courses).

ASSESSMENT OBJECTIVES AND WEIGHTINGS

ASSESSMENT OBJECTIVE	DESCRIPTION	% IN INTERNATIONAL GCSE
AO1	Knowledge and understanding of biology	38%–42%
AO2	Application of knowledge and understanding, analysis and evaluation of biology	38%–42%
AO3	Experimental skills, analysis and evaluation of data and methods in biology	19%–21%

EXPERIMENTAL SKILLS

In the assessment of experimental skills, students may be tested on their ability to:

- solve problems set in a practical context

- apply scientific knowledge and understanding in questions with a practical context

- devise and plan investigations, using scientific knowledge and understanding when selecting appropriate techniques

- demonstrate or describe appropriate experimental and investigative methods, including safe and skilful practical techniques

- make observations and measurements with appropriate precision, record these methodically and present them in appropriate ways

- identify independent, dependent and control variables

- use scientific knowledge and understanding to analyse and interpret data to draw conclusions from experimental activities that are consistent with the evidence

- communicate the findings from experimental activities, using appropriate technical language, relevant calculations and graphs

- assess the reliability of an experimental activity

- evaluate data and methods taking into account factors that affect accuracy and validity.

CALCULATORS

Students are permitted to take a suitable calculator into the examinations. Calculators with QWERTY keyboards or that can retrieve text or formulae will not be permitted.

UNIT 1
ORGANISMS AND LIFE PROCESSES

All living organisms are composed of microscopic units known as cells. These building blocks of life have a number of features in common, which allow them to grow, reproduce, and generate more organisms. In Chapter 1 we start by looking at the structure and function of cells, and the essential life processes that go on within them. Despite the fact that cells are similar in structure, there are many millions of different species of organisms. Chapter 2 looks at the diversity of living things and how we can classify them into groups on the basis of the features that they show.

1 LIFE PROCESSES

There are structural features that are common to the cells of all living organisms. In this chapter you will find out about these features and look at some of the processes that keep cells alive.

LEARNING OBJECTIVES

- Understand the characteristics shared by living organisms
- Describe cell structures and their functions, including the nucleus, cytoplasm, cell membrane, cell wall, mitochondria, chloroplasts, ribosomes and vacuole
- Know the similarities and differences in the structures of plant and animal cells
- Understand the role of enzymes as biological catalysts in metabolic reactions
- Understand how temperature changes can affect enzyme function, including changes to the shape of the active site
- Understand how enzyme function can be affected by changes in pH altering the active site
- Investigate how enzyme activity can be affected by changes in temperature

BIOLOGY ONLY
- Investigate how enzyme activity can be affected by changes in pH

- Describe the differences between aerobic and anaerobic respiration
- Understand how the process of respiration produces ATP in living organisms

- Know that ATP provides energy for cells
- Know the word equation and balanced chemical symbol equation for aerobic respiration
- Know the word equations for anaerobic respiration
- Investigate the evolution of carbon dioxide and heat from respiring seeds or other suitable living organisms
- Understand the processes of diffusion, osmosis and active transport by which substances move into and out of cells
- Understand how factors affect the rate of movement of substances into and out of cells
- Investigate diffusion in a non-living system (agar jelly)

BIOLOGY ONLY
- Explain the importance of cell differentiation in the development of specialised cells

- Describe the levels of organisation within organisms – organelles, cells, tissues, organ systems

BIOLOGY ONLY
- Understand the advantages and disadvantages of using stem cells in medicine.

All living organisms are composed of units called **cells**. The simplest organisms are made from single cells (Figure 1.1) but more complex plants and animals are composed of millions of cells. In many-celled (**multicellular**) organisms, there may be hundreds of different types of cells with different structures. They are specialised so that they can carry out particular functions in the animal or plant. Despite all the differences, there are basic features that are the same in all cells.

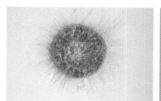   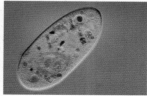

▲ Figure 1.1 Many simple organisms have 'bodies' made from single cells. Here are four examples.

There are eight life processes which take place in most living things. Organisms:

■ require nutrition – plants make their own food, animals eat other organisms

■ respire – release energy from their food

■ excrete – get rid of waste products

■ respond to stimuli – are sensitive to changes in their surroundings

■ move – by the action of muscles in animals, and slow growth movements in plants

■ control their internal conditions – maintain a steady state inside the body

■ reproduce – produce offspring

■ grow and develop – increase in size and complexity, using materials from their food.

CELL STRUCTURE

This part of the book describes the cell structure of 'higher' organisms such as animals, plants and fungi. The cells of bacteria are simpler in structure and will be described in Chapter 2.

Most cells contain certain parts such as the nucleus, cytoplasm and cell membrane. Some cells have structures missing, for instance red blood cells are unusual in that they have no nucleus. The first chapter in a biology textbook will usually present diagrams of 'typical' plant and animal cells. In fact, there is really no such thing as a 'typical' cell. Humans, for example, are composed of hundreds of different kinds of cells – from nerve cells to blood cells, skin cells to liver cells. What we really mean by a 'typical' cell is a general diagram that shows all the features that you will find in most cells (Figure 1.2). However, not all these are present in *all* cells – for example the cells in the parts of a plant that are not green do not contain chloroplasts.

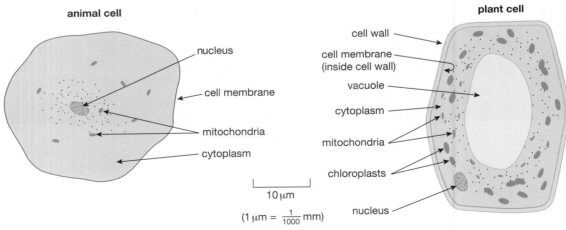

▲ Figure 1.2 The structure of a 'typical' animal and plant cell.

The living material that makes up a cell is called **cytoplasm**. It has a texture rather like sloppy jelly, in other words somewhere between a solid and a liquid. Unlike a jelly, it is not made of one substance but is a complex material made of many different structures. You can't see many of these structures under an ordinary light microscope. An electron microscope has a much higher magnification and can show the details of these structures, which are called **organelles** (Figure 1.3).

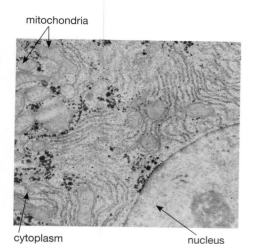
mitochondria

cytoplasm   nucleus

▲ Figure 1.3 The organelles in a cell can be seen using an electron micropscope.

The largest organelle in the cell is the **nucleus**. Nearly all cells have a nucleus. The few types that don't are usually dead (e.g. the xylem vessels in a stem, Chapter 11) or don't live for very long (e.g. red blood cells, Chapter 5). The nucleus controls the activities of the cell. It contains **chromosomes** (46 in human cells) which carry the genetic material, or **genes**. Genes control the activities in the cell by determining which proteins the cell can make. The DNA remains in the nucleus, but the instructions for making proteins are carried out of the nucleus to the cytoplasm, where the proteins are assembled on tiny structures called **ribosomes**. A cell contains thousands of ribosomes, but they are too small to be seen through a light microscope.

One very important group of proteins found in cells are **enzymes.** Enzymes control the chemical reactions that take place in the cytoplasm.

All cells are surrounded by a **cell membrane**, sometimes called the cell *surface* membrane to distinguish it from other membranes inside the cell. This is a thin layer like a 'skin' on the surface of the cell. It forms a boundary between the cytoplasm of the cell and the outside. However, it is not a complete barrier. Some chemicals can pass into the cell and others can pass out. We say that the membrane is **partially permeable**. The membrane can go further than this and actually *control* the movement of some substances – it is **selectively permeable**.

One organelle that is found in the cytoplasm of all living cells is the **mitochondrion** (plural mitochondria). In cells that need a lot of energy such as muscle or nerve cells, there are many mitochondria. This gives us a clue to their function. They carry out some of the reactions of **respiration** (see page 12) releasing energy that the cell can use. Most of the energy from respiration is released in the mitochondria.

PLANT CELLS

All of the structures you have seen so far are found in both animal and plant cells. However, some structures are only ever found in plant cells. There are three in particular – the cell wall, a permanent vacuole and chloroplasts.

The **cell wall** is a layer of non-living material that is found outside the cell membrane of plant cells. It is made mainly of a carbohydrate called **cellulose**, although other chemicals may be added to the wall in some cells. Cellulose is a tough material that helps the cell keep its shape and is one reason why the 'body' of a plant has a fixed shape. Animal cells do not have a cell wall and tend to be more variable in shape. Plant cells absorb water, producing an internal pressure that pushes against adjacent cells, giving the plant support (see Chapter 11). Without a cell wall strong enough to resist these pressures, this method of support would be impossible. The cell wall is porous, so it is not a barrier to water or dissolved substances. We call it *freely permeable*.

Mature (fully grown) plant cells often have a large central space surrounded by a membrane, called a **vacuole**. This vacuole is a permanent feature of the cell. It is filled with a watery liquid called cell sap, which is a store of dissolved sugars, mineral ions and other solutes. Animal cells do contain vacuoles, but they are only small, temporary structures.

Cells of the green parts of plants, especially the leaves, contain another very important organelle, the **chloroplast**. Chloroplasts absorb light energy to make food in the process of photosynthesis (see Chapter 10). They contain a green pigment called **chlorophyll**. Cells from the parts of a plant that are not green, such as the flowers, roots and woody stems, have no chloroplasts.

KEY POINT

Nearly all cells contain cytoplasm, a nucleus, a cell membrane and mitochondria. As well as these structures, plant cells have a cell wall and a permanent vacuole, and plant cells that photosynthesise contain chloroplasts.

Figure 1.4 shows some animal and plant cells seen through the light microscope.

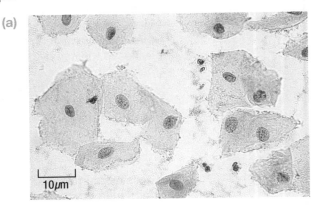

(a)

10μm

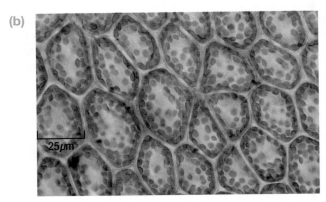

(b)

25μm

▲ Figure 1.4 (a) Cells from the lining of a human cheek. (b) Cells from the photosynthetic tissue of a leaf.

ENZYMES: CONTROLLING REACTIONS IN THE CELL

The chemical reactions that take place in a cell are controlled by a group of proteins called enzymes. Enzymes are biological **catalysts**. A catalyst is a chemical which speeds up a reaction without being used up itself. It takes part in the reaction, but afterwards is unchanged and free to catalyse more reactions. Cells contain hundreds of different enzymes, each catalysing a different reaction. This is how the activities of a cell are controlled – the nucleus contains the genes, which control the production of enzymes, which then catalyse reactions in the cytoplasm:

genes → proteins (enzymes) → catalyse reactions

Everything a cell does depends on which enzymes it can make, which in turn depends on which genes in its nucleus are working.

What hasn't been mentioned is why enzymes are needed at all. They are necessary because the temperatures inside organisms are low (e.g. the human body temperature is about 37 °C) and without catalysts, most of the reactions that happen in cells would be far too slow to allow life to go on. The reactions can only take place quickly enough when enzymes are present to speed them up.

It is possible for there to be thousands of different sorts of enzymes because they are proteins, and protein molecules have an enormous range of structures and shapes (see Chapter 4).

The molecule that an enzyme acts on is called its **substrate**. Each enzyme has a small area on its surface called the **active site**. The substrate attaches to the active site of the enzyme. The reaction then takes place and products are formed. When the substrate joins up with the active site it lowers the energy needed for the reaction to start, allowing the products to be formed more easily.

Enzymes also catalyse reactions where large molecules are built up from smaller ones. In this case, several substrate molecules attach to the active site, the reaction takes place and the larger product molecule is formed. The product then leaves the active site.

The substrate fits into the active site of the enzyme like a key fitting into a lock. Just as a key will only fit one lock, a substrate will only fit into the active site of a particular enzyme. This is known as the **lock and key model** of enzyme action. It is the reason why enzymes are *specific*, i.e. an enzyme will only catalyse one reaction.

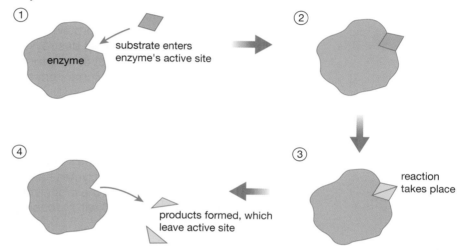

▲ Figure 1.5 Enzymes catalyse reactions at their active site. This acts like a 'lock' to the substrate 'key'. The substrate fits into the active site, and products are formed. This happens more easily than without the enzyme – so enzymes act as catalysts.

After an enzyme molecule has catalysed a reaction, the product is released from the active site, and the enzyme is free to act on more substrate molecules.

FACTORS AFFECTING ENZYMES

KEY POINT

'Optimum' temperature means the 'best' temperature, in other words the temperature at which the reaction takes place most rapidly.

DID YOU KNOW?
Kinetic energy is the energy an object has because of its movement. The molecules of enzyme and substrate are moving faster, so they have more kinetic energy.

A number of factors affect the activity of enzymes. The rate of reaction may be increased by raising the concentration of the enzyme or the substrate. Two other factors that affect enzymes are temperature and pH.

TEMPERATURE

The effect of temperature on the action of an enzyme is easiest to see as a graph, where we plot the rate of the reaction against temperature (Figure 1.6).

Enzymes in the human body have evolved to work best at body temperature (37 °C). The graph in Figure 1.6 shows a peak on the curve at this temperature, which is called the *optimum temperature* for the enzyme.

As the enzyme is heated up to the optimum temperature, the rise in temperature increases the rate of reaction. This is because higher temperatures give the molecules of the enzyme and the substrate more kinetic energy, so they collide more often. More collisions means that the reaction will take place more frequently.

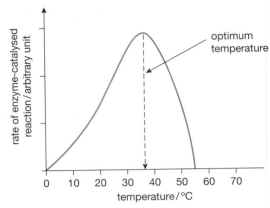

▲ Figure 1.6 Effect of temperature on the action of an enzyme.

However, above the optimum, temperature starts to have another effect. Enzymes are made of protein, and proteins are broken down by heat. From 40 °C upwards, the heat destroys the enzyme. We say that it is **denatured**. You can see the effect of denaturing when you boil an egg. The egg white is made of protein, and turns from a clear runny liquid into a white solid as the heat denatures the protein. Denaturing changes the shape of the active site so that the substrate will no longer fit into it. Denaturing is permanent – the enzyme molecules will no longer catalyse the reaction.

Not all enzymes have an optimum temperature near 37 °C, only those of animals such as mammals and birds, which all have body temperatures close to this value. Enzymes have evolved to work best at the normal body temperature of the organism. Bacteria that always live at an average temperature of 10 °C will probably have enzymes with an optimum temperature near 10 °C.

pH

The pH around the enzyme is also important. The pH inside cells is neutral (pH 7) and most enzymes have evolved to work best at this pH. At extremes of pH either side of neutral, the enzyme activity decreases, as shown in Figure 1.7. The pH at which the enzyme works best is called its *optimum pH*. Either side of the optimum, the pH affects the structure of the enzyme molecule and changes the shape of its active site, so that the substrate will not fit into it so well.

KEY POINT

Although most enzymes work best at a neutral pH, a few have an optimum below or above pH 7. The stomach produces hydrochloric acid, which makes its contents very acidic (see Chapter 4). Most enzymes stop working at a low pH, but the stomach makes an enzyme called pepsin which has an optimum pH of about 2, so that it is adapted to work well in these unusually acidic surroundings.

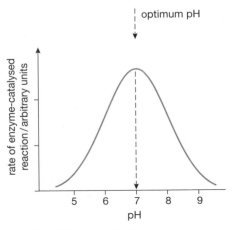

▲ Figure 1.7 Most enzymes work best at a neutral pH.

ACTIVITY 1

▼ PRACTICAL: AN INVESTIGATION INTO THE EFFECT OF TEMPERATURE ON THE ACTIVITY OF AMYLASE

The digestive enzyme amylase breaks down starch into the sugar maltose. If the speed at which the starch disappears is recorded, this is a measure of the activity of the amylase.

Figure 1.8 shows apparatus which can be used to record how quickly the starch is used up.

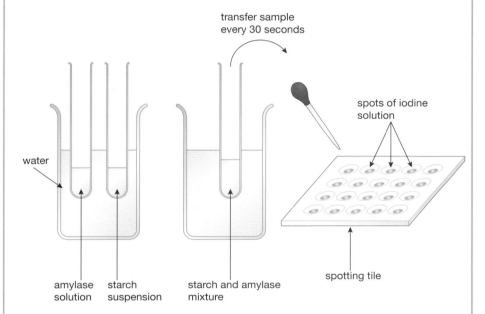

▲ Figure 1.8 Investigating the breakdown of starch by amylase at different temperatures.

Spots of iodine solution are placed into the dips on the spotting tile. Using a syringe, 5 cm³ of starch suspension is placed in one boiling tube, and 5 cm³ of amylase solution in another tube, using a different syringe. The beaker is filled with water at 20 °C. Both boiling tubes are placed in the beaker of water for 5 minutes, and the temperature recorded.

The amylase solution is then poured into the starch suspension, leaving the tube containing the mixture in the water bath. Immediately, a small sample of the mixture is removed from the tube with a pipette and added to the first drop of iodine solution on the spotting tile. The colour of the iodine solution is recorded.

A sample of the mixture is taken every 30 seconds for 10 minutes and tested for starch as above, until the iodine solution remains yellow, showing that all the starch is used up.

The experiment is repeated, maintaining the water bath at different temperatures between 20 °C and 60 °C. A set of results is shown in the table below.

Time / min	Colour of mixture at different temperatures / (°C)				
	20	30	40	50	60
0.0	Blue-black	Blue-black	Blue-black	Blue-black	Blue-black
0.5	Blue-black	Blue-black	Brown	Blue-black	Blue-black
1.0	Blue-black	Blue-black	Yellow	Blue-black	Blue-black
1.5	Blue-black	Blue-black	Yellow	Blue-black	Blue-black
2.0	Blue-black	Blue-black	Yellow	Brown	Blue-black
2.5	Blue-black	Blue-black	Yellow	Brown	Blue-black
3.0	Blue-black	Blue-black	Yellow	Brown	Blue-black
3.5	Blue-black	Blue-black	Yellow	Yellow	Blue-black
4.0	Blue-black	Blue-black	Yellow	Yellow	Blue-black
5.5	Blue-black	Blue-black	Yellow	Yellow	Blue-black
6.0	Blue-black	Brown	Yellow	Yellow	Blue-black
6.5	Blue-black	Brown	Yellow	Yellow	Blue-black
7.0	Blue-black	Yellow	Yellow	Yellow	Blue-black
7.5	Blue-black	Yellow	Yellow	Yellow	Brown
8.0	Blue-black	Yellow	Yellow	Yellow	Brown
8.5	Brown	Yellow	Yellow	Yellow	Yellow
9.0	Brown	Yellow	Yellow	Yellow	Yellow
9.5	Yellow	Yellow	Yellow	Yellow	Yellow
10.0	Yellow	Yellow	Yellow	Yellow	Yellow

The rate of reaction can be calculated from the time taken for the starch to be fully broken down, as shown by the colour change from blue-black to yellow. For example, at 50 °C the starch had all been digested after 3.5 minutes. The rate is found by dividing the volume of the starch (5 cm³) by the time:

$$\text{Rate} = \frac{5.0\,cm^3}{3.5\,min} = 1.4\ cm^3 \text{ per min}$$

Plotting a graph of rate against temperature should produce a curve similar to the one shown in Figure 1.6. Try this, using the results in the table. Better still, you may be able to do this experiment and provide your own results.

If the curve doesn't turn out quite like the one in Figure 1.6, can you suggest why this is? How could you improve the experiment to get more reliable results?

BIOLOGY ONLY

ACTIVITY 2

▼ PRACTICAL: AN INVESTIGATION INTO THE EFFECT OF pH ON THE ACTIVITY OF CATALASE

Buffer solutions are solutions of salts that resist changes in pH. Different buffer solutions can be prepared for maintaining different values of pH. Buffer solutions are useful for finding the effect of pH on enzyme activity.

Hydrogen peroxide (H_2O_2) is a product of metabolism. Hydrogen peroxide is toxic (poisonous), so it must not be allowed to build up in cells. The enzyme catalase protects cells by breaking down hydrogen peroxide into the harmless products water and oxygen:

$$2H_2O_2 \rightarrow 2H_2O + O_2$$

Potato cells contain a high concentration of catalase. A large potato is chopped into small pieces and placed in a blender with an equal volume of distilled water. The blender is switched on to mince up the potato tissue and release the catalase from the cells. The potato debris is allowed to settle to the bottom and the liquid extract above the debris removed.

The extract is tested for catalase activity at different values of pH. Using a graduated syringe, 5 cm³ of extract is placed in a boiling tube and 5 cm³ of pH 7 buffer solution added from another syringe. The tube is shaken gently to mix the buffer with the potato extract. The mixture is left for 5 minutes, then 5 cm³ of 5% hydrogen peroxide solution is added to the tube from a third syringe. A bung and delivery tube is quickly inserted in the boiling tube and the end of the delivery tube placed in a beaker of water (Figure 1.9).

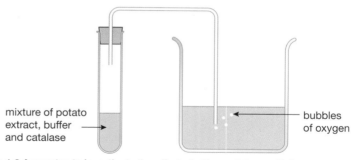

mixture of potato extract, buffer and catalase

bubbles of oxygen

▲ Figure 1.9 Apparatus to investigate the effect of pH on catalase activity.

The bubbles of oxygen gas produced in the first minute after adding the hydrogen peroxide are counted. The number of bubbles per minute is a measure of the initial reaction rate.

The experiment is repeated, using different buffers. Some results are shown in the table below.

pH	Rate of reaction / bubbles per minute
5	6
6	39
7	47
8	14

Plot a bar chart of the results.

Which pH gives the fastest rate of reaction?

How could you modify this experiment to find a more precise value of the optimum pH for the enzyme catalase?

Think – are four values of pH enough? Is there a better way than counting bubbles of gas?

How could you control the temperature during the experiment?

How could you modify the experiment to make the results more reliable?

Think – is one set of results enough?

END OF BIOLOGY ONLY

HOW THE CELL GETS ITS ENERGY

A cell needs a source of energy in order to be able to carry out all the processes needed for life. It gets this energy by breaking down food molecules to release the stored chemical energy that they contain. This process is called respiration. Many people think that respiration means the same as 'breathing', but although there are links between the two processes, the biological meaning of respiration is very different.

KEY POINT

Respiration is called an *oxidation* reaction, because oxygen is used to break down food molecules.

Respiration happens in all the cells of our body. Oxygen is used to oxidise food, and carbon dioxide (and water) are released as waste products. The main food oxidised is a sugar called **glucose**. Glucose contains stored chemical energy that can be converted into other forms of energy that the cell can use. It is rather like burning a fuel to get the energy out of it, except that burning releases most of the energy as heat. Respiration releases some heat energy, but most is used to make a substance called **ATP** (see below). The energy stored in the ATP molecules can then be used for a variety of purposes, such as:

■ contraction of muscle cells, producing movement

■ active transport of molecules and ions (see page 18)

■ building large molecules, such as proteins

■ cell division.

The energy released as heat is also used to maintain a constant body temperature in mammals and birds (see Chapter 8).

The overall reaction for respiration is:

$$\text{glucose} + \text{oxygen} \rightarrow \text{carbon dioxide} + \text{water} \quad (+ \text{energy})$$
$$C_6H_{12}O_6 + 6O_2 \rightarrow 6CO_2 + 6H_2O \quad (+ \text{energy})$$

DID YOU KNOW?

Carbon from the respired glucose passes out into the atmosphere as carbon dioxide. The carbon can be traced through this pathway using a radioactive form of carbon called carbon-14.

This is called **aerobic respiration**, because it uses oxygen. Aerobic respiration happens in the cells of humans and those of animals, plants and many other organisms. It is important to realise that the equation above is only a *summary* of the process. It actually takes place gradually, as a sequence of small steps, which release the energy of the glucose in small amounts. Each step in the process is catalysed by a different enzyme. The later steps in the process are the aerobic ones, and these release the most energy. They happen in the mitochondria of the cell.

ATP – THE ENERGEY 'CURRENCY' OF THE CELL

Respiration releases energy while other cell processes use it up. Cells have a way of passing the energy from respiration to the other processes that need it. They do this using a chemical called **adenosine triphosphate** or **ATP**. ATP is present in all living cells.

ATP is composed of an organic molecule called adenosine attached to three phosphate groups. In a cell, ATP can be broken down losing one phosphate group and forming adenosine diphosphate or ADP (Figure 1.10 (a)).

(a) When energy is needed ATP is broken down into ADP and phosphate (P):

(b) During respiration ATP is made from ADP and phosphate:

▲ Figure 1.10 ATP is the energy 'currency' of the cell.

When this reaction takes place, chemical energy is released and can be used to drive metabolic processes that need it.

During respiration the opposite happens – energy from the oxidation of glucose is used to drive the reverse reaction and a phosphate is added onto ADP (Figure 1.10 (b)).

ATP is often described as the energy 'currency' of the cell. It transfers energy between the process that releases it (respiration) and the processes in a cell that use it up.

ANAEROBIC RESPIRATION

There are some situations where cells can respire *without* using oxygen. This is called **anaerobic respiration**. In anaerobic respiration, glucose is not completely broken down, so less energy is released. The advantage of anaerobic respiration is that it can occur in situations where oxygen is in short supply. Two important examples of this are in yeast cells and muscle cells.

Yeasts are single-celled fungi. They are used in processes such as baking bread. When yeast cells are prevented from getting enough oxygen, they stop respiring aerobically and start to respire anaerobically instead. The glucose is partly broken down into ethanol (alcohol) and carbon dioxide:

glucose → ethanol + carbon dioxide (+ some energy)

This process is described in more detail in Chapter 21.

Think about the properties of ethanol – it makes a good fuel and will burn to produce a lot of heat, so it still has a lot of chemical energy 'stored' in it.

Muscle cells can also respire anaerobically when they are short of oxygen. If muscles are overworked, the blood cannot reach them fast enough to deliver enough oxygen for aerobic respiration. This happens when a person does a 'burst' of activity, such as a sprint, or quickly lifting a heavy weight. This time the glucose is broken down into a substance called **lactate**:

glucose → lactate (+ some energy)

Anaerobic respiration provides enough energy to keep the overworked muscles going for a short period. During the exercise, the level of lactate rises in the muscle cells and bloodstream.

After the exercise the lactate is respired aerobically in the mitochondria. The volume of oxygen needed to completely oxidise the lactate that builds up in the body during anaerobic respiration is called the **oxygen debt**.

ACTIVITY 3

▼ PRACTICAL: DEMONSTRATION OF THE PRODUCTION OF CARBON DIOXIDE BY SMALL LIVING ORGANISMS

Hydrogen carbonate indicator solution is normally orange, but turns yellow if carbon dioxide is added to it. The indicator is sensitive to small changes in carbon dioxide concentration, and can be used to show production of carbon dioxide by small organisms such as woodlice, maggots (fly larvae) or germinating seeds.

The organisms are placed in a stoppered boiling tube with the indicator, as shown in Figure 1.11. The gauze platform supports the organisms above the hydrogen carbonate indicator solution and stops them from coming into contact with the chemical.

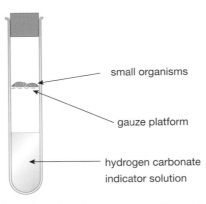

small organisms

gauze platform

hydrogen carbonate indicator solution

▲ Figure 1.11 Testing for carbon dioxide production by small organisms.

Of the three species of organisms mentioned above, which do you think would change the colour of the indicator most quickly? If you are able to observe each of the organisms, this might help with your prediction.

When you have made your prediction (called a 'hypothesis'), plan an investigation to test it. Take care to consider the variables that need to be controlled, and don't forget to include a description of a fourth tube that you would need to set up as the experimental Control (see Appendix A for an explanation of these terms).

It may be possible for you to carry out the investigation using similar apparatus and organisms.

ACTIVITY 4

▼ PRACTICAL: DEMONSTRATION THAT HEAT IS PRODUCED BY RESPIRATION

Some peas are soaked in water for 24 hours, so that they start to germinate. A second batch of peas is boiled, to kill them. Each set of peas is washed in a 1% bleach solution, which acts as a disinfectant, killing any bacteria present on the surface of the peas. The peas are then rinsed twice in distilled water to remove any traces of bleach.

Each batch of peas is placed in an inverted vacuum flask as shown in Figure 1.12, leaving some air in each flask. A vacuum flask insulates its contents, so that any small temperature change inside the flask can be measured.

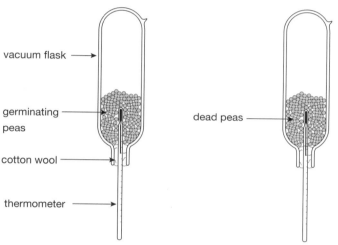

▲ Figure 1.12 Experiment to show that heat is produced during respiration in germinating peas.

The seeds produce carbon dioxide gas, which is denser than air. The inverted flasks and cotton wool allow this to escape. It might otherwise kill the peas.

The apparatus is left set up for a couple of days, and the temperature inside each flask measured at the start and end of the experiment.

The following results were obtained from this experiment:

Temperature in both flasks at the start = 21 °C

Temperature in flask with dead peas at the end = 21 °C

Temperature in flask with living peas at the end = 24 °C

Can you explain these results? Why is it necessary to kill any microorganisms on the surface of the peas? Explain the importance of the flask containing dead peas.

MOVEMENT OF MATERIALS IN AND OUT OF CELLS

Cell respiration shows the need for cells to be able to take in certain substances from their surroundings, such as glucose and oxygen, and get rid of others, such as carbon dioxide and water. As you have seen, the cell surface membrane can control which chemicals can pass in and out – it is described as selectively permeable.

There are three main ways that molecules and ions can move through the membrane. They are diffusion, active transport and osmosis.

DIFFUSION

Many substances can pass through the membrane by **diffusion**. Diffusion happens when a substance is more concentrated in one place than another. For example, if the cell is making carbon dioxide by respiration, the concentration of carbon dioxide inside the cell will be higher than outside. This difference in concentration is called a concentration gradient. The molecules of carbon dioxide are constantly moving about because of their kinetic energy. The cell membrane is permeable to carbon dioxide, so the molecules can move in either direction through it. Since there is a higher concentration of carbon dioxide molecules inside the cell than outside, over time, more molecules will move from inside to outside than move in the other direction. We say that there is a *net* movement of the molecules out of the cell (Figure 1.13).

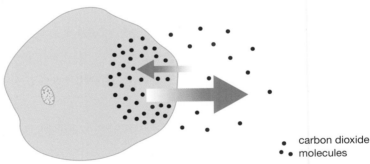

carbon dioxide molecules

▲ Figure 1.13 Carbon dioxide is produced by respiration, so its concentration builds up inside the cell. Although the carbon dioxide molecules diffuse in both directions across the cell membrane, the overall (net) movement is out of the cell, down the concentration gradient.

The opposite happens with oxygen. Respiration uses up oxygen, so there is a concentration gradient of oxygen from outside to inside the cell. There is therefore a net movement of oxygen *into* the cell by diffusion.

The rate of diffusion is affected by various factors.

- The concentration gradient. Diffusion happens more quickly when there is a steep concentration gradient (i.e. a big difference in concentrations between two areas).

- The surface area to volume ratio. A larger surface area in proportion to the volume will increase the rate.

- The distance. The rate is decreased if the distance over which diffusion has to take place is greater.

- The temperature. The rate is greater at higher temperatures. This is because a high temperature provides the particles with more kinetic energy.

KEY POINT

Diffusion is the net movement of particles (molecules or ions) from a region of high concentration to a region of low concentration, i.e. down a concentration gradient.

Safety Note: Wear eye protection and avoid all skin contact with the acid and the dyed agar blocks.

ACTIVITY 5

▼ PRACTICAL: DEMONSTRATION OF DIFFUSION IN A JELLY

Agar is a jelly that is used for growing cultures of bacteria. It has a consistency similar to the cytoplasm of a cell. Like cytoplasm, it has a high water content. Agar can be used to show how substances diffuse through a cell.

This demonstration uses the reaction between hydrochloric acid and potassium permanganate solution. When hydrochloric acid comes into contact with potassium permanganate, the purple colour of the permanganate disappears.

A Petri dish is prepared which contains a 2 cm deep layer of agar jelly, dyed purple with potassium permanganate. Three cubes of different sizes are cut out of the jelly, with side lengths 2 cm, 1 cm and 0.5 cm. The cubes have different volumes and total surface areas. They also have a different surface area to volume ratio, as shown in the table below.

Length of side of cube / cm	Volume of cube / cm³ (length × width × height)	Surface area of cube / cm² (length × width of one side) × 6	Ratio of surface area to volume of cube (surface area divided by volume)
2	(2 × 2 × 2) = 8	(2 × 2) × 6 = 24	24/8 = 3
1	(1 × 1 × 1) = 1	(1 × 1) × 6 = 6	6/1 = 6
0.5	(0.5 × 0.5 × 0.5) = 0.125	(0.5 × 0.5) × 6 = 1.5	1.5/0.125 = 12

Notice that the smallest cube has the largest surface area to volume ratio. The same is true of cells – a small cell has a larger surface area to volume ratio than a large cell.

The cubes are carefully dropped, at the same time, into a beaker of dilute hydrochloric acid (Figure 1.14).

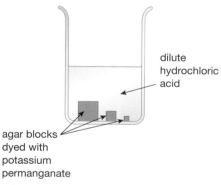

dilute hydrochloric acid

agar blocks dyed with potassium permanganate

▶ Fig 1.14 Investigating diffusion in a jelly.

The time taken for each cube to turn colourless is noted.

Which cube would be the first to turn colourless and which the last? Explain the reasoning behind your prediction.

If the three cubes represented cells of different sizes, which cell would have the most difficulty in obtaining substances by diffusion?

It may be possible for you to try this experiment, using similar apparatus.

ACTIVE TRANSPORT

Diffusion happens because of the kinetic energy of the particles. It does not need an 'extra' source of energy from respiration. However, sometimes a cell needs to take in a substance when there is very little of that substance outside the cell, in other words *against* a concentration gradient. It can do this by another process, called **active transport**.

During active transport a cell uses energy from respiration to take up substances, rather like a pump uses energy to move a liquid from one place to another. In fact, biologists speak of the cell 'pumping' ions or molecules in or out. The pumps are large protein molecules located in the cell membrane, and they are driven by the breakdown of ATP. An example of a place where this happens is in the human small intestine, where some glucose in the gut is absorbed into the cells lining the intestine by active transport. The roots of plants also take up certain mineral ions in this way. Cells use active transport to control the uptake of many substances.

KEY POINT

Active transport is the movement of substances against a concentration gradient, using energy from respiration.

OSMOSIS

Water moves across cell membranes by a special sort of diffusion, called **osmosis**. Osmosis happens when the total concentrations of all dissolved substances inside and outside the cell are different. Water will move across the membrane from the more dilute solution to the more concentrated one. Notice that this is still obeying the rules of diffusion – the water moves from where there is a higher concentration of *water* molecules to a lower concentration of *water* molecules. Osmosis can only happen if the membrane is permeable to water but not to some other solutes. We say that it is **partially permeable**.

Osmosis is important for moving water from cell to cell, for example in plant roots. You will find out more about osmosis in Chapter 11.

KEY POINT

Osmosis in cells is the net movement of water from a dilute solution to a more concentrated solution across the partially permeable cell membrane.

SPECIALISED EXCHANGE SURFACES

All cells exchange substances with their surroundings, but some parts of animals or plants are specially adapted for the exchange of materials because they have a very large surface area in proportion to their volume. In animals, two examples are the alveoli (air sacs) of the lungs (Chapter 3) and the villi of the small intestine (Chapter 4). Diffusion is a slow process, and organs that rely on diffusion need a large surface over which it can take place. The alveoli allow the exchange of oxygen and carbon dioxide to take place between the air and the blood during breathing. The villi of the small intestine provide a large surface area for the absorption of digested food. In plants, exchange surfaces are also adapted by having a large surface area, such as the spongy mesophyll of the leaf (Chapter 10) and the root hairs (Chapter 11).

KEY POINT

'Adapted' or 'an **adaptation**' means that the structure of a cell or an organism is suited to its function. It is a word that is very commonly used in biology, and will appear again in many of the chapters of this book. We also use it when we say that organisms are adapted to their environment.

BIOLOGY ONLY

CELL DIVISION AND DIFFERENTIATION

Multicellular organisms like animals and plants begin life as a single fertilised egg cell, called a **zygote**. This divides into two cells, then four, then eight and so on, until the adult body contains countless millions of cells (Figure 1.15).

This type of cell division is called **mitosis** and is under the control of the genes. You can read a full account of mitosis in Chapter 17, but it is worthwhile considering an outline of the process now. First of all the chromosomes in the nucleus are copied, then the nucleus splits into two, so that the genetic information is shared equally between the two 'daughter' cells. The cytoplasm then divides (or in plant cells a new cell wall develops) forming two smaller cells. These then take in food substances to supply energy and building materials so that they can grow to full size. The process is repeated, but as the developing embryo grows, cells become specialised to carry out particular roles. This specialisation is also under the control of the genes, and is called **differentiation**. Different kinds of cells develop depending on where they are located in the embryo, for example a nerve cell in the spinal cord, or an epidermal cell in the outer layer of the skin (Figure 1.16). Throughout this book you will read about cells that have a structure adapted for a particular function.

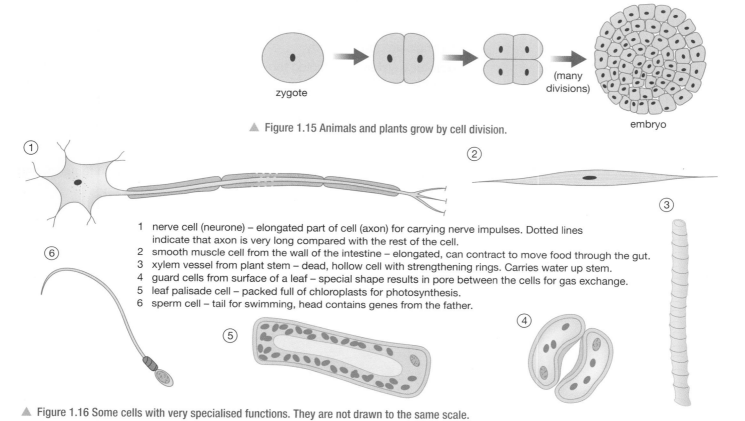

▲ Figure 1.15 Animals and plants grow by cell division.

1 nerve cell (neurone) – elongated part of cell (axon) for carrying nerve impulses. Dotted lines indicate that axon is very long compared with the rest of the cell.
2 smooth muscle cell from the wall of the intestine – elongated, can contract to move food through the gut.
3 xylem vessel from plant stem – dead, hollow cell with strengthening rings. Carries water up stem.
4 guard cells from surface of a leaf – special shape results in pore between the cells for gas exchange.
5 leaf palisade cell – packed full of chloroplasts for photosynthesis.
6 sperm cell – tail for swimming, head contains genes from the father.

▲ Figure 1.16 Some cells with very specialised functions. They are not drawn to the same scale.

What is hard to understand about this process is that through mitosis all the cells of the body have the *same* genes. For cells to function differently, they must produce different proteins, and different genes code for the production of these different proteins. How is it that some genes are 'switched on' and others are 'switched off' to produce different cells? The answer to this question is very complicated, and scientists are only just beginning to work it out.

END OF BIOLOGY ONLY

CELLS, TISSUES AND ORGANS

Cells with a similar function are grouped together as **tissues**. For example, the muscle of your arm contains millions of similar muscle cells, all specialised for one function – contraction to move the arm bones. This is muscle tissue. However, a muscle also contains other tissues, such as blood, nervous tissue and epithelium (lining tissue). A collection of several tissues carrying out a particular function is called an **organ**. The main organs of the human body are shown in Figure 1.17. Plants also have tissues and organs. Leaves, roots, stems and flowers are all plant organs.

In animals, jobs are usually carried out by several different organs working together. This is called an **organ system**. For example, the digestive system consists of the gut, along with glands such as the pancreas and gall bladder. The function of the whole system is to digest food and absorb the digested products into the blood. There are seven main systems in the human body. These are the:

- digestive system
- gas exchange system – including the lungs, which exchange oxygen and carbon dioxide
- circulatory system – including the heart and blood vessels, which transport materials around the body
- excretory system – including the kidneys, which filter toxic waste materials from the blood
- nervous system – consisting of the brain, spinal cord and nerves, which coordinate the body's actions
- endocrine system – glands secreting hormones, which act as chemical messengers
- reproductive system – producing sperm in males and eggs in females, and allowing the development of the embryo.

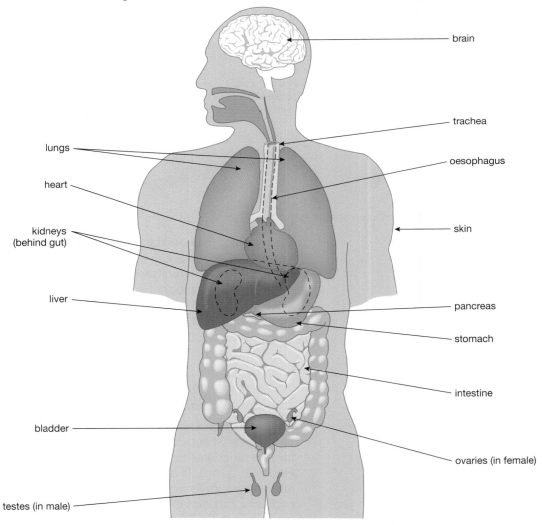

▲ Figure 1.17 Some of the main organs of the human body.

BIOLOGY ONLY

STEM CELLS

A **stem cell** is a cell that has the ability to divide many times by mitosis while remaining undifferentiated. Later, it can differentiate into specialised cells such as muscle or nerve cells. In humans there are two main types of stem cells:

- *Embryonic stem cells* are found in the early stage of development of the embryo. They can differentiate into any type of cell.

- *Adult stem cells* are found in certain adult tissues such as bone marrow, skin, and the lining of the intestine. They have lost the ability to differentiate into any type of cell but can form a number of specialised tissues. For example, bone marrow cells can divide many times but are only able to produce different types of red and white blood cells.

The use of stem cells to treat (or prevent) a disease, or to repair damaged tissues is called *stem cell therapy*. At present, the most common form of stem cell therapy is the use of bone marrow transplants. Bone marrow transplants are used to treat patients with conditions such as leukaemia (a type of blood cancer). Some cancer treatments use chemicals that kill cancer cells (chemotherapy) but this also destroys healthy cells. Bone marrow transplants supply stem cells that can divide and differentiate, replacing cells lost from the body during chemotherapy. Bone marrow transplants are now a routine procedure and have been used successfully for over 30 years. Bone marrow and other adult stem cells are readily available, but they have limited ability to differentiate into other types of cell.

Scientists are able to isolate and culture embryonic stem cells (Figure 1.18). These are obtained from fertility clinics where parents choose to donate their unused embryos for research. In the future, it is hoped that we will be able to use embryonic stem cells to treat many diseases such as diabetes along with brain disorders such as Parkinson's disease. Stem cells could also be used to repair damaged nerve tissue. Until now, treatments using embryonic stem cells have not progressed beyond the experimental stage. Stem cell research can also present problems. Many people object morally to using cells from embryos for medical purposes despite the fact that they might one day be used to cure many diseases.

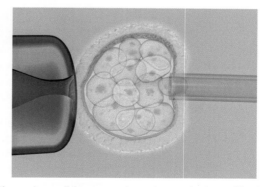

▲ Figure 1.18 Extracting a stem cell from an embryo at an early stage of its development. The embryo consists of a ball of about 20 cells. A single cell is removed by drawing it into a fine glass capillary tube.

END OF BIOLOGY ONLY

LOOKING AHEAD – MEMBRANES IN CELLS

If you continue to study biology beyond International GCSE, you will learn more about the structure and function of cells. You might like to look on the Internet for some electron micrographs and carry out some further research into cells.

Electron micrographs allow us to see cells at a much greater magnification than by using a light microscope. They also reveal more detail. The image produced by a light microscope can only distinguish features about the size of a mitochondrion. The electron microscope has a much greater *resolution*. Resolution is the ability to distinguish two points in an image as being separate. The maximum resolution of a light microscope is about 200 nanometres (nm), whilst with an electron microscope we can distinguish structures less than 1 nm in size. That is why ribosomes are only visible using an electron microscope – they are about 25 nm in diameter. A nanometre (nm) is 10^{-9} m, or one millionth of a milimetre.

Electron microscopy reveals that much of the cytoplasm is made up of membranes. As well as the cell surface membrane, there are membranes around organelles such as the nucleus, mitochondria and chloroplasts. In addition, there is an extensive system of membranes running throughout the cytoplasm, called the endoplasmic reticulum (ER). Some ER is covered in ribosomes, and is called rough ER (Figure 1.19).

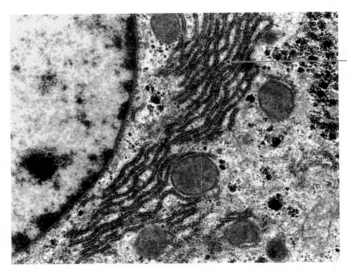

rough endoplasmic reticulum

▲ Figure 1.19 Rough endoplasmic reticulum is a system of membranes extending throughout the cytoplasm of a cell. It is covered with ribosomes the (tiny dots). Ribosomes are the site of protein synthesis.

There are thousands of different chemical reactions that take place inside cells. A key function of a cell membrane is to separate cell functions into different compartments so they don't take place together. For example, the reactions and enzymes of aerobic respiration are kept inside the mitochondria, separate from the rest of the cytoplasm.

CHAPTER QUESTIONS

More questions on life processes can be found at the end of Unit 1 on page 32.

SKILLS   CRITICAL THINKING

1 Which of the following comparisons of animal and plant cells is *not* true?

	Animal cells	Plant cells
A	do not have chloroplasts	have chloroplasts
B	have mitochondria	do not have mitochondria
C	have temporary vacuoles	have permanent vacuoles
D	do not have cellulose cell walls	have cellulose cell walls

2 Which of the following descriptions is correct?

   A The cell wall is freely permeable and the cell membrane is partially permeable

   B The cell wall is partially permeable and the cell membrane is freely permeable

   C Both the cell wall and the cell membrane are freely permeable

   D Both the cell wall and the cell membrane are partially permeable

3 What are the products of anaerobic respiration in yeast?

   A ethanol and carbon dioxide   **B** lactate and carbon dioxide

   C carbon dioxide and water   **D** ethanol and water

BIOLOGY ONLY

4 Which of the following is the best definition of 'differentiation'?

   A The organisation of the body into cells, tissues and organs

   B A type of cell division resulting in the growth of an embryo

   C The adaptation of a cell for its function

   D The process by which the structure of a cell becomes specialised for its function

END OF BIOLOGY ONLY

SKILLS   INTERPRETATION

SKILLS   CRITICAL THINKING

SKILLS   INTERPRETATION

5 a Draw a diagram of a plant cell. Label all of the parts. Alongside each label write the function of that part.

   b Write down three differences between the cell you have drawn and a 'typical' animal cell.

6 Write a short description of the nature and function of enzymes. Include in your description:

   ◼ a definition of an enzyme

   ◼ a description of the 'lock and key' model of enzyme action

   ◼ an explanation of the difference between intracellular and extracellular enzymes.

   Your description should be about a page in length, including a labelled diagram.

7 The graph shows the effect of temperature on an enzyme. The enzyme was extracted from microorganism that lives in hot mineral springs near a volcano.

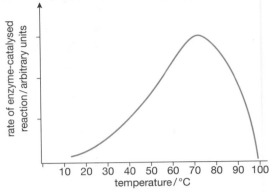

a What is the optimum temperature of this enzyme?

b Explain why the activity of the enzyme is greater at 60 °C than at 30 °C.

c The optimum temperature of enzymes in the human body is about 37 °C. Explain why this enzyme is different.

d What happens to the enzyme at 90 °C?

8 Explain the differences between diffusion and active transport.

9 The nerve cell called a **motor neurone** (page 19) and a **palisade** cell of a leaf (page 19) are both very specialised cells. Read about each of these and explain very briefly (three or four lines) how each is adapted to its function.

10 The diagram shows a cell from the lining of a human kidney tubule. A major role of the cell is to absorb glucose from the fluid passing along the tubule and pass it into the blood, as shown by the arrow on the diagram.

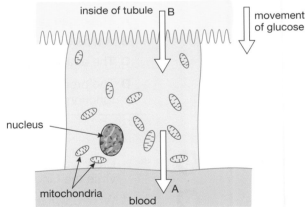

a What is the function of the mitochondria?

b The tubule cell contains a large number of mitochondria. They are needed for the cell to transport glucose across the cell membrane into the blood at 'A'. Suggest the method that the cell uses to do this and explain your answer.

c The mitochondria are *not* needed to transport the glucose into the cell from the tubule at 'B'. Name the process by which the ions move across the membrane at 'B' and explain your answer.

d The surface membrane of the tubule cell at 'B' is greatly folded. Suggest how this adaptation helps the cell to carry out its function.

2 THE VARIETY OF LIVING ORGANISMS

There is an enormous variety of living organisms. Biologists put them into groups according to their structure and function. The members of each group have certain features in common.

LEARNING OBJECTIVES

- Understand the difference between eukaryotic and prokaryotic organisms

- Describe the features common to plants and recognise examples of flowering plants such as maize, peas and beans

- Describe the features common to animals and recognise examples such as mammals and insects

- Describe the features common to fungi and recognise examples such as *Mucor* and yeast

- Describe the features common to protoctists and recognise examples such as *Amoeba*, *Chlorella* and *Plasmodium*

- Describe the features common to bacteria and recognise examples such as *Lactobacillus bulgaricus* and *Pneumococcus*

- Describe the features common to viruses and recognise examples such as the influenza virus, the HIV virus and the tobacco mosaic virus

- Understand the term 'pathogen' and know that pathogens may include fungi, bacteria, protoctists or viruses.

There are more than ten million species of organisms alive on Earth today, and many more that once lived on Earth but are now extinct. In order to make sense of this enormous variety biologists classify organisms, putting them into groups. Members of each group are related – they are descended from a common ancestor by the process of evolution (see Chapter 19). This common ancestry is reflected in the similarities of structure and function of the members of a group.

The five major groups of living organisms are plants, animals, fungi, protoctists and bacteria.

▲ Figure 2.1 (a) A pea plant. Its leaves and stem cells contain chloroplasts, giving them their green colour. The white flowers are pollinated by insects. (b) Maize plants are pollinated by wind. These are the male flowers, which make the pollen. (c) The female maize flowers produce seeds after pollination.

PLANTS

You will be familiar with flowering plants, such as those shown in Figure 2.1. This group, or **kingdom**, also contains simpler plants, such as mosses and ferns. All plants are **multicellular,** which means that their 'bodies' are made up of many cells. Their main distinguishing feature is that their cells contain chloroplasts and they carry out photosynthesis – the process that uses light energy to convert simple inorganic molecules such as water and carbon dioxide into complex organic compounds (see Chapter 10). One of these organic compounds is the carbohydrate **cellulose**, and all plants have cell walls made of this material.

Plants can make many other organic compounds as a result of photosynthesis. One of the first to be made is the storage carbohydrate **starch**, which is often found inside plant cells. Another is the sugar **sucrose**, which is transported around the plant and is sometimes stored in fruits and other plant organs. The structure and function of flowering plants is dealt with in Unit 3 of this book.

ANIMALS

You will be even more familiar with this kingdom, since it contains the species *Homo sapiens*, i.e. humans! The variety of the animal kingdom is also enormous, including organisms such as sponges, molluscs, worms, starfish, insects and crustaceans, through to larger animals such as fish, amphibians, reptiles, birds and mammals (Figure 2.2). The last five groups are all **vertebrates**, which means that they have a vertebral column, or backbone. All other animals lack this feature, and are called **invertebrates**.

(a)

(b)

(c)

▲ Figure 2.2 (a) A housefly. (b) A mosquito, feeding on human blood. Houseflies and mosquitoes are both insects, which make up the largest sub-group of all the animals. About 60% of all animal species are insects. (c) This high jumper's movement is coordinated by a complex nervous system.

Animals are also multicellular organisms. Their cells never contain chloroplasts, so they are unable to carry out photosynthesis. Instead, they gain their nutrition by feeding on other animals or plants. Animal cells also lack cell walls, which allows their cells to change shape, an important feature for organisms that need to move from place to place. Movement in animals is achieved in various ways, but often involves coordination by a nervous system (see Chapter 6). Another feature common to most animals is that they store carbohydrate in their cells as a compound called **glycogen** (see Chapter 4). The structure and function of animals is dealt with in Unit 2 of this book.

FUNGI

Fungi include mushrooms and toadstools, as well as moulds. These groups of fungi are multicellular. Another group of fungi is the yeasts, which are **unicellular** (made of single cells). Different species of yeasts live everywhere – on the surface of fruits, in soil, water, and even on dust in the air. The yeast powder used for baking contains millions of yeast cells (Figure 2.3). The cells of fungi never contain chloroplasts, so they cannot photosynthesise. Their cells have cell walls, but they are not composed of cellulose (Figure 2.4).

EXTENSION WORK

Because fungi have cell walls, they were once thought to be plants that had lost their chlorophyll. We now know that their cell wall is not made of cellulose as in plants, but of a different chemical called **chitin** (the same material that makes up the outside skeleton of insects). There are many ways that fungi are very different from plants (the most obvious is that fungi do not photosynthesise) and they are not closely related to plants at all.

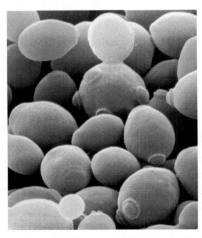

▲ Figure 2.3 Yeast cells, highly magnified

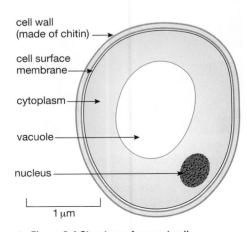

cell wall (made of chitin)

cell surface membrane

cytoplasm

vacuole

nucleus

1 μm

▲ Figure 2.4 Structure of a yeast cell

The singular of hyphae is hypha.

▲ Figure 2.5 Toadstools growing on a rotting tree trunk

▲ Figure 2.6 The 'pin mould' *Mucor* growing on a piece of bread. The dark spots are structures that produce spores for reproduction.

A mushroom or toadstool is the reproductive structure of the organism, called a fruiting body (Figure 2.5). Under the soil, the mushroom has many fine thread-like filaments called **hyphae** (pronounced high-fee). A mould is rather like a mushroom without the fruiting body. It just consists of the network of hyphae (Figure 2.6). The whole network is called a **mycelium** (pronounced my-sea-lee-um). Moulds feed by absorbing nutrients from dead (or sometimes living) material, so they are found wherever this is present, for example, in soil, rotting leaves or decaying fruit.

If you leave a piece of bread or fruit exposed to the air for a few days, it will soon become mouldy. Mould spores carried in the air have landed on the food and grown into a mycelium of hyphae (Figure 2.7).

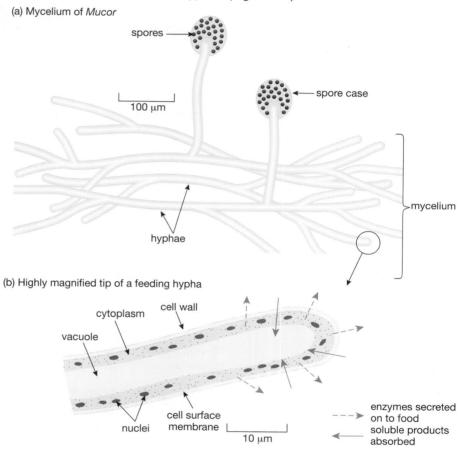

(a) Mycelium of *Mucor*

spores

spore case

100 µm

mycelium

hyphae

(b) Highly magnified tip of a feeding hypha

cytoplasm

cell wall

vacuole

nuclei

cell surface membrane

10 µm

enzymes secreted on to food

soluble products absorbed

▲ Figure 2.7 The structure of a typical mould fungus, the 'pin mould' *Mucor*.

The thread-like hyphae of *Mucor* have cell walls surrounding their cytoplasm. The cytoplasm contains many nuclei. In other words the hyphae are not divided up into separate cells.

When a spore from *Mucor* lands on the food, a hypha grows out from it. The hypha grows and branches again and again, until the mycelium covers the surface of the food. The hyphae secrete digestive enzymes on to the food, breaking it down into soluble substances such as sugars, which are then absorbed by the mould. Eventually, the food is used up and the mould must infect another source of food by producing more spores.

When an organism feeds on dead organic material in this way, and digestion takes place outside of the organism, this is called **saprotrophic** nutrition. Enzymes that are secreted out of cells for this purpose are called *extracellular* enzymes (see Chapter 1).

PROTOCTISTS

Protoctists are sometimes called the 'dustbin kingdom', because they are a mixed group of organisms that don't fit into the plants, animals or fungi. Most protoctists are microscopic single-celled organisms (Figure 2.8). Some look like animal cells, such as *Amoeba*, which lives in pond water. These are known as **protozoa**. Other protoctists have chloroplasts and carry out photosynthesis, so are more like plants. These are called **algae**. Most algae are unicellular, but some species such as seaweeds are multicellular and can grow to a great size. Some protoctists are the agents of disease, such as *Plasmodium*, the organism that causes malaria.

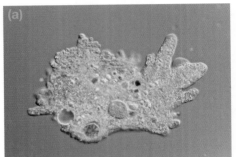 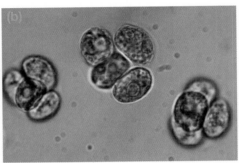 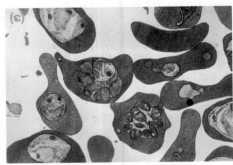

▲ Figure 2.8 (a) *Amoeba*, a protozoan that lives in ponds (b) *Chlorella*, a unicellular freshwater alga (c) Blood cells containing the protoctist parasite *Plasmodium*, the organism responsible for causing malaria

EUKARYOTIC AND PROKARYOTIC ORGANISMS

All the organisms described so far are composed of **eukaryotic** cells and are known as eukaryotic organisms. 'Eukaryotic' means 'having a nucleus' – their cells contain a nucleus surrounded by a membrane, along with other membrane bound organelles, such as mitochondria and chloroplasts.

There are also organisms made of simpler cells, which have no nucleus, mitochondria or chloroplasts. These are called **prokaryotic** cells. 'Prokaryotic' means 'before nucleus'. The main forms of prokaryotic organisms are the bacteria.

BACTERIA

Bacteria are small single-celled organisms. Their cells are much smaller than those of eukaryotic organisms and have a much simpler structure. To give you some idea of their size, a typical animal cell might be 10 to 50 μm in diameter (1 μm, or one micrometre, is a millionth of a metre). Compared with this, a typical bacterium is only 1 to 5 μm in length (Figure 2.9) and its volume is thousands of times smaller than that of the animal cell.

There are three basic shapes of bacteria: spheres, rods and spirals, but they all have a similar internal structure (Figure 2.10).

All bacteria are surrounded by a cell wall, which protects the bacterium and keeps the shape of the cell. Bacterial cell walls are not made of cellulose but a complex compound of sugars and proteins called peptidoglycan. Some species have another layer outside this wall, called a **capsule** or slime layer. Both give the bacterium extra protection. Underneath the cell wall is the cell membrane, as in other cells. The middle of the cell is made of cytoplasm. Since it is a prokaryotic cell, the bacterium has no nucleus. Instead, its genetic material (DNA) is in a single chromosome, loose in the cytoplasm, forming a circular loop.

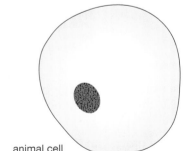

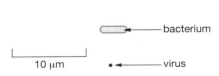

animal cell

bacterium

10 μm

virus

▲ Figure 2.9 A bacterium is much smaller than an animal cell. The relative size of a virus is also shown.

(a) **Some different bacterial shapes**

spheres:
singles, pairs, chains
or groups

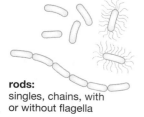

rods:
singles, chains, with
or without flagella

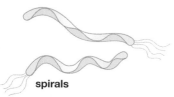

spirals

▲ Figure 2.11 The bacterium *Lactobacillus
bulgaricus*, used in the production of
yoghurt.

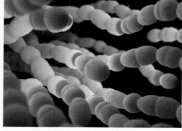

▲ Figure 2.12 Rounded cells of the bacterium
Pneumococcus, one cause of pneumonia.

(b) **Internal structure of a bacterium**

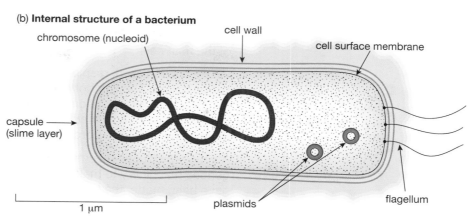

◀ ▲ Figure 2.10 Structure of bacteria

Some bacteria can swim, and are propelled through water by corkscrew-like
movements of structures called **flagella** (a single one of these is called a
flagellum). However, many bacteria do not have flagella and cannot move by
themselves. Other structures present in the cytoplasm include the **plasmids**.
These are small circular rings of DNA, carrying some of the bacterium's genes.
Not all bacteria contain plasmids, although about three-quarters of all known
species do. Plasmids have very important uses in genetic engineering (see
Chapter 22).

Some bacteria contain a form of chlorophyll in their cytoplasm, and can carry
out photosynthesis. However, most bacteria feed off other living or dead
organisms. Along with the fungi, many bacteria are important **decomposers**
(see Chapter 14), recycling dead organisms and waste products in the soil
and elsewhere. Some bacteria are used by humans to make food, such as
Lactobacillus bulgaricus, a rod-shaped species used in the production of
yoghurt from milk (Figure 2.11). Other species are **pathogens**, which means
that they cause disease (Figure 2.12).

Despite the relatively simple structure of the bacterial cell, it is still a living
cell that carries out the normal processes of life, such as respiration, feeding,
excretion, growth and reproduction. As you have seen, some bacteria can
move, and they can also respond to a range of stimuli. For example, they may
move towards a source of food, or away from a poisonous chemical. You
should think about these features when you compare bacteria with the next
group, the much simpler viruses.

VIRUSES

All **viruses** are parasites, and can only reproduce inside living cells. The cell
in which the virus lives is called the host. There are many different types of
viruses. Some live in the cells of animals or plants, and there are even viruses
which infect bacteria. Viruses are much smaller than bacterial cells: most are
between 0.01 and 0.1 μm in diameter (Figure 2.9).

Viruses are not made of cells. A virus particle is very simple. It has no nucleus
or cytoplasm, and is composed of a core of genetic material surrounded by a
protein coat (Figure 2.13). The genetic material can be either **DNA**, or a similar
chemical called **RNA** (see Chapter 16). In either case, the genetic material
makes up just a few genes – all that is needed for the virus to reproduce inside
its host cell.

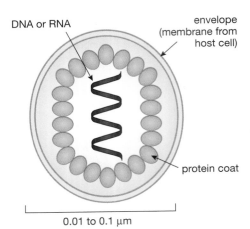

DNA or RNA

envelope (membrane from host cell)

protein coat

0.01 to 0.1 μm

▲ Figure 2.13 The structure of a typical virus, such as the type causing influenza (flu).

EXTENSION WORK

AIDS is not actually a disease but a 'syndrome'. A syndrome is a set of symptoms caused by a medical condition. In the case of HIV, the virus severely damages the person's immune system, so they are more likely to get other diseases, such as tuberculosis. They may also develop some unusual types of cancer. This collection of different symptoms is referred to as AIDS.

▲ Figure 2.15 Discoloration of the leaves of a tobacco plant, caused by infection with tobacco mosaic virus.

Sometimes a membrane called an envelope may surround a virus particle, but the virus does not make this. Instead it is 'stolen' from the surface membrane of the host cell.

Viruses do not feed, respire, excrete, move, grow or respond to their surroundings. They do not carry out any of the normal 'characteristics' of living things except reproduction, and they can only do this parasitically. This is why biologists do not consider viruses to be living organisms. You can think of them as being on the border between an organism and a non-living chemical.

A virus reproduces by entering the host cell and taking over the host's genetic machinery to make more virus particles. After many virus particles have been made, the host cell dies and the particles are released to infect more cells. Many human diseases are caused in this way, such as influenza ('flu'). Other examples include colds, measles, mumps, polio and rubella ('German measles'). Of course, the reproduction process does not continue forever. Usually, the body's immune system destroys the virus and the person recovers. Sometimes, however, a virus cannot be destroyed by the immune system quickly enough, and it may cause permanent damage or death. With other infections, the virus may attack cells of the immune system itself. This is the case with HIV (the Human Immunodeficiency Virus), which causes the illness called AIDS (Acquired Immune Deficiency Syndrome).

Viruses don't just parasitise animal cells. Some infect plant cells, such as the tobacco mosaic virus (Figure 2.14), which interferes with the ability of the tobacco plant to make chloroplasts, causing mottled patches to develop on the leaves (Figure 2.15).

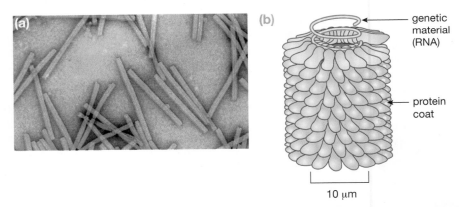

(a)

(b)

genetic material (RNA)

protein coat

10 μm

▲ Figure 2.14 (a) Tobacco mosaic virus (TMV), seen through an electron microscope. (b) Structure of part of a TMV particle, magnified 1.25 million times.

CHAPTER QUESTIONS

More questions on the variety of living organisms can be found at the end of Unit 1 on page 32.

SKILLS CRITICAL THINKING

1 Which of the following is *not* a characteristic of plants?

    A cells contain chloroplasts

    B cell wall made of cellulose

    C bodies are multicellular

    D store carbohydrate as glycogen

2 Fungi carry out *saprotrophic nutrition*. What is the meaning of this term?

A extracellular digestion of dead organic matter

B feeding on other living organisms

C making organic molecules by photosynthesis

D secreting digestive enzymes

3 Below are three groups of organisms.

1. viruses

2. bacteria

3. yeasts

Which of these organisms are prokaryotic?

A 1 only

B 2 only

C 1 and 2

D 1, 2 and 3

4 Which of the following diseases is *not* caused by a virus?

A influenza

B measles

C malaria

D AIDS

5 **a** Name the kingdom to which each of the following organisms belongs:

i mushroom

ii *Chlorella*

iii moss

iv *Lactobacillus*

b The diagram shows a species of protoctist called *Euglena*. Use the diagram to explain why *Euglena* is classified as a protoctist and not as an animal or plant.

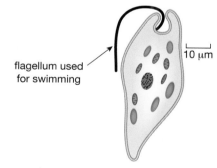

flagellum used for swimming

10 μm

6 **a** Draw a diagram to show the structure of a typical virus particle.

b Is a virus a living organism? Explain your answer.

c Explain the statement 'viruses are all parasites'.

7 Explain the meanings of the following terms:

a invertebrate

b hyphae

c saprotrophic

UNIT QUESTIONS

SKILLS CRITICAL THINKING   These three organelles are found in cells: nucleus, chloroplast and mitochondrion.

    a Which of the above organelles would be found in:

        i a cell from a human muscle? **(1)**

        ii a palisade cell from a leaf? **(1)**

        iii a cell from the root of a plant? **(1)**

SKILLS REASONING   **b** Explain fully why the answers to ii) and iii) above are different. **(1)**

SKILLS INTERPRETATION   **c** What is the function of each organelle? **(3)**

                                                      Total 7 marks

SKILLS INTERPRETATION, REASONING In multicellular organisms, cells are organised into tissues, organs and organ systems.

    a The diagram shows a section through an artery and a capillary.

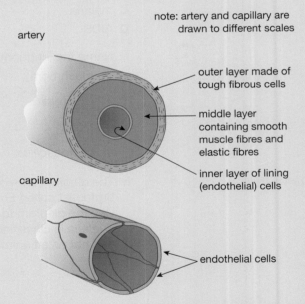

note: artery and capillary are drawn to different scales

artery

outer layer made of tough fibrous cells

middle layer containing smooth muscle fibres and elastic fibres

inner layer of lining (endothelial) cells

capillary

endothelial cells

    Explain why an artery can be considered to be an organ whereas a capillary cannot. **(2)**

    b Organ systems contain two or more organs whose functions are linked. The digestive system is one human organ system. (See Chapter 4.)

SKILLS CRITICAL THINKING   **i** What does the digestive system do? **(2)**

         ii Name three organs in the human digestive system. Explain what each organ does as part of the digestive system. **(6)**

        iii Name two other human organ systems and, for each system, name two organs that are part of the system. **(6)**

                                                        Total 16 marks

SKILLS   ANALYSIS

3 Catalase is an enzyme found in many plant and animal cells. It catalyses the breakdown of hydrogen peroxide into water and oxygen.

$$\text{hydrogen peroxide} \xrightarrow{\text{catalase}} \text{water} + \text{oxygen}$$

a In an investigation into the action of catalase in potato, 20 g of potato tissue was put into a small beaker containing hydrogen peroxide weighing 80 g in total. The temperature was maintained at 20 °C throughout the investigation. As soon as the potato was added, the mass of the beaker and its contents was recorded until there was no further change in mass. The results are shown in the graph.

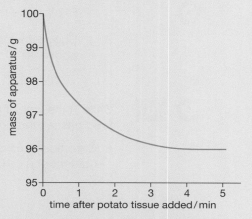

5   i How much oxygen was formed in this investigation? Explain your answer. **(2)**

6   ii Estimate the time by which half this mass of oxygen had been formed. **(2)**

SKILLS   REASONING   **8**

iii Explain, in terms of collisions between enzyme and substrate molecules, why the rate of reaction changes during the course of the investigation. **(2)**

b The students repeated the investigation at 30 °C. What difference, if any, would you expect in:

SKILLS   DECISION MAKING   **9**

i the mass of oxygen formed?

ii the time taken to form this mass of oxygen?

Explain your answers. **(4)**

Total 10 marks

SKILLS   CRITICAL THINKING   **7**

4 Different particles move across cell membranes using different processes.

a The table below shows some ways in which active transport, osmosis and diffusion are similar and some ways in which they are different.
Copy and complete the table with ticks and crosses. **(3)**

Feature	Active transport	Osmosis	Diffusion
particles must have kinetic energy			
requires energy from respiration			
particles move down a concentration gradient			

b The graph shows the results of an investigation into the rate of diffusion of sodium ions across the membranes of potato cells.

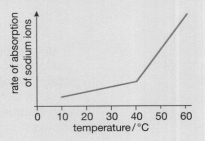

SKILLS ANALYSIS, REASONING 6

i Explain the increase in the rate of diffusion up to 40 °C. **(2)**

ii Suggest why the rate of increase is much steeper at temperatures above 40 °C. **(2)**

Total 7 marks

SKILLS REASONING 8

5 Cells in the wall of the small intestine divide by mitosis to replace cells lost as food passes through.

a Chromosomes contain DNA. The graph shows the changes in the DNA content of a cell in the wall of the small intestine as it divides by mitosis.

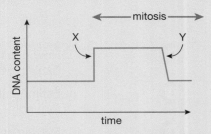

i Why is it essential that the DNA content is doubled (X) before mitosis begins? **(2)**

SKILLS ANALYSIS 8

ii What do you think happens to the cell at point Y? **(1)**

b The diagram shows a cell in the wall of a villus in the small intestine. Some of the processes involved in the absorption of glucose are also shown.

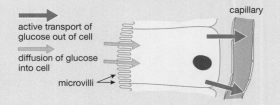

SKILLS INTERPRETATION 6

i What is the importance of the small intestine having villi? **(1)**

7

ii Suggest how the microvilli adapt this cell to its function of absorbing glucose. **(1)**

8

iii Suggest how the active transport of glucose out of the cell and into the blood stream helps with the absorption of glucose from the small intestine. **(2)**

Total 7 marks

SKILLS ▶ REASONING　 **6**

! Safety Note: Eye protection should be worn when setting up the apparatus as sodium hydroxide is very hazardous to the eyes.

A respirometer is used to measure the rate of respiration. The diagram shows a simple respirometer. The sodium hydroxide solution in the apparatus absorbs carbon dioxide. Some results from the investigation are also shown.

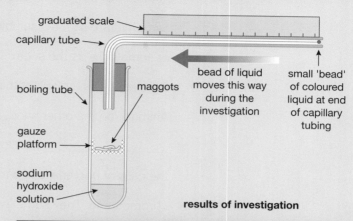

graduated scale
capillary tube
bead of liquid moves this way during the investigation
small 'bead' of coloured liquid at end of capillary tubing
boiling tube
maggots
gauze platform
sodium hydroxide solution

results of investigation

Experiment	Distance moved by bead / mm
1	20
2	3
3	18

a Assume that the maggots in the apparatus respire aerobically.

　i Write the symbol equation for aerobic respiration. (4)

　ii From the equation, what can you assume about the amount of oxygen taken in and carbon dioxide given off by the maggots? Explain your answer. (3)

　iii Result 2 is significantly different from the other two results. Suggest a reason for this. (2)

　iv How would the results be different if the organisms under investigation respired anaerobically? (2)

Total 11 marks

SKILLS ▶ INTERPRETATION　5　**7**

The table below shows some features of different groups of organisms. Copy and complete the table by putting a tick in the box if the organism has that feature, or a cross if it lacks the feature.

Feature	Type of organism		
	Plant	Fungus	Virus
they are all parasites			
they are made up of a mycelium of hyphae			
they can only reproduce inside living cells			
they feed by extracellular digestion by enzymes			
they store carbohydrate as starch			

Total 5 marks

SKILLS CRITICAL THINKING **8**

Copy and complete the following account.

Plants have cell walls made of _____ . They store

carbohydrate as the insoluble compound called _____

or sometimes as the sugar _____ . Plants make these

substances as a result of the process called _____ .

Animals, on the other hand, store carbohydrate as the compound

_____ . Both animals' and plants' cells have nuclei, but

the cells of bacteria lack a true nucleus, having their DNA in a circular

chromosome. They sometimes also contain small rings of DNA called

_____ , which are used in genetic engineering. Bacteria

and fungi break down organic matter in the soil. They are known as

_____ . Some bacteria are pathogens, which means that they

_____ .

Total 8 marks

SKILLS DECISION MAKING **9**

The diagram shows the apparatus used to investigate germination of pea
seeds. At the start of the experiment the temperature in both flasks was 19 °C

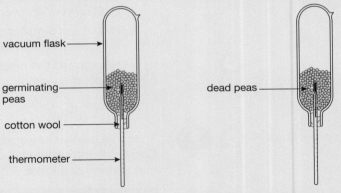

The apparatus was left for 24 hours. At the end of this time the temperature in
the flask with germinating peas was 22 °C, while the temperature in the flask
with dead peas was 19 °C.

a Explain the biological reason for this difference in temperature. (2)

b The seeds in both flasks were washed in disinfectant before the
experiment. Explain why this was done. (1)

c Cotton wool was used to hold the seeds in the flasks.
Suggest why cotton wool was used instead of a rubber bung. (1)

SKILLS EXECUTIVE FUNCTION

d Pea seeds were used in both flasks. State another variable that
should have been controlled in the experiment. (1)

Total 5 marks

SKILLS   EXECUTIVE
         FUNCTION

7   **10**   A piece of meat is a tissue composed of muscle fibres. Muscle fibres use ATP when they contract. Describe an investigation to find out if a solution of ATP will cause the contraction of muscle fibres.

Total 6 marks

UNIT 2
ANIMAL PHYSIOLOGY

Physiology is the branch of biology that looks at how living things function. It studies the workings of an organism at different levels from cells, tissues and organs through to the whole organism. In this unit we look at animal physiology and in particular how the human body works. It is important to study physiology, not least because knowledge of the body is essential in understanding how to treat it when it goes wrong. This is reflected in the fact that one of the six categories of Nobel Prize is awarded for 'physiology or medicine'.

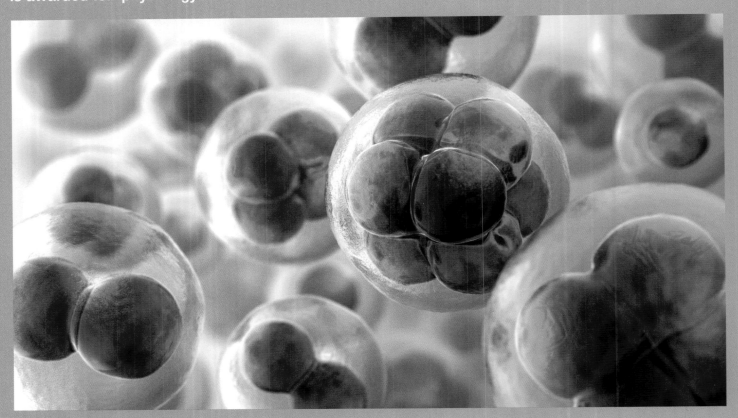

3 BREATHING AND GAS EXCHANGE

When we breathe, air is moved in and out of the lungs so that gas exchange can take place between the air and the blood. This chapter looks at these processes, and also deals with some ways that smoking can damage the lungs and stop these vital organs from working properly.

LEARNING OBJECTIVES

- Describe the structure of the thorax, including the ribs, intercostal muscles, diaphragm, trachea, bronchi, bronchioles, alveoli and pleural membranes

- Understand the role of the intercostal muscles and the diaphragm in ventilation

- Explain how alveoli are adapted for gas exchange by diffusion between air in the lungs and blood in capillaries

- Investigate breathing in humans, including the release of carbon dioxide and the effect of exercise

- Understand the biological consequences of smoking in relation to the lungs and circulatory system, including coronary heart disease.

Cells get their energy by oxidising foods such as glucose, during the process called **respiration**. If cells are to respire aerobically, they need a continuous supply of oxygen from the blood. In addition, carbon dioxide from respiration needs to be removed from the body. In humans, these gases are exchanged between the blood and the air in the lungs.

RESPIRATION AND BREATHING

You need to understand the difference between respiration and breathing. Respiration is the oxidation reaction that releases energy from foods such as glucose (Chapter 1). Breathing is the mechanism that moves air into and out of the lungs, allowing gas exchange to take place. The lungs and associated structures are often called the 'respiratory system' but this can be confusing. It is better to call them the gas exchange system and this is the term we use in this book.

THE STRUCTURE OF THE GAS EXCHANGE SYSTEM

The lungs are enclosed in the chest or **thorax** by the ribcage and a muscular sheet of tissue called the **diaphragm** (Figure 3.1). As you will see, the actions of these two structures bring about the movements of air into and out of the lungs. Joining each rib to the next are two sets of muscles called **intercostal muscles** ('costals' are rib bones). The diaphragm separates the contents of the thorax from the abdomen. It is not flat, but a shallow dome shape, with a fibrous middle part forming the 'roof' of the dome, and muscular edges forming the walls.

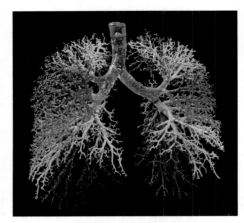

▲ Figure 3.2 This cast of the human lungs was made by injecting a pair of lungs with a liquid plastic. The plastic was allowed to set, then the lung tissue was dissolved away with acid.

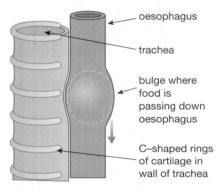

oesophagus

trachea

bulge where food is passing down oesophagus

C–shaped rings of cartilage in wall of trachea

▲ Figure 3.3 C-shaped cartilage rings in the trachea.

EXTENSION WORK

In the bronchi, the cartilage forms complete, circular rings. In the trachea, the rings are incomplete, and shaped like a letter 'C'. The open part of the ring is at the back of the trachea, next to where the oesophagus (gullet) lies as it passes through the thorax. When food passes along the oesophagus by peristalsis (see Chapter 4) the gaps in the rings allow the lumps of food to pass through more easily, without the peristaltic wave 'catching' on the rings (Figure 3.3).

KEEPING THE AIRWAYS CLEAN

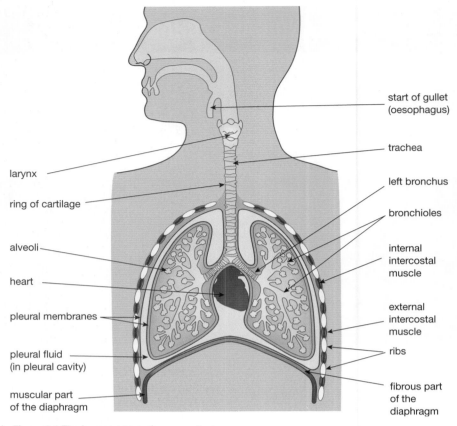

start of gullet (oesophagus)

trachea

left bronchus

bronchioles

internal intercostal muscle

external intercostal muscle

ribs

fibrous part of the diaphragm

larynx

ring of cartilage

alveoli

heart

pleural membranes

pleural fluid (in pleural cavity)

muscular part of the diaphragm

▲ Figure 3.1 The human gas exchange system

The air passages of the lungs form a highly branching network (Figure 3.2). This is why it is sometimes called the **bronchial tree**.

When we breathe in, air enters our nose or mouth and passes down the windpipe or **trachea**. The trachea splits into two tubes called the **bronchi** (singular **bronchus**), one leading to each lung. Each bronchus divides into smaller and smaller tubes called **bronchioles**, eventually ending at microscopic air sacs, called **alveoli** (singular **alveolus**). It is here that gas exchange with the blood takes place.

The walls of trachea and bronchi contain rings of gristle or **cartilage**. These support the airways and keep them open when we breathe in. They are rather like the rings in a vacuum cleaner hose – without them the hose would squash flat when the cleaner sucks air in.

The inside of the thorax is separated from the lungs by two thin, moist membranes called the **pleural membranes**. They make up a continuous envelope around the lungs, forming an airtight seal. Between the two membranes is a space called the **pleural cavity**, filled with a thin layer of liquid called **pleural fluid**. This acts as lubrication, so that the surfaces of the lungs don't stick to the inside of the chest wall when we breathe.

The trachea and larger airways are lined with a layer of cells that have an important role in keeping the airways clean. Some cells in this lining secrete a sticky liquid called **mucus**, which traps particles of dirt or bacteria that are breathed in. Other cells are covered with tiny hair-like structures called **cilia** (Figure 3.4). The cilia beat backwards and forwards, sweeping the mucus and trapped particles out towards the mouth. In this way, dirt and bacteria are

▲ Figure 3.4 This electron microscope picture shows cilia from the lining of the trachea.

prevented from entering the lungs, where they might cause an infection. As you will see, one of the effects of smoking is that it destroys the cilia and stops this protection mechanism from working properly.

VENTILATION OF THE LUNGS

Ventilation means moving air in and out of the lungs. This requires a difference in air pressure – the air moves from a place where the pressure is high to one where it is low. Ventilation depends on the fact that the thorax is an airtight cavity. When we breathe, we change the volume of our thorax, which alters the pressure inside it. This causes air to move in or out of the lungs.

There are two movements that bring about ventilation: those of the ribs and the diaphragm. If you put your hands on your chest and breathe in deeply, you can feel your ribs move upwards and outwards. They are moved by the intercostal muscles (Figure 3.5). The outer (external) intercostals contract, pulling the ribs up. At the same time, the muscles of the diaphragm contract, pulling the diaphragm down into a more flattened shape (Figure 3.6a). Both these movements increase the volume of the chest and cause a slight drop in pressure inside the thorax compared with the air outside. Air then enters the lungs (inhalation).

The opposite happens when you breathe out deeply. The external intercostals relax, and the internal intercostals contract, pulling the ribs down and in. At the same time, the diaphragm muscles relax and the diaphragm goes back to its normal dome shape. The volume of the thorax decreases, and the pressure in the thorax is raised slightly above atmospheric pressure. This time the difference in pressure forces air out of the lungs (Figure 3.6b). Exhalation is helped by the fact that the lungs are elastic, so that they have a tendency to collapse and empty like a balloon.

EXTENSION WORK

During normal (shallow) breathing, the elasticity of the lungs and the weight of the ribs acting downwards is enough to cause exhalation. The internal intercostals are only really used for deep (forced) breathing out, for instance when we are exercising.

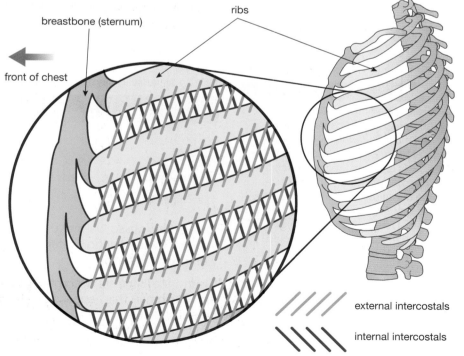

breastbone (sternum)

ribs

front of chest

external intercostals

internal intercostals

▲ Figure 3.5 Side view of the chest wall, showing the ribs. The diagram shows how the two sets of intercostal muscles run between the ribs. When the external intercostals contract, they move the ribs upwards. When the internal intercostals contract, the ribs are moved downwards.

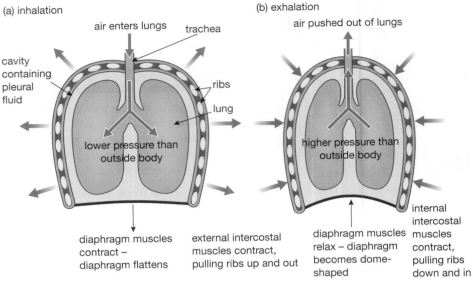

▲ Figure 3.6 Changes in the position of the ribs and diaphragm during breathing. (a) Breathing in (inhalation). (b) Breathing out (exhalation).

KEY POINT

It is important that you remember the changes in volume and pressure during ventilation. If you have trouble understanding these, think of what happens when you use a bicycle pump. If you push the pump handle, the air in the pump is squashed, its pressure rises and it is forced out of the pump. If you pull on the handle, the air pressure inside the pump falls a little, and air is drawn in from outside. This is similar to what happens in the lungs. In exams, students sometimes talk about the lungs *forcing* the air in and out – they don't!

GAS EXCHANGE IN THE ALVEOLI

You can tell what is happening during gas exchange if you compare the amounts of different gases in atmospheric air with the air breathed out (Table 3.1).

Table 3.1 Approximate percentage volume of gases in atmospheric (inhaled) and exhaled air.

Gas	Atmospheric air / %	Exhaled air / %
nitrogen	78	79
oxygen	21	16
carbon dioxide	0.04	4
other gases (mainly argon)	1	1

HINT

Be careful when interpreting percentages! The *percentage* of a gas in a mixture can vary, even if the actual *amount* of the gas stays the same. This is easiest to understand from an example. Imagine you have a bottle containing a mixture of 20% oxygen and 80% nitrogen. If you used a chemical to absorb all the oxygen in the bottle, the nitrogen left would now be 100% of the gas in the bottle, despite the fact that the *amount* of nitrogen would still be the same. That is why the percentage of nitrogen in inhaled and exhaled air is slightly different.

Exhaled air is also warmer than atmospheric air, and is saturated with water vapour. The amount of water vapour in the atmosphere varies depending on weather conditions.

Clearly, the lungs are absorbing oxygen into the blood and removing carbon dioxide from it. This happens in the alveoli. To do this efficiently, the alveoli must have a structure which brings the air and blood very close together, over a very large surface area. There are enormous numbers of alveoli. It has been calculated that the two lungs contain about 700 000 000 of these tiny air sacs, giving a total surface area of 60 m². That's bigger than the floor area of an average classroom! Viewed through a high-powered microscope, the alveoli look rather like bunches of grapes, and are covered with tiny blood capillaries (Figure 3.7).

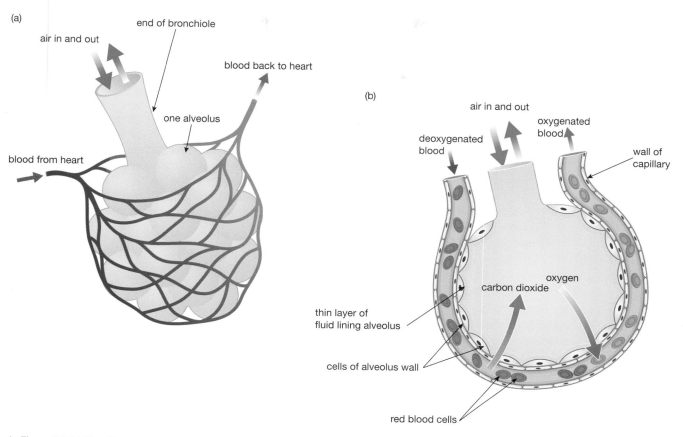

(a)

air in and out

end of bronchiole

blood back to heart

one alveolus

blood from heart

(b)

air in and out

deoxygenated blood

oxygenated blood

wall of capillary

oxygen

carbon dioxide

thin layer of fluid lining alveolus

cells of alveolus wall

red blood cells

▲ Figure 3.7 (a) Alveoli and the surrounding capillary network. (b) Diffusion of oxygen and carbon dioxide takes place between the air in the alveolus and the blood in the capillaries.

HINT

Be careful – students sometimes write 'The alveolus has *cell walls*'. This statement is not correct – a cell wall is part of a plant cell! The correct way to describe the structure is: 'The alveolus has *a wall made of cells*'.

Deoxygenated blood is pumped from the heart to the lungs and passes through the capillaries surrounding the alveoli. The blood has come from the respiring tissues of the body, where it has given up some of its oxygen to the cells, and gained carbon dioxide. Around the lungs, the blood is separated from the air inside each alveolus by only two cell layers; the cells making up the wall of the alveolus, and the capillary wall itself. This is a distance of less than a thousandth of a millimetre.

EXTENSION WORK

The thin layer of fluid lining the inside of the alveoli comes from the blood. The capillaries and cells of the alveolar wall are 'leaky' and the blood pressure pushes fluid out from the blood plasma into the alveolus. Oxygen dissolves in this moist surface before it passes through the alveolar wall into the blood.

Because the air in the alveolus has a higher concentration of oxygen than the blood entering the capillary network, oxygen diffuses from the air, across the wall of the alveolus and into the blood. At the same time there is more carbon dioxide in the blood than there is in the air in the lungs. This means that there is a diffusion gradient for carbon dioxide in the other direction, so carbon dioxide diffuses the other way, out of the blood and into the alveolus. The result is that the blood which leaves the capillaries and flows back to the heart has gained oxygen and lost carbon dioxide. The heart then pumps the oxygenated blood around the body again, to supply the respiring cells (see Chapter 5).

Safety Note: Wear eye protection and breathe gently; do not blow. A clean mouthpiece must be used for each person.

ACTIVITY 1

▼ PRACTICAL: COMPARING THE CARBON DIOXIDE CONTENT OF INHALED AND EXHALED AIR

The apparatus in Figure 3.8 can be used to compare the amount of carbon dioxide in inhaled and exhaled air. A person breathes gently in and out through the middle tube. Exhaled air passes out through one tube of indicator solution and inhaled air is drawn in through the other tube. If limewater is used, the limewater in the 'exhaled' tube will turn cloudy before the limewater in the 'inhaled' tube. (If hydrogen carbonate indicator solution is used instead, it changes from red to yellow.)

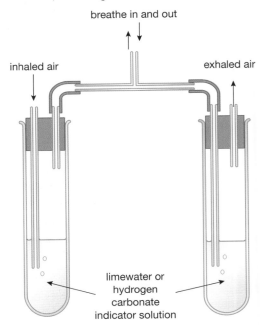

breathe in and out

inhaled air

exhaled air

limewater or hydrogen carbonate indicator solution

▲ Figure 3.8 Apparatus for Experiment 6.

ACTIVITY 2

▼ PRACTICAL: AN INVESTIGATION INTO THE EFFECT OF EXERCISE ON BREATHING RATE

It is easy to show the effect of exercise on a person's breathing rate. They sit quietly for five minutes, making sure that they are completely relaxed. They then count the number of breaths they take in one minute, recording their results in a table. They wait a minute, and then count their breaths again, recording the result, and repeating if necessary until they get a steady value for the 'resting rate'.

The person then carries out some vigorous exercise, such as running on the spot for three minutes. Immediately after they finish the exercise, they sit down and record the breathing rate as before. They then continue to record their breaths per minute, every minute, until they return to their normal resting rate.

Safety Note: Wear suitable footwear for exercising and if doing step-ups use a sturdy secure low box or a PE bench.

The table shows the results from an investigation into the breathing rate of two girls, A and B, before and after exercise.

Time from start of experiment (min)	Breathing rate / breaths per min	
	A	B
1	13	13
2	14	12
3	14	12
Rate after 3 min vigorous exercise:		
7	28	17
8	24	13
9	17	12
10	14	12

Plot a line graph of these results, using the same axes for both subjects. Join the data points using straight lines, and leave a gap during the period of exercise, when no readings were taken.

Why does breathing rate need to rise during exercise? Explain as fully as possible. Why does the rate not return to normal as soon as a subject finishes the exercise? (see Chapter 1).

Describe the difference in the breathing rates of the two girls (A and B) after exercise. Which girl is more fit? Explain your reasoning.

THE EFFECTS OF SMOKING

In order for the lungs to exchange gases properly, the air passages need to be clear, the alveoli need to be free from dirt particles and bacteria, and they must have as big a surface area as possible in contact with the blood. There is one habit that can upset all of these conditions – smoking.

Links between smoking and diseases of the lungs are now a proven fact. Smoking is associated with lung cancer, bronchitis and emphysema. It is also a major contributing factor to other conditions, such as coronary heart disease and ulcers of the stomach and intestine. Pregnant women who smoke are more likely to give birth to underweight babies.

Coronary heart disease will be described in Chapter 5 after you have studied the structure of the heart. Here we will look at a number of other medical conditions that are caused by smoking.

EFFECTS OF SMOKE ON THE LINING OF THE AIR PASSAGES

You saw above how the lungs are kept free of particles of dirt and bacteria by the action of mucus and cilia. In the trachea and bronchi of a smoker, the cilia are destroyed by the chemicals in cigarette smoke.

The reduced numbers of cilia mean that the mucus is not swept away from the lungs, but remains to block the air passages. This is made worse by the fact that the smoke irritates the lining of the airways, stimulating the cells to secrete more mucus. The sticky mucus blocking the airways is the source of 'smoker's cough'. Irritation of the bronchial tree, along with infections from bacteria in the mucus, can cause the lung disease **bronchitis**. Bronchitis blocks normal air flow, so the sufferer has difficulty breathing properly.

EMPHYSEMA

Emphysema is another lung disease that kills about 20 000 people in Britain every year. Smoking is the cause of one type of emphysema. Smoke damages the walls of the alveoli, which break down and fuse together again, forming enlarged, irregular air spaces (Figure 3.9).

This greatly reduces the surface area for gas exchange, which becomes very inefficient. The blood of a person with emphysema carries less oxygen. In serious cases, this leads to the sufferer being unable to carry out even mild exercise, such as walking. Emphysema patients often have to have a supply of oxygen nearby at all times (Figure 3.10). There is no cure for emphysema, and usually the sufferer dies after a long and distressing illness.

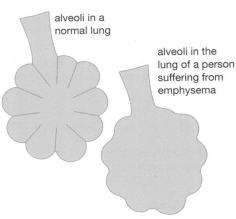

alveoli in a normal lung

alveoli in the lung of a person suffering from emphysema

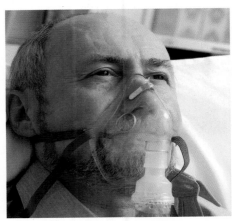

EXTENSION WORK

A person who has chronic (long-term) bronchitis and emphysema is said to be suffering from chronic obstructive pulmonary disease, or COPD. COPD is a progressive disease for which there is no cure.

▲ Figure 3.9 The alveoli of a person suffering from emphysema have a greatly reduced surface area and inefficient gas exchange.

▲ Figure 3.10 Patients with emphysema often need to breathe air enriched with oxygen in order to stay alive.

LUNG CANCER

Evidence of the link between smoking and lung cancer first appeared in the 1950s. In one study, a number of patients in hospital were given a series of questions about their lifestyles. They were asked about their work, hobbies, housing and so on, including a question about how many cigarettes they smoked. The same questionnaire was given to two groups of patients. The first group were all suffering from lung cancer. The second (**Control**) group were in hospital with various other illnesses, but not lung cancer. To make it a fair comparison, the Control patients were matched with the lung cancer patients for sex, age and so on.

When the results were compared, one difference stood out (Table 3.2). A greater proportion of the lung cancer patients were smokers than in the Control patients. There seemed to be a connection between smoking and getting lung cancer.

Table 3.2 Comparison of the smoking habits of lung cancer patients and other patients.

	Percentage of patients who were non-smokers	Percentage of patients who smoked more than 15 cigarettes a day
lung cancer patients	0.5	25
Control patients (with illnesses other than lung cancer)	4.5	13

Although the results didn't prove that smoking caused lung cancer, there was a statistically significant link between smoking and the disease: this is called a 'correlation'.

Over 20 similar investigations in nine countries have revealed the same findings. In 1962 a report called 'Smoking and health' was published by the Royal College of Physicians of London, which warned the public about the dangers of smoking. Not surprisingly, the first people to take the findings seriously were doctors, many of whom stopped smoking. This was reflected in their death rates from lung cancer. In ten years, while deaths among the general male population had risen by 7%, the deaths of male doctors from the disease had *fallen* by 38%.

Cigarette smoke contains a strongly addictive drug – **nicotine**. Smoke contains over 7000 chemicals, including; carbon monoxide, arsenic, ammonia, formaldehyde, cyanide, benzene, and toluene. More than 60 of the chemicals are known to cause cancer. These chemicals are called **carcinogens**, and are contained in the tar that collects in a smoker's lungs. Cancer happens when cells mutate and start to divide uncontrollably, forming a **tumour** (Figure 3.11). If a lung cancer patient is lucky, they may have the tumour removed by an operation before the cancer cells spread to other tissues of the body. Unfortunately tumours in the lungs usually cause no pain, so they are not discovered until it is too late – it may be inoperable, or tumours may have developed elsewhere.

If you smoke you are not *bound* to get lung cancer, but the risk that you will get it is much greater. In fact, the more cigarettes you smoke, the more the risk increases (Figure 3.12).

▲ Figure 3.11 This lung is from a patient with lung cancer.

DID YOU KNOW?
People often talk about 'yellow nicotine stains'. In fact it is the *tar* that stains a smoker's fingers and teeth. Nicotine is a colourless, odourless chemical.

DID YOU KNOW?
Studies have shown that the type of cigarette smoked makes very little difference to the smoker's risk of getting lung cancer. Filtered and 'low tar' cigarettes only reduce the risk slightly.

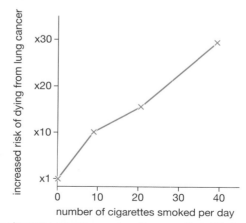

▲ Figure 3.12 The more cigarettes a person smokes, the more likely it is they will die of lung cancer. For example, smoking 20 cigarettes a day increases the risk by about 15 times.

The obvious thing to do is not to start smoking. However, if you are a smoker, giving up the habit soon improves your chance of survival (Figure 3.14). After a few years, the likelihood of your dying from a smoking-related disease is almost back to the level of a non-smoker.

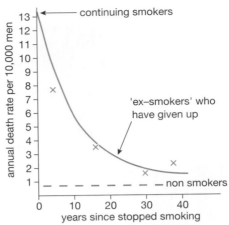

▲ Figure 3.13 Death rates from lung cancer for smokers, non-smokers and ex-smokers.

CARBON MONOXIDE IN SMOKE

One of the harmful chemicals in cigarette smoke is the poisonous gas **carbon monoxide**. When this gas is breathed in with the smoke, it enters the bloodstream and interferes with the ability of the blood to carry oxygen. Oxygen is carried around in the blood in the red blood cells, attached to a chemical called **haemoglobin** (see Chapter 5). Carbon monoxide can combine with the haemoglobin much more tightly than oxygen can, forming a compound called **carboxyhaemoglobin**. The haemoglobin will combine with carbon monoxide in preference to oxygen. When this happens, the blood carries much less oxygen around the body. Carbon monoxide from smoking is also a major cause of heart disease (Chapter 5).

If a pregnant woman smokes, she will be depriving her unborn **fetus** of oxygen (Figure 3.14). This has an effect on its growth and development, and leads to the mass of the baby at birth being lower, on average, than the mass of babies born to non-smokers.

SOME SMOKING STATISTICS

- It is estimated that there are over 1 billion smokers worldwide. In 2014 they consumed 5.8 *trillion* cigarettes.

- Every year nearly 6 million people are killed by tobacco-related illnesses. If the current trend continues, by 2030 this will rise to 8 million deaths per year and 80% of these premature deaths will be in developing countries.

- Smoking causes almost 80% of deaths from lung cancer, 80% of deaths from bronchitis and emphysema, and 14% of deaths from heart disease.

- More than a quarter of all cancer deaths are attributable to smoking. These include cancer of the lung, mouth, lip, throat, bladder, kidney, pancreas, stomach, liver and cervix.

- While demand for tobacco has steadily fallen in developed countries like the UK, cigarette consumption is being increasingly concentrated in the developing world.

- 9.6 million adults in the UK smoke cigarettes, 20% of men and 17% of women. However, 22% of women and 30% of men in the UK are now ex-smokers. Surveys show that about two-thirds of current smokers would like to stop smoking.

- It is estimated that worldwide, 31% of men and 8% of women are smokers. Consumption varies widely between different countries, but generally the areas of the world where there has been no change in consumption, or an increase, are southern and central Asia, Eastern Europe and Africa.

■ In China alone there are about 350 million smokers, who consume about one-third of all cigarettes smoked worldwide. Large multinational tobacco companies have long been keen to enter the Chinese market.

■ In China there are over a million deaths a year from smoking-related diseases. This figure is expected to double by 2025.

■ In developing countries, smoking has a greater economic impact. Poorer smokers spend significant amounts of their income on cigarettes rather than necessities like food, healthcare and education.

■ Tobacco farming uses up land that could be used for growing food crops. In 2012, 7.5 million tonnes of tobacco leaf were grown on almost 4.3 million hectares of land (an area larger than Switzerland).

Sources: Action on Smoking and Health (ASH) fact sheets (2015- 2016); ASH research reports (2014-2016)]

GIVING UP SMOKING

Most smokers admit that they would like to find a way to give up the habit. The trouble is that the nicotine in tobacco is a very addictive drug, and causes withdrawal symptoms when people stop smoking. These include cravings for a cigarette, restlessness and a tendency to put on weight (nicotine depresses the appetite).

There are various ways that smokers can be helped to give up their habit. One method is 'vaping', which involves inhaling a vapour containing nicotine from an electronic cigarette or e-cigarette (Figure 3.15). Other methods use nicotine patches (Figure 3.16) or nicotine chewing gum. They all work in a similar way, providing the smoker with a source of nicotine without the harmful tar from cigarettes. The nicotine is absorbed by the body and reduces the craving for a cigarette. Gradually, the patient reduces the nicotine dose until they are weaned off the habit.

EXTENSION WORK

You could carry out an Internet search to find out about the different methods people use to help them give up smoking. Which methods have the highest success rate? Is there any evidence that suggests e-cigarettes are not safe?

There are several other ways that people use to help them give up smoking, including the use of drugs that reduce withdrawal symptoms, acupuncture and even hypnotism.

CHAPTER QUESTIONS

More questions on breathing can be found at the end of Unit 2 on page 130.

SKILLS   CRITICAL THINKING

1 The structures below are found in the human bronchial tree

1. alveoli   3. bronchioles
2. trachea   4. bronchi

Which of the following shows the route taken by air after it is breathed in through the mouth?

A 2 → 3 → 4 → 1   **C** 2 → 4 → 3 → 1
B 1 → 4 → 3 → 2   **D** 4 → 1 → 2 → 3

2 Which of the following is *not* a feature of an efficient gas exchange surface?

A thick walls   **C** close proximity to blood capillaries
B moist lining   **D** large surface area

3 Which row in the table shows the correct percentage of oxygen in atmospheric and exhaled air?

	Atmospheric air / %	Exhaled air / %
A	78	21
B	21	16
C	16	4
D	4	0.04

4 Chemicals in cigarette smoke lead to the breakdown of the walls of the alveoli. What is the name given to this disease?

A bronchitis

B emphysema

C coronary heart disease

D lung cancer

5 Copy and complete the table, which shows what happens in the thorax during ventilation of the lungs. Two boxes have been completed for you.

	Action during inhalation	Action during exhalation
external intercostal muscles	contract	
internal intercostal muscles		
ribs		move down and in
diaphragm		
volume of thorax		
pressure in thorax		
volume of air in lungs		

6 A student wrote the following about the lungs.

When we breathe in, our lungs inflate, sucking air in and pushing the ribs up and out, and forcing the diaphragm down. This is called respiration. In the air sacs of the lungs, the air enters the blood. The blood then takes the air around the body, where it is used by the cells. The blood returns to the lungs to be cleaned. When we breathe out, our lungs deflate, pulling the diaphragm up and the ribs down. The stale air is pushed out of the lungs.

The student did not have a good understanding of the workings of the lungs. Re-write their description, using correct biological words and ideas.

7 Sometimes, people injured in an accident such as a car crash suffer from a *pneumothorax*. This is an injury where the chest wall is punctured, allowing air to enter the pleural cavity (see Figure 3.1). A patient was brought to the casualty department of a hospital, suffering from a pneumothorax on the left side of his chest. His left lung had collapsed, but he was able to breathe normally with his right lung.

a Explain why a pneumothorax caused the left lung to collapse.

b Explain why the right lung was not affected.

c If a patient's lung is injured or infected, a surgeon can sometimes 'rest' it by performing an operation called an *artificial pneumothorax*. What do you think might be involved in this operation?

8 Briefly explain the importance of the following.

   a The trachea wall contains C-shaped rings of cartilage.

   b The distance between the air in an alveolus and the blood in an alveolar capillary is less than 1/1000th of a millimetre.

   c The lining of the trachea contains mucus-secreting cells and cells with cilia.

   d Smokers have a lower concentration of oxygen in their blood than non-smokers.

   e Nicotine patches and nicotine chewing gum can help someone give up smoking.

   f The lungs have a surface area of about 60 m² and a good blood supply.

9 Explain the differences between the lung diseases bronchitis and emphysema.

10 A long-term investigation was carried out into the link between smoking and lung cancer. The smoking habits of male doctors aged 35 or over were determined while they were still alive, then the number and causes of deaths among them were monitored over a number of years. (Note that this survey was carried out in the 1950s – very few doctors smoke these days!) The results are shown in the graph.

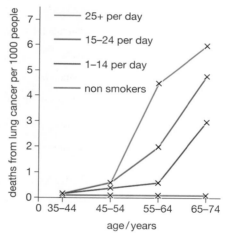

   a Write a paragraph to explain what the researchers found out from the investigation.

   b How many deaths from lung cancer would be expected for men aged 55 who smoked 25 cigarettes a day up until their death? How many deaths from lung cancer would be expected for men in the same age group smoking 10 a day?

   c Table 3.2 (page 47) shows the findings of another study linking lung cancer with smoking. Which do you think is the more convincing evidence of the link, this investigation or the findings illustrated in Table 3.2?

11 Design and make a hard-hitting leaflet explaining the link between smoking and lung cancer. It should be aimed at encouraging an adult smoker to give up the habit. You could use suitable computer software to produce your design. Include some smoking statistics, perhaps from an Internet search. However don't use too many, or they may put the person off reading the leaflet!

4 FOOD AND DIGESTION

Food is essential for life. The nutrients obtained from it are used in many different ways by the body. This chapter looks at the different kinds of food, and how the food is broken down by the digestive system and absorbed into the blood, so that it can be carried to all the tissues of the body.

LEARNING OBJECTIVES

- Identify the chemical elements present in carbohydrates, proteins and lipids (fats and oils)

- Describe the structure of carbohydrates, proteins and lipids as large molecules made up from smaller basic units – starch and glycogen from simple sugars, protein from amino acids, and lipids from fatty acids and glycerol

- Investigate food samples for the presence of glucose, starch, protein and fat

- Understand that a balanced diet should include appropriate proportions of carbohydrate, protein, lipid, vitamins, minerals, water and dietary fibre

- Identify the sources and describe the functions of carbohydrate, lipid, protein, vitamins A, C and D, the mineral ions calcium and iron, water, and dietary fibre as components of the diet

- Understand how energy requirements vary with activity levels, age and pregnancy

- Describe the structure and function of the human alimentary canal, including the mouth, oesophagus, stomach, small intestine (duodenum and ileum), large intestine (colon and rectum) and pancreas

- Understand how food is moved through the gut by peristalsis

- Understand the role of digestive enzymes, including the digestion of starch to glucose by amylase and maltase, the digestion of proteins to amino acids by proteases and the digestion of lipids to fatty acids and glycerol by lipases

- Understand that bile is produced by the liver and stored in the gall bladder, and understand the role of bile in neutralising stomach acid and emulsifying lipids

- Understand how the small intestine is adapted for absorption, including the structure of a villus

BIOLOGY ONLY

- Investigate the energy content in a food sample

We need food for three main reasons:

- to supply us with a 'fuel' for energy

- to provide materials for growth and repair of tissues

- to help fight disease and keep our bodies healthy.

A BALANCED DIET

▲ Figure 4.1 A balanced diet contains all the types of food the body needs, in just the right amounts.

The food that we eat is called our diet. No matter what you like to eat, your diet must include the following five groups of food substances if your body is to work properly and stay healthy – **carbohydrates**, **lipids**, **proteins**, **minerals** and **vitamins** – along with **dietary fibre** and water. Food should provide you with all of these substances, but they must also be present in the *right* amounts. A diet that provides enough of these substances and in the correct proportions to keep you healthy is called a **balanced diet** (Figure 4.1). We will look at each type of food in turn, to find out about its chemistry and the role that it plays in the body.

CARBOHYDRATES

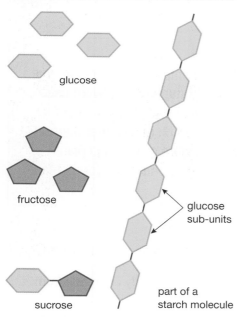

▲ Figure 4.2 Glucose and fructose are 'single sugar' molecules. A molecule of glucose joined to a molecule of fructose forms the 'double sugar' called sucrose. Starch is a polymer of many glucose sub-units.

Carbohydrates only make up about 1% of the mass of the human body, but they have a very important role. They are the body's main 'fuel' for supplying cells with energy. Cells release this energy by oxidising a sugar called **glucose**, in the process called cell respiration (see Chapter 1). Glucose and other sugars belong to one group of carbohydrates.

Glucose is found naturally in many sweet-tasting foods, such as fruits and vegetables. Other foods contain different sugars, such as the fruit sugar called **fructose**, and the milk sugar, **lactose**. Ordinary table sugar, the sort some people put in their tea or coffee, is called **sucrose**. Sucrose is the main sugar that is transported through plant stems. This is why we can extract it from sugar cane, which is the stem of a large grass-like plant. Sugars have two physical properties that you will probably know: they all taste sweet, and they are all soluble in water.

We can get all the sugar we need from natural foods such as fruits and vegetables, and from the **digestion** of starch. Many processed foods contain large amounts of *added* sugar. For example, a typical can of cola can contain up to seven teaspoons (27 g) of sugar! There is hidden sugar in many other foods. A tin of baked beans contains about 10 g of added sugar. This is on top of all the food that we eat with a more obvious sugar content, such as cakes, biscuits and sweets.

In fact, we get most of the carbohydrate in our diet not from sugars, but from **starch**. Starch is a large, *insoluble* molecule. Because it does not dissolve, it is found as a storage carbohydrate in many plants, such as potato, rice, wheat and millet. The 'staple diets' of people from around the world are starchy foods like rice, potatoes, bread and pasta. Starch is a polymer of glucose – it is made of long chains of hundreds of glucose molecules joined together (Figure 4.2).

Starch is only found in plant tissues, but animal cells sometimes contain a very similar carbohydrate called **glycogen**. This is also a polymer of glucose, and is found in tissues such as liver and muscle, where it acts as a store of energy for these organs.

As you will see, large carbohydrates such as starch and glycogen have to be broken down into simple sugars during digestion, so that they can be absorbed into the blood.

Another carbohydrate that is a polymer of glucose is **cellulose**, the material that makes up plant cell walls. Humans are *not* able to digest cellulose, because our gut doesn't make the enzyme needed to break down the cellulose molecule. This means that we are not able to use cellulose as a source of energy. However, it still has a vitally important function in our diet. It forms **dietary fibre** or 'roughage', which gives the muscles of the gut something to push against as the food is moved through the intestine. This keeps the gut contents moving, avoiding constipation and helping to prevent serious diseases of the intestine, such as colitis and bowel cancer.

EXTENSION WORK

'Single' sugars such as glucose and fructose are called **monosaccharides**. Sucrose molecules are made of two monosaccharides (glucose and fructose) joined together, so sucrose is called a **disaccharide**. Lactose is also a disaccharide, made of glucose joined to another monosaccharide called galactose. Polymers of sugars, such as starch, glycogen and cellulose, are called polysaccharides.

LIPIDS (FATS AND OILS)

Lipids contain the same three elements as carbohydrates – carbon, hydrogen and oxygen – but the proportion of oxygen in a lipid is much lower than in a carbohydrate. For example, beef and lamb both contain a fat called tristearin, which has the formula $C_{51}H_{98}O_6$. This fat, like other animal fats, is a solid at room

▲ Figure 4.3 These foods are all rich in lipids.

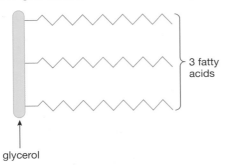

3 fatty acids

glycerol

▲ Figure 4.4 Lipids are made up of a molecule of glycerol joined to three fatty acids. The many different fatty acids form the variable part of the molecule.

temperature, but melts if you warm it up. On the other hand, plant lipids are usually liquid at room temperature, and are called oils. Meat, butter, cheese, milk, eggs and oily fish are all rich in animal fats, as well as foods fried in animal fat. Vegetable oils include many types used for cooking, such as olive oil, corn oil and rapeseed oil, as well as products made from oils, such as margarine (Figure 4.3).

Lipids make up about 10% of our body's mass. They form an essential part of the structure of all cells, and fat is deposited in certain parts of the body as a long-term store of energy, for example under the skin and around the heart and kidneys. The fat layer under the skin acts as insulation, reducing heat loss through the surface of the body. Fat around organs such as the kidneys also helps to protect them from mechanical damage.

The chemical 'building blocks' of lipids are two types of molecule called **glycerol** and **fatty acids**. Glycerol is an oily liquid. It is also known as glycerine, and is used in many types of cosmetics. In lipids, a molecule of glycerol is joined to three fatty acid molecules. There are many different fatty acid molecules, which give us the many different kinds of lipid found in food (Figure 4.4).

Although lipids are an essential part of our diet, too much lipid is unhealthy, especially a type called *saturated* fat, and a lipid compound called **cholesterol**. These substances have been linked to heart disease (see Chapter 5).

DID YOU KNOW?

Saturated lipids (saturated fats) are more common in food from animal sources, such as meat and dairy products. 'Saturated' is a word used in chemistry, which means that the fatty acids of the lipids contain no double bonds. Other lipids are *unsaturated*, which means that their fatty acids contain double bonds. These are more common in plant oils. There is evidence that unsaturated lipids are healthier for us than saturated ones.

PROTEINS

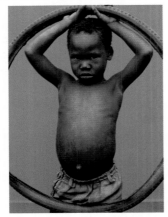

▲ Figure 4.5 This child is suffering from a lack of protein in his diet, a disease called kwashiorkor. His swollen belly is not due to a full stomach, but is caused by fluid collecting in the tissues. Other symptoms include loss of weight, poor muscle growth, general weakness and flaky skin.

Proteins make up about 18% of the mass of the body. This is the second largest percentage after water. All cells contain protein, so we need it for growth and repair of tissues. Many compounds in the body are made from protein, including enzymes.

Most foods contain some protein, but certain foods such as meat, fish, cheese and eggs are particularly rich in it. You will notice that these foods are animal products. Plant material generally contains less protein, but some foods, especially beans, peas and nuts, are richer in protein than others.

However, we don't need much protein in our diet to stay healthy. Doctors recommend a maximum daily intake of about 70g. In more economically developed countries, people often eat far more protein than they need, whereas in many poorer countries a protein-deficiency disease called **kwashiorkor** is common (Figure 4.5).

Like starch, proteins are also polymers, but whereas starch is made from a single molecular building block (glucose), proteins are made from 20 different sub-units called **amino acids**. All amino acids contain four chemical elements: carbon, hydrogen and oxygen (as in carbohydrates and fats) along with nitrogen. Two amino acids also contain sulfur. The amino acids are linked together in long chains, which are usually folded up or twisted into spirals, with cross-links holding the chains together (Figure 4.6).

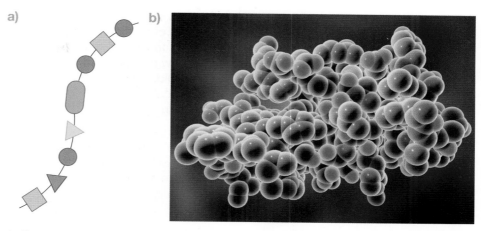

a)   b)

▲ Figure 4.6 (a) A chain of amino acids forming part of a protein molecule. Each shape represents a different amino acid. (b) A computer model of the protein insulin. This substance, like all proteins, is made of a long chain of amino acids arranged in a particular order and folded into a specific shape.

The *shape* of a protein is very important in allowing it to carry out its function, and the *order* of amino acids in the protein decides its shape. Because there are 20 different amino acids, and they can be arranged in any order, the number of different protein structures that can be made is enormous. As a result, there are thousands of different kinds of proteins in organisms, from structural proteins such as collagen and keratin in skin and nails, to proteins with more specific functions, such as enzymes and haemoglobin.

MINERALS

All the foods you have read about so far are made from just five chemical elements: carbon, hydrogen, oxygen, nitrogen and sulfur. Our bodies contain many other elements that we get from our food as 'minerals' or 'mineral ions'. Some are present in large amounts in the body, for example calcium, which is used for making teeth and bones. Others are present in much smaller amounts, but still have essential jobs to do. For instance our bodies contain about 3 g of iron, but without it our blood would not be able to carry oxygen. Table 4.1 shows just a few of these minerals and the reasons they are needed.

Table 4.1 Some examples of minerals needed by the body.

Mineral	Approximate mass in an adult body / g	Location or role in body	Examples of foods rich in minerals
calcium	1000	making teeth and bones	dairy products, fish, bread, vegetables
phosphorus	650	making teeth and bones; part of many chemicals, e.g. DNA and ATP	most foods
sodium	100	in body fluids, e.g. blood	common salt, most foods
chlorine	100	in body fluids, e.g. blood	common salt, most foods
magnesium	30	making bones; found inside cells	green vegetables
iron	3	part of haemoglobin in red blood cells, helps carry oxygen	red meat, liver, eggs, some vegetables, e.g. spinach

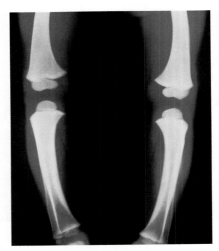

▲ Figure 4.7 An x-ray of the legs of a child showing the symptoms of rickets.

If a person doesn't get enough of a mineral from their diet, they will show the symptoms of a 'mineral deficiency disease'. For example, a one-year-old child needs to consume about 0.6 g (600 mg) of calcium every day, to make the bones grow properly and harden. Anything less than this over a prolonged period could result in poor bone development. The bones become deformed, a disease called rickets (Figure 4.7). Rickets can also be caused by lack of vitamin D in the diet (see below).

Similarly, 16-year-olds need about 12 mg of iron in their daily food intake. If they don't get this amount, they can't make enough haemoglobin for their red blood cells (see Chapter 5). This causes a condition called anaemia. People who are anaemic become tired and lack energy, because their blood doesn't carry enough oxygen.

VITAMINS

During the early part of the twentieth century, experiments were carried out that identified another class of food substances. When young laboratory rats were fed a diet of pure carbohydrate, lipid and protein, they all became ill and died. If they were fed on the same pure foods with a little added milk, they grew normally. The milk contained chemicals that the rats needed in small amounts to stay healthy. These chemicals are called vitamins. The results of one of these experiments are shown in Figure 4.8.

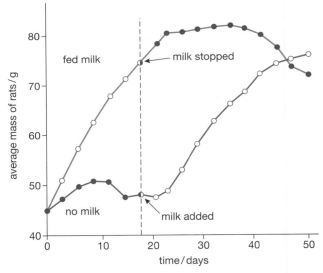

▲ Figure 4.8 Rats were fed a diet of pure carbohydrate, lipid and protein, with and without added milk. Vitamins in the milk had a dramatic effect on their growth.

At first, the chemical nature of vitamins was not known, and they were given letters to distinguish between them, such as vitamin A, vitamin B and so on. Each was identified by the effect a lack of the vitamin (vitamin deficiency) had on the body. For example, vitamin D is needed for growing bones to take up calcium salts. A deficiency of this vitamin can result in rickets (Figure 4.7), just as a lack of calcium can.

We now know the chemical structure of the vitamins and the exact ways in which they work in the body. As with vitamin D, each has a particular function. Vitamin A is needed to make a light-sensitive chemical in the retina of the eye (see Chapter 6). A lack of this vitamin causes night blindness, where the

The cure for scurvy was discovered as long ago as 1753. Sailors on long voyages often got scurvy because they ate very little fresh fruit and vegetables (the main source of vitamin C). A ship's doctor called James Lind wrote an account of how the disease could quickly be cured by eating fresh oranges and lemons. The famous explorer Captain Cook, on his world voyages in 1772 and 1775, kept his sailors healthy by making sure that they ate fresh fruit. By 1804, all British sailors were made to drink lime juice to prevent scurvy. This is how they came to be called 'limeys', a word that was later used by Americans for all British people.

person finds it difficult to see in dim light. Vitamin C is needed to make fibres of a material called connective tissue. This acts as a 'glue', bonding cells together in a tissue. It is found in the walls of blood vessels and in the skin and lining surfaces of the body. Vitamin C deficiency leads to a disease called **scurvy**, where wounds fail to heal, and bleeding occurs in various places in the body. This is especially noticeable in the gums (Figure 4.9).

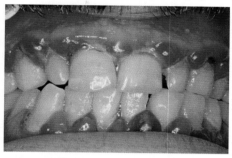

▲ Figure 4.9 Vitamin C helps lining cells such as those in the mouth and gums stick to each other. Lack of vitamin C causes scurvy, where the mouth and gums become damaged and bleed.

Vitamin B is not a single substance, but a collection of many different substances called the vitamin B group. It includes vitamins B1 (thiamine), B2 (riboflavin) and B3 (niacin). These compounds are involved in the process of cell respiration. Different deficiency diseases result if any of them are missing from the diet. For example, lack of vitamin B1 results in the weakening of the muscles and paralysis, a disease called **beri-beri**.

The main vitamins, their role in the body and some foods which are good sources of each, are summarised in Table 4.2.

Notice that the amounts of vitamins that we need are very small, but we cannot stay healthy without them.

Table 4.2 Summary of the main vitamins. Note that you only need remember the sources and functions of vitamins A, C and D.

Vitamin	Recommended daily amount in diet[1]	Use in the body	Effect of deficiency	Some foods that are a good source of the vitamin
A	0.8 mg	making a chemical in the retina; also protects the surface of the eye	night blindness, damaged cornea of eye	fish liver oils, liver, butter, margarine, carrots
B1	1.1 mg	helps with cell respiration	beri-beri	yeast extract, cereals
B2	1.4 mg	helps with cell respiration	poor growth, dry skin	green vegetables, eggs, fish
B3	16 mg	helps with cell respiration	pellagra (dry red skin, poor growth, and digestive disorders)	liver, meat, fish.
C	80 mg	sticks together cells lining surfaces such as the mouth	scurvy	fresh fruit and vegetables
D	5 μg	helps bones absorb calcium and phosphate	rickets, poor teeth	fish liver oils; also made in skin in sunlight

[1]Figures are the European Union's recommended daily intake for an adult (2012). 'mg' stands for milligram (a thousandth of a gram) and 'μg' for microgram (a millionth of a gram).

FOOD TESTS

It is possible to carry out simple chemical tests to find out if a food contains starch, glucose, protein or lipid. Practical 8 uses pure substances for the tests, but it is possible to do them on normal foods too. Unless the food is a liquid like milk, it needs to be cut up into small pieces and ground with a pestle and mortar, then shaken with some water in a test tube. This is done to extract the components of the food and dissolve any soluble substances such as sugars.

▲ Figure 4.10a Testing for starch using iodine

▲ Figure 4.10b Glucose with Benedict's solution, before and after heating

ACTIVITY 3

▼ PRACTICAL: TEST FOR STARCH

A little starch is placed on a spotting tile. A drop of yellow-brown iodine solution is added to the starch. The iodine reacts with the starch, forming a very dark blue, or 'blue-black' colour (Figure 4.10 (a)). Starch is insoluble, but this test will work on a solid sample of food, such as potato, or a suspension of starch in water.

▼ PRACTICAL: TEST FOR GLUCOSE

Glucose is called a reducing sugar. This is because the test for glucose involves reducing an alkaline solution of copper (II) sulfate to copper (I) oxide.

A small spatula measure of glucose is placed in a test tube and a little water added (about 2 cm deep). The tube is shaken to dissolve the glucose. Several drops of Benedict's solution are added to the tube, enough to colour the mixture blue (Figure 4.10 (b)).

A water bath is prepared by half-filling a beaker with water and heating it on a tripod and gauze. The test tube is placed in the beaker and the water allowed to boil (using a water bath is safer than heating the tube directly in the Bunsen burner). After a few seconds the clear blue solution gradually changes colour, forming a cloudy orange or 'brick red' precipitate of copper (I) oxide (Figure 4.10 (b)).

All other 'single' sugars (monosaccharides), such as fructose, are reducing sugars, as well as some 'double' sugars (disaccharides), such as the milk sugar, lactose. However, ordinary table sugar (sucrose) is not. If sucrose is boiled with Benedict's solution it will stay a clear blue colour.

▼ PRACTICAL: TEST FOR PROTEIN

The test for protein is sometimes called the 'biuret' test, after the coloured compound that is formed.

A little protein, such as powdered egg white (albumen), is placed in a test tube and about 2 cm depth of water added. The tube is shaken to mix the powder with the water. An equal volume of dilute (5%) potassium hydroxide solution is added and the tube shaken again. Finally two drops of 1% copper sulfate solution are added. A purple colour develops. (Sometimes these two solutions are supplied already mixed together as 'biuret solution'.)

▼ PRACTICAL: TEST FOR LIPID

Fats and oils are insoluble in water, but will dissolve in ethanol (alcohol). The test for lipid uses this fact.

A pipette is used to place one drop of olive oil in the bottom of a test tube. About 2 cm depth of ethanol is added, and the tube is shaken to dissolve the oil. The solution is poured into a test tube that is about three-quarters full with cold water. A white cloudy layer forms on the top of the water. The white layer is caused by the ethanol dissolving in the water and leaving the lipid behind as a suspension of tiny droplets, called an emulsion.

Baked Beans in tomato sauce

Nutritional information

Typical values	Amount per 100 g	Amount per serving (207 g)
Energy	313 kJ / 76 kcal	646 kJ / 155 kcal
Protein	4.7 g	9.7 g
Carbohydrate (of which sugars)	13.6 g (6.0 g)	28.2 g (12.4 g)
Fat (of which saturates)	0.2 g (Trace)	0.4 g (0.1 g)
Fibre	3.7 g	7.7 g
Sodium	0.5 g	1.0 g

HIGH IN FIBRE

210g e

▲ Figure 4.11 Food packaging is labelled with the proportions of different food types that it contains, along with its energy content. The energy in units called kilocalories (kcal) is also shown, but scientists no longer use this old-fashioned unit.

ENERGY FROM FOOD

Some foods contain more energy than others. It depends on the proportions of carbohydrate, lipid and protein that they contain. Their energy content is measured in kilojoules (kJ). If a gram of carbohydrate is fully oxidised, it produces about 17 kJ, whereas a gram of lipid yields over twice as much as this (39 kJ). Protein can produce about 18 kJ per gram. If you look on a food label, it usually shows the energy content of the food, along with the amounts of different nutrients that it contains (Figure 4.11).

Foods with a high percentage of lipid, such as butter or nuts, contain a large amount of energy. Others, like fruits and vegetables, which are mainly composed of water, have a much lower energy content (Table 4.3).

Table 4.3 Energy content of some common foods

Food	kJ per 100g	Food	kJ per 100g
margarine	3200	white bread	1060
butter	3120	chips	990
peanuts	2400	grilled beef steak	930
samosa	2400	fried cod	850
chocolate	2300	roast chicken	770
Cheddar cheese	1700	boiled potatoes	340
table sugar	1650	milk	270
cornflakes	1530	baked beans	270
rice	1500	yoghurt	200
spaghetti	1450	boiled cabbage	60
fried beefburger	1100	lettuce	40

Food scientists measure the amount of energy in a sample of food by burning it in a calorimeter (Figure 4.12). The calorimeter is filled with oxygen, to make sure that the food will burn easily. A heating filament carrying an electrical current ignites the food. The energy given out by the burning food is measured by using it to heat up water flowing through a coil in the calorimeter.

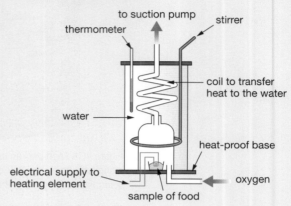

▶ Figure 4.12 A food calorimeter

If you have samples of food that will easily burn in air, you can measure the energy in them by a similar method, using the heat from the burning food to warm up water in a test tube (see Activity 9).

Even while you are asleep you need a supply of energy – in order to keep warm, for your heart to keep beating, to allow messages to be sent through your nerves, and for other body functions. However, the energy you need at other times depends on the physical work you do. The total amount of energy that a person needs to keep healthy depends on their age and body size, and also on the amount of activity they do. Table 4.4 shows some examples of how much energy is needed each day by people of different age, sex and occupation.

Remember that these are approximate figures, and they are averages. Generally, the greater a person's weight, the more energy that person needs. This is why men, with a greater average body mass, need more energy than women. The energy needs of a pregnant woman are increased, mainly because of the extra weight that she has to carry. A heavy manual worker, such as a labourer, needs extra energy for increased muscle activity.

It is not only the recommended energy requirements that vary with age, sex and pregnancy, but also the *content* of the diet. For instance, during pregnancy a woman may need extra iron or calcium in her diet, for the growth of the fetus. In younger women, the blood loss during menstruation (periods) can result in anaemia, producing a need for extra iron in the diet.

Table 4.4 The daily energy needs of different types of people.

Age/sex/occupation of person	Energy needed per day / kJ
newborn baby	2000
child aged 2	5000
child aged 6	7500
girl aged 12–14	9000
boy aged 12–14	11000
girl aged 15–17	9000
boy aged 15–17	12000
female office worker	9500
male office worker	10500
heavy manual worker	15000
pregnant woman	10000
breast-feeding woman	11300

BIOLOGY ONLY

ACTIVITY 4

▼ PRACTICAL: MEASURING THE ENERGY CONTENT OF A FOOD

If a sample of food will burn well in air, its energy content can be measured using a simplified version of the food calorimeter (Figure 4.13). Suitable foods are dry pasta, crispbread or biscuits. It is not advisable to use nuts, since some people are allergic to them.

Safety Note: Wear eye protection and do not taste the food. Do not burn nuts. Do not hold the food samples in your fingers when 'spearing' with the needle; push down on to a board.

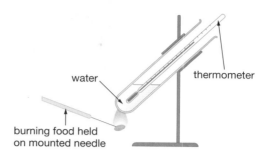

▲ Figure 4.13 Measuring the energy content of a food sample

First of all the mass of the food sample is found, by weighing it on a balance. A measured volume of water (20 cm³) is placed in a boiling tube, and the tube supported in a clamp on a stand as shown in Figure 4.13. The temperature of the water is recorded.

The food is speared on the end of a mounted needle, and then held in a Bunsen burner flame until it catches fire (this may take 30 seconds or so). When the food is alight, the mounted needle is used to hold the burning food underneath the boiling tube of water so that the flame heats up the water. This is continued, relighting the food if it goes out, until the food will no longer burn.

The final temperature of the water is measured, using the thermometer to stir the water gently, to make sure that the heat is evenly distributed.

Two facts are needed to calculate the energy content of the food.

- The energy needed to raise the temperature of one gram of water by 1 °C is 4.2 joules.
- A volume of 1 cm³ of water has a mass of 1 g.

Multiplying the rise in temperature of the water by the mass of the water and then by 4.2 gives the number of joules of energy that were transferred to the water. Dividing this by the mass of the food gives the energy per gram:

Energy in joules per gram (J per g)

$$= \frac{(\text{final temperature} - \text{temperature at start}) \times 20 \,(g) \times 4.2 \,(\text{J per g per °C})}{\text{mass of food (g)}}$$

For example, imagine you had a piece of pasta weighing 0.55 g. The starting temperature of the 20 g of water was 21 °C. After using the burning pasta to heat up the water, the temperature of the water was 43 °C.

$$\text{Energy content of the pasta} = \frac{(43 - 21) \times 20 \times 4.2}{0.55}$$

$$= 3360 \text{ J per g } (3400 \text{ J per g to 2 significant figures})$$

Comparison of the energy content of different foods

You may be able to use a similar method to find the energy content of suitable foods that will burn easily. Suggest a hypothesis that you could test about the energy content of the foods, and design an experiment to test your hypothesis. Explain how you will make sure that your results are reliable.

END OF BIOLOGY ONLY

DIGESTION

Food, such as a piece of bread, contains carbohydrates, lipids and proteins, but they are not the same carbohydrates, lipids and proteins as in our tissues. The components of the bread must first be broken down into their 'building blocks' before they can be absorbed through the wall of the gut. This process is called **digestion**. The digested molecules – sugars, fatty acids, glycerol and amino acids – along with minerals, vitamins and water, can then be carried around the body in the blood. When they reach the tissues they are reassembled into the molecules that make up our cells.

Digestion is speeded up by **enzymes**, which are biological catalysts (see Chapter 1). Although most enzymes stay inside cells, the digestive enzymes are made by the tissues and glands in the gut and pass out of cells – on to the gut contents where they act on the food. This *chemical* digestion is helped by *mechanical* digestion. Mechanical digestion is the physical breakdown of food. The most obvious place where this happens is in the mouth, where the teeth bite and chew the food, cutting it into smaller pieces that have a larger surface area. This means that enzymes can act on the food more quickly. Other parts of the gut also help with mechanical digestion. For example, muscles in the wall of the stomach contract to churn up the food while it is being chemically digested.

PERISTALSIS

Muscles are also responsible for moving the food along the gut. The walls of the intestine contain two layers of muscles. One layer has fibres arranged in rings around the gut. This is the circular muscle layer. The other has fibres running along the length of the gut, and is called the longitudinal muscle layer. Together these two layers act to push the food along. When the circular muscles contract and the longitudinal muscles relax, the gut is made narrower. When the opposite happens, i.e. the longitudinal muscles contract and the circular muscles relax, the gut becomes wider. Waves of muscle contraction like this pass along the gut, pushing the food along, rather like squeezing toothpaste from a tube. This is called **peristalsis** (Figure 4.14). It means that movement of food in the gut doesn't depend on gravity – we can still eat standing on our heads!

> **HINT**
>
> A good definition of digestion is: 'Digestion is the chemical and mechanical breakdown of food. It converts large insoluble molecules into small soluble molecules, which can be absorbed into the blood.'

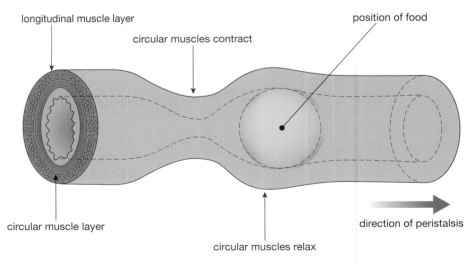

longitudinal muscle layer

circular muscles contract

position of food

circular muscle layer

direction of peristalsis

circular muscles relax

▲ Figure 4.14 Peristalsis: contraction of circular muscles behind the food narrows the gut, pushing the food along. When the circular muscles are contracted, the longitudinal ones are relaxed, and vice versa.

THE DIGESTIVE SYSTEM

Figure 4.15 shows a simplified diagram of the human digestive system. It is simplified so that you can see the order of the organs along the gut. The real gut is much longer than this, and coiled up so that it fills the whole space of the abdomen. Overall, its length in an adult is about 8 m. This gives plenty of time for the food to be broken down and absorbed as it passes through the gut.

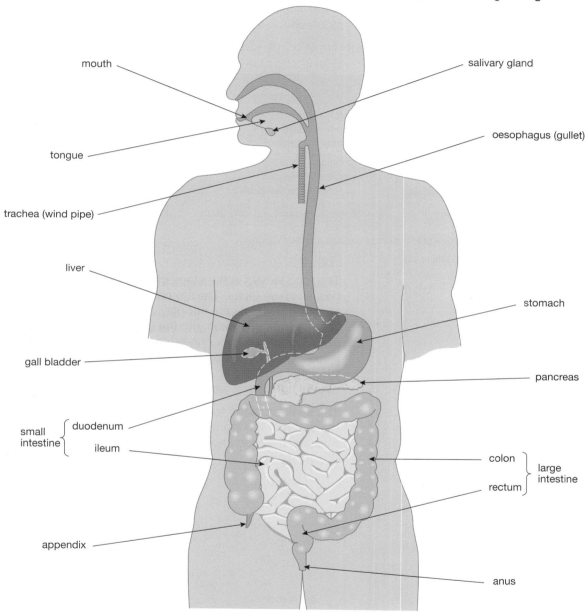

▲ Figure 4.15 The human digestive system

The mouth, stomach and the first part of the small intestine (called the **duodenum**) all break down the food using enzymes, either made in the gut wall itself, or by glands such as the **pancreas**. Digestion continues in the last part of the small intestine (the **ileum**) and it is here that the digested food is absorbed. The last part of the gut, the large intestine, is mainly concerned with absorbing water out of the remains, and storing the waste products (**faeces**) before they are removed from the body.

The three main classes of food are broken down by three classes of enzymes. Carbohydrates are digested by enzymes called **carbohydrases**. Proteins are acted upon by **proteases**, and enzymes called **lipases** break down lipids. Some of the places in the gut where these enzymes are made are shown in Table 4.5.

Digestion begins in the mouth. **Saliva** helps moisten the food and contains the enzyme **amylase**, which starts the breakdown of starch. The chewed lump of food, mixed with saliva, then passes along the **oesophagus** (gullet) to the stomach.

KEY POINT

Amylase digests starch into maltose. Amylase is the enzyme, starch is the substrate and maltose is the product.

Table 4.5 Some of the enzymes that digest food in the human gut. The substances shown in bold are the end products of digestion that can be absorbed from the gut into the blood.

Class of enzyme	Examples	Digestive action	Source of enzyme	Where it acts in the gut
carbohydrases	amylase amylase maltase	starch → maltose[1] starch → maltose maltose → **glucose**	salivary glands pancreas wall of small intestine	mouth small intestine small intestine
proteases	pepsin trypsin peptidases	proteins → peptides[2] proteins → peptides peptides → **amino acids**	stomach wall pancreas wall of small intestine	stomach small intestine small intestine
lipases	lipase	lipids → **glycerol** and **fatty acids**	pancreas	small intestine

[1]Maltose is a disaccharide made of two glucose molecules joined together.

[2]Peptides are short chains of amino acids.

The food is held in the stomach for several hours, while initial digestion of protein takes place. The stomach wall secretes hydrochloric acid, so the stomach contents are strongly acidic. This has a very important function. It kills bacteria that are taken into the gut along with the food, helping to protect us from food poisoning. The protease enzyme that is made in the stomach, called **pepsin**, has to be able to work in these acidic conditions, and has an optimum pH value of about 2. This is unusually low – most enzymes work best at near neutral conditions (see Chapter 1).

The semi-digested food is held back in the stomach by a ring of muscle at the outlet of the stomach, called a **sphincter muscle**. When this relaxes, it releases the food into the first part of the small intestine, called the duodenum (Figure 4.16).

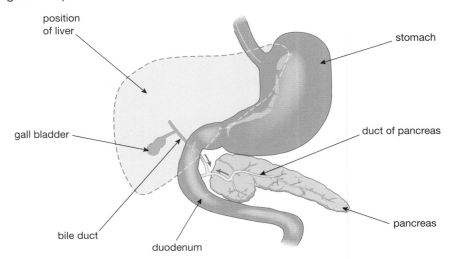

▲ Figure 4.16 The first part of the small intestine, the duodenum, receives digestive juices from the liver and pancreas through tubes called ducts.

▲ Figure 4.17 Bile turns fats into an emulsion of tiny droplets for easier digestion.

Several digestive enzymes are added to the food in the duodenum. These are made by the pancreas, and digest starch, proteins and lipids (Table 4.5). As well as this, the **liver** makes a digestive juice called **bile**. Bile is a green liquid that is stored in the **gall bladder** and passes down the **bile duct** on to the food. Bile does not contain enzymes, but has another important function. It turns any large lipid globules in the food into an emulsion of tiny droplets (Figure 4.17). This increases the surface area of the lipid, so that lipase enzymes can break it down more easily.

Bile and pancreatic juice have another function. They are both alkaline. The mixture of semi-digested food and enzymes coming from the stomach is acidic, and needs to be neutralised by the addition of alkali before it continues on its way through the gut.

As the food continues along the intestine, more enzymes are added, until the parts of the food that can be digested have been fully broken down into soluble end products, which can be absorbed. This is the role of the last part of the small intestine, the ileum.

ABSORPTION IN THE ILEUM

The ileum is highly adapted to absorb the digested food. The lining of the ileum has a very large surface area, which means that it can quickly and efficiently absorb the soluble products of digestion into the blood. The length of the intestine helps to provide a large surface area, and this is aided by folds in its lining, but the greatest increase in area is due to tiny projections from the lining, called **villi** (Figure 4.18).

(a)

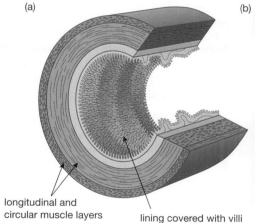

longitudinal and circular muscle layers

lining covered with villi

(b)

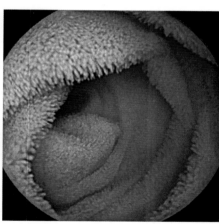

▲ Figure 4.18 (a) The inside lining of the ileum is adapted to absorb digested food by the presence of millions of tiny villi. (b) Photo of the inside a patient's ileum, taken using a camera attached to an endoscope. You can see thousands of tiny villi covering the lining.

The singular of villi is 'villus'. Each villus is only about 1–2 mm long, but there are millions of them, so that the total area of the lining is thought to be about 300 m². This provides a massive area in contact with the digested food. As well as this, high-powered microscopy has revealed that the surface cells of each villus themselves have hundreds of minute projections, called **microvilli**, which increase the surface area for absorption even more (Figure 4.19).

Each villus contains a network of blood capillaries. Most of the digested food enters these blood vessels, but the products of fat digestion, as well as tiny fat droplets, enter a tube in the middle of the villus, called a **lacteal**. The lacteals form part of the body's lymphatic system, which transports a liquid called lymph. This lymph eventually drains into the blood system too.

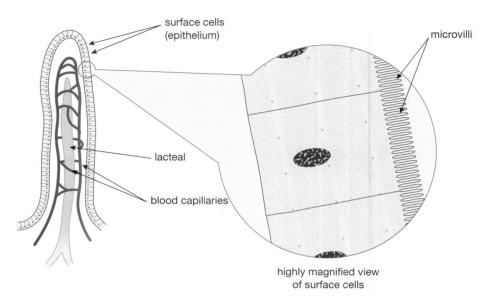

▲ Figure 4.19 Each villus contains blood vessels and a lacteal, which absorb the products of digestion. The surface cells of the villus are covered with microvilli, which further increase the surface area for absorption.

The surface of a villus is made of a single layer of cells called an epithelium. This means that there is only a short distance between the digested food in the ileum and the blood capillaries making it easier for the products of digestion to diffuse through and enter the blood. The epithelium cells contain many mitochondria, which supply the energy needed for active transport of some substances.

In addition each villus contains muscle fibres which contract to move the villus. The villi are in constant motion, keeping them in contact with the contents of the ileum and maintaining a steep concentration gradient for diffusion of the products of digestion.

The blood vessels from the ileum join up to form a large blood vessel called the **hepatic portal vein**, which leads to the liver (see Chapter 5). The liver acts rather like a food processing factory, breaking some molecules down, and building up and storing others. For example, glucose from carbohydrate digestion is converted into glycogen and stored in the liver. Later, the glycogen can be converted back into glucose when the body needs it (see Chapter 7).

The digested food molecules are distributed around the body by the blood system (see Chapter 5). The soluble food molecules are absorbed from the blood into cells of tissues, and are used to build new parts of cells. This is called assimilation.

KEY POINT

Removal of faeces by the body is sometimes incorrectly called excretion. Excretion is a word that only applies to materials that are the waste products of cells of the body, such as carbon dioxide. Faeces are not products of cell metabolism – they consist of waste that has passed through the gut and left the body via the anus, without entering the cells. The correct name for this process is egestion.

THE LARGE INTESTINE – ELIMINATION OF WASTE

By the time that the contents of the gut have reached the end of the small intestine, most of the digested food, as well as most of the water, has been absorbed. The waste material consists mainly of cellulose (fibre) and other indigestible remains, water, dead and living bacteria and cells lost from the lining of the gut. The function of the first part of the large intestine, called the **colon**, is to absorb most of the remaining water from the contents, leaving a semi-solid waste material called faeces. This is stored in the **rectum**, until expelled out of the body through the **anus**.

CHAPTER QUESTIONS

More questions on food and digestion can be found at the end of Unit 2 on page 130.

SKILLS CRITICAL THINKING

1 Which of the following organic molecules contains carbon, hydrogen, oxygen and nitrogen?

A glycogen

B lipid

C cellulose

D protein

2 Which of the following substances would give a positive test when boiled with Benedict's solution?

A fructose

B lipid

C starch

D protein

3 Which of the following statements about digestion is *not* correct?

A Digestion produces fatty acids and glycerol

B Digestion converts insoluble molecules into soluble molecules

C Digestion changes proteins into amino acids

D Digestion releases energy from food

4 Which of the following organs does *not* produce digestive enzymes?

A salivary gland

B gall bladder

C stomach

D pancreas

SKILLS EXECUTIVE FUNCTION

5 The diagram shows an experiment that was set up as a model to show why food needs to be digested.

The Visking tubing acts as a model of the small intestine because it has tiny holes in it that some molecules can pass through. The tubing was left in the boiling tube for an hour, then the water in the tube was tested for starch and glucose.

a Describe how you would test the water for starch, and for glucose. What would the results be for a 'positive' test in each case?

SKILLS REASONING

b The tests showed that glucose was present in the water, but starch was not. Explain why.

c If the tubing takes the place of the intestine, what part of the body does the water in the boiling tube represent?

SKILLS CRITICAL THINKING

d What does 'digested' mean?

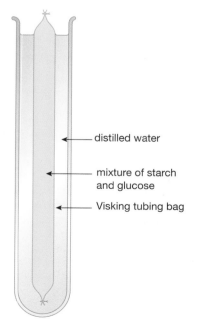

— distilled water

— mixture of starch and glucose

— Visking tubing bag

6 A student carried out an experiment to find out the best conditions for the enzyme pepsin to digest protein. For the protein, she used egg white powder, which forms a cloudy white suspension in water. The table below shows how the four tubes were set up.

Tube	Contents
A	5 cm³ egg white suspension, 2 cm³ pepsin, 3 drops of dilute acid. Tube kept at 37 °C
B	5 cm³ egg white suspension, 2 cm³ distilled water, 3 drops of dilute acid. Tube kept at 37 °C
C	5 cm³ egg white suspension, 2 cm³ pepsin, 3 drops of dilute acid. Tube kept at 20 °C
D	5 cm³ egg white suspension, 2 cm³ pepsin, 3 drops of dilute alkali. Tube kept at 37 °C

The tubes were left for 2 hours and the results were then observed. Tubes B, C and D were still cloudy. Tube A had gone clear.

a Three tubes were kept at 37 °C. Why was this temperature chosen?

b Explain what had happened to the protein in tube A.

c Why did tube D stay cloudy?

d Tube B is called a **Control**. Explain what this means.

e Tube C was left for another 3 hours. Gradually it started to clear. Explain why digestion of the protein happened more slowly in this tube.

f The lining of the stomach secretes hydrochloric acid. Explain the function of this.

g When the stomach contents pass into the duodenum, they are still acidic. How are they neutralised?

7 Copy and complete the following table of digestive enzymes.

Enzyme	Food on which it acts	Products
amylase		
trypsin		
		fatty acids and glycerol

8 Describe four adaptations of the small intestine (ileum) that allow it to absorb digested food efficiently.

9 Bread is made mainly of starch, protein and lipid. Imagine a piece of bread about to start its journey through the human gut. Describe what happens to the bread as it passes through the mouth, stomach, duodenum, ileum and colon. Explain how the bread is moved along the gut. Your description should be illustrated by two or three simplified diagrams.

SKILLS PROBLEM SOLVING

10 The diagram shows a method that can be used to measure the energy content of some types of food. A student placed 20 cm^3 of water in a boiling tube and measured its temperature. He weighed a small piece of pasta, and then held it in a Bunsen burner flame until it caught alight. He then used the burning pasta to heat the boiling tube of water, until the pasta had finished burning. Finally, he measured the temperature of the water at the end of the experiment.

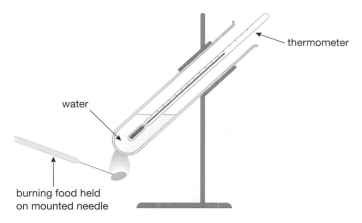

To answer the questions that follow, use the following information.

- The density of water is 1 g per cm^3.

- The pasta weighed 0.22 g.

- The water temperature at the start was 21 °C and at the end was 39 °C.

- The heat energy supplied to the water can be found from the formula:

  energy (in joules) = mass of water × temperature change × 4.2

  a Calculate the energy supplied to the water in the boiling tube in joules (J). Convert this to kilojoules (kJ) by dividing by 1000.

  b Calculate the energy released from the pasta as kilojoules per gram of pasta (kJ per g).

SKILLS REASONING

  c The correct figure for the energy content of pasta is 14.5 kJ per g. The student's result is an underestimate. Write down three reasons why he may have got a lower than expected result. (Hint: think about how the design of the apparatus might introduce errors.)

SKILLS DECISION MAKING

  d Suggest one way the apparatus could be modified to reduce these errors.

SKILLS REASONING

  e The energy in a peanut was measured using the method described above. The peanut was found to contain about twice as much energy per gram as the pasta. Explain why this is the case.

5 BLOOD AND CIRCULATION

Large, multicellular animals need a circulatory system to transport substances to and from the cells of the body. This chapter looks at the structure and function of the circulatory systems of humans and other animals, the composition of mammalian blood, and disorders associated with the heart and circulation.

LEARNING OBJECTIVES

- Understand why simple unicellular organisms can rely on diffusion for movement of substances in and out of the cell

- Understand the need for a transport system in multicellular organisms

- Understand the general structure of the circulation system, including the blood vessels to and from the heart and lungs, liver and kidneys

- Describe the structure of the heart and how it functions

- Understand how factors may increase the risk of developing coronary heart disease

- Explain how the heart rate changes during exercise and under the influence of adrenaline

- Understand how the structures of arteries, veins and capillaries relate to their functions

- Describe the composition of blood: red blood cells, white blood cells, platelets and plasma

- Understand the role of plasma in the transport of carbon dioxide, digested food, urea, hormones and heat energy

- Understand how the adaptations of red blood cells make them suitable for the transport of oxygen, including shape, the absence of a nucleus and the presence of haemoglobin

- Understand how the immune system responds to infection using white blood cells, illustrated by phagocytes ingesting pathogens and lymphocytes releasing antibodies specific to the pathogen

BIOLOGY ONLY

- Understand how vaccination results in the manufacture of memory cells, which enable future antibody production to the pathogen to occur sooner, faster and in greater quantity

- Understand how platelets are involved in blood clotting, which prevents blood loss and the entry of microorganisms

THE NEED FOR CIRCULATORY SYSTEMS

Figure 5.1 shows the circulatory system of a mammal.

Blood is pumped around a closed circuit made up of the heart and blood vessels. As it travels around the body, it collects materials from some places and unloads them in others. In mammals, blood transports:

- oxygen from the lungs to all other parts of the body
- carbon dioxide from all parts of the body to the lungs
- nutrients from the gut to all parts of the body
- urea from the liver to the kidneys.

Hormones, antibodies and many other substances are also transported by the blood. It also distributes heat around the body.

Single-celled organisms, like the ones shown in Figure 5.2, do not have circulatory systems.

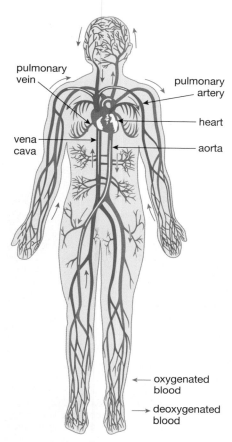

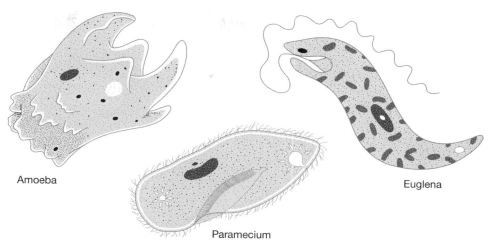

▲ Figure 5.2 Unicellular organisms do not have circulatory systems.

▲ Figure 5.1 The human circulatory system.

There is no circulatory system to carry materials around the very small 'bodies' of these single-celled organisms. Materials can easily move around the cell without a special system. There is no need for lungs or gills to obtain oxygen from the environment either. Single-celled organisms obtain oxygen by diffusion through the surface membrane of the cell. The rest of the cell then uses the oxygen. The area of the cell's surface determines how much oxygen the organism can get (the supply rate), and the volume of the cell determines how much oxygen the organism uses (the demand rate).

The ratio of supply to demand can be written as: $\dfrac{\text{surface area}}{\text{volume}}$

This is called the 'surface area to volume ratio' and it is affected by the size of an organism (see Chapter 1, Activity 5). Single-celled organisms have a high surface area to volume ratio. Their cell surface membrane has a large enough area to supply all the oxygen that their volume demands. In larger animals, the surface area to volume ratio is lower.

Large animals cannot get all the oxygen they need through their surface (even if the body surface would allow it to pass through) – there just isn't enough surface to supply all that volume. To overcome this problem, large organisms have evolved special gas exchange organs and circulatory systems. The gills of fish and the lungs of mammals are linked to a circulatory system that carries oxygen to all parts of the body. The same idea applies to obtaining nutrients – the gut obtains nutrients from food and the circulatory system distributes the nutrients around the body.

THE CIRCULATORY SYSTEMS OF DIFFERENT ANIMALS

One of the main functions of a circulatory system in animals is to transport oxygen. Blood is pumped to a gas exchange organ to load oxygen. It is then pumped to other parts of the body where it unloads the oxygen. There are two main types of circulatory systems in animals.

- In a **single circulatory** system the blood is pumped from the heart to the gas exchange organ and then directly to the rest of the body.

- In a **double circulatory** system the blood is pumped from the heart to the gas exchange organ, back to the heart and then to the rest of the body.

Figure 5.3 shows the difference between these systems.

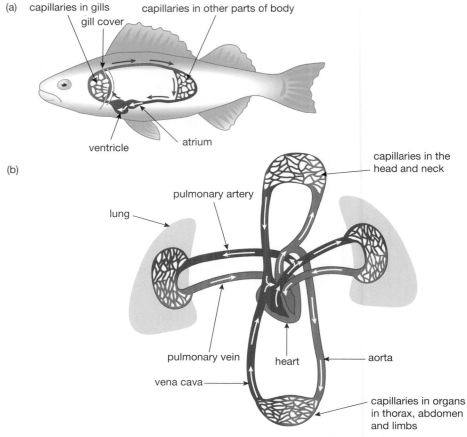

▲ Figure 5.3 (a) The single circulatory system of a fish. The blood passes through the heart only once in a complete circuit of the body. (b) The double circulatory system of a human (and other mammals). The blood passes through the heart twice in one complete circuit of the body.

There are two parts to a double circulatory system:

- The **pulmonary circulation**. Deoxygenated blood leaves the heart through the pulmonary arteries, and is circulated through the lungs, where it becomes oxygenated. The oxygenated blood returns to the heart through the pulmonary veins.

- The **systemic circulation**. Oxygenated blood leaves the heart through the aorta and is circulated through all other parts of the body, where it unloads its oxygen. Deoxygenated blood returns to the heart through the vena cava.

A double circulatory system is more efficient than a single circulatory system. The heart pumps the blood twice, so higher pressures can be maintained. The blood travels more quickly to organs. In the single circulatory system of a fish, blood loses pressure as it passes through the gills. It then travels more slowly to the other organs.

The human circulatory system comprises:

- the heart – this is a pump

- blood vessels – these carry the blood around the body; **arteries** carry blood away from the heart and towards other organs, **veins** carry blood towards the heart and away from other organs and **capillaries** carry blood through organs, linking the arteries and veins

- blood – the transport medium.

Figure 5.4 shows the main blood vessels in the human circulatory system.

DID YOU KNOW?

There are actually two vena cavae (the plural of vena cava). One brings blood back from the head and arms and the other returns blood from the rest of the body (see Figure 5.4).

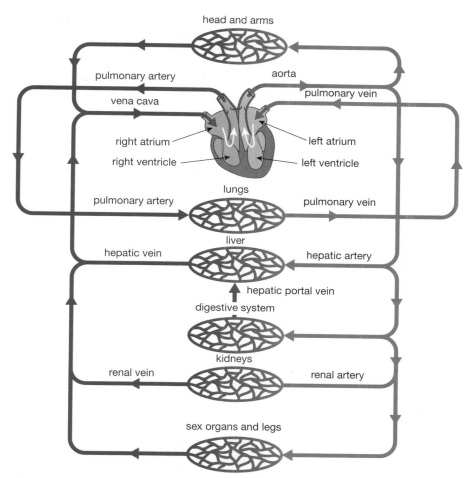

▲ Figure 5.4 The main components of the human circulatory system

THE STRUCTURE AND FUNCTION OF THE HUMAN HEART

The human heart is a pump (Figure 5.5). It pumps blood around the body at different speeds and at different pressures according to the body's needs.

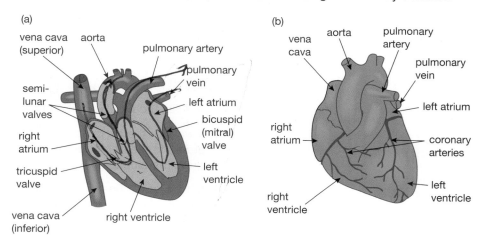

▲ Figure 5.5 The human heart: (a) vertical section; (b) external view

Blood is moved through the heart by a series of contractions and relaxations of the muscle in the walls of the four chambers. These events form the **cardiac cycle**. The main stages are illustrated in Figure 5.6.

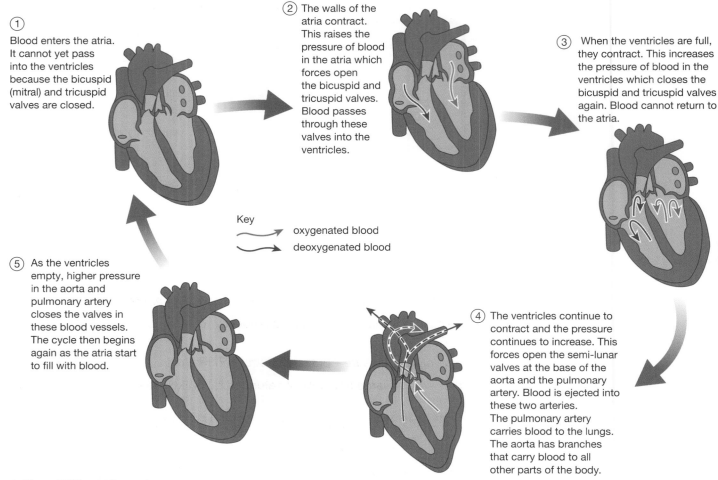

① Blood enters the atria. It cannot yet pass into the ventricles because the bicuspid (mitral) and tricuspid valves are closed.

② The walls of the atria contract. This raises the pressure of blood in the atria which forces open the bicuspid and tricuspid valves. Blood passes through these valves into the ventricles.

③ When the ventricles are full, they contract. This increases the pressure of blood in the ventricles which closes the bicuspid and tricuspid valves again. Blood cannot return to the atria.

Key
→ oxygenated blood
→ deoxygenated blood

⑤ As the ventricles empty, higher pressure in the aorta and pulmonary artery closes the valves in these blood vessels. The cycle then begins again as the atria start to fill with blood.

④ The ventricles continue to contract and the pressure continues to increase. This forces open the semi-lunar valves at the base of the aorta and the pulmonary artery. Blood is ejected into these two arteries. The pulmonary artery carries blood to the lungs. The aorta has branches that carry blood to all other parts of the body.

▲ Figure 5.6 The cardiac cycle.

HINT

Note that during the cardiac cycle both atria contract at the same time (and then relax). After this both ventricles contract at the same time (and then relax). Students are sometimes confused about this, and think that one ventricle contracts, followed by the other.

KEY POINT

Cardiac muscle is unlike any other muscle in our bodies. It never gets fatigued ('tired') like skeletal muscle. On average, cardiac muscle fibres contract and then relax again about 70 times a minute. In a lifetime of 70 years, this special muscle will contract over two billion times – without taking a rest!

The structure of the heart is adapted to its function in several ways:

■ It is divided into a left side and a right side by a wall of muscle called the septum. The right **ventricle** pumps blood only to the lungs while the left **ventricle** pumps blood to all other parts of the body. This requires much more pressure, which is why the wall of the left ventricle is much thicker than that of the right ventricle.

■ Valves ensure that blood can flow only in one direction through the heart.

■ The walls of the **atria** are thin. They can be stretched to receive blood as it returns to the heart but can contract with enough force to push blood through the bicuspid and tricuspid valves into the ventricles.

■ The walls of the heart are made of **cardiac muscle**, which can contract and then relax continuously without becoming fatigued.

■ The cardiac muscle has its own blood supply – the coronary circulation. Blood reaches the muscle via **coronary arteries**. These carry blood to capillaries that supply the heart muscle with oxygen and nutrients. Blood is returned to the right atrium via **coronary veins**.

CORONARY HEART DISEASE

The coronary arteries are among the narrowest in the body. They are easily blocked by a build-up of fatty substances (including **cholesterol**) in their walls. This can cut off the blood supply to an area of cardiac muscle. The affected muscle can no longer receive oxygen and glucose, so it cannot respire and release energy. This means it is unable to contract, resulting in a heart attack. This is called **coronary heart disease** (CHD). It can lead to severe health problems and is often fatal.

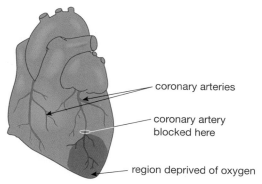

coronary arteries

coronary artery blocked here

region deprived of oxygen

▲ Figure 5.7 A blockage of a coronary artery cuts off the blood supply to part of the heart muscle.

A number of factors make coronary heart disease more likely:

- heredity – some people inherit a tendency to develop coronary heart disease

- high blood pressure – puts more strain on the heart

- diet – eating large amounts of saturated fat is likely to raise cholesterol levels

- smoking – raises blood pressure and makes blood clots more likely to form

- stress – raises blood pressure

- lack of exercise – regular exercise helps to reduce blood pressure and strengthens the heart.

HEART RATE

Normally the heart beats about 70 times a minute, but this can change according to the needs of the body. When we exercise, muscles must release more energy. They need an increased supply of oxygen for aerobic respiration (see Chapter 1). To deliver the extra oxygen, both the number of beats per minute (heart rate) and the volume of blood pumped with each beat (called stroke volume) increase.

When we are angry or afraid our heart rate again increases. The increased output supplies extra blood to the muscles, enabling them to release extra energy through aerobic respiration. This allows us to fight or run away and is called the 'fight or flight' response. It is triggered by **secretion** of the hormone **adrenaline** from the adrenal glands (see Chapter 7).

When we sleep, our heart rate decreases as all our organs are working more slowly. They need to release less energy and so need less oxygen.

These changes in the heart rate are controlled by nerve impulses from a part of the brain called the **medulla** (Figure 5.8). When we start to

DID YOU KNOW?

Have you noticed a 'hollow' or 'fluttering' feeling in your stomach when you are anxious? It happens because blood that would normally flow to your stomach and intestines has been diverted to the muscles to allow the 'fight or flight' response.

exercise, our muscles produce more carbon dioxide in aerobic respiration. Receptors in the aorta and the carotid artery (the artery leading to the head) detect this increase. They send electrical signals called nerve impulses through the sensory nerve to the medulla. The medulla responds by sending nerve impulses along the accelerator nerve. When carbon dioxide production returns to normal, the medulla receives fewer impulses. It responds by sending nerve impulses along a decelerator nerve.

(not to scale)

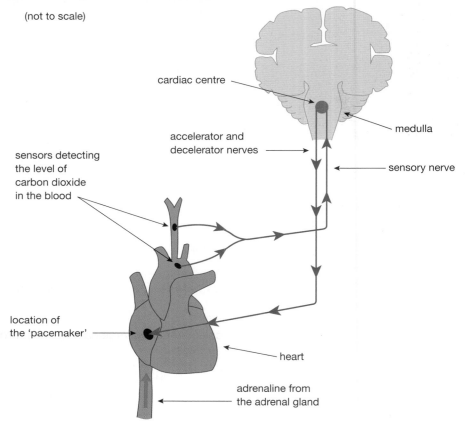

cardiac centre

medulla

accelerator and decelerator nerves

sensory nerve

sensors detecting the level of carbon dioxide in the blood

location of the 'pacemaker'

heart

adrenaline from the adrenal gland

▲ Figure 5.8 How the heart rate is controlled.

The accelerator nerve increases the heart rate. It also causes the heart to beat with more force and so increases blood pressure. The decelerator nerve decreases the heart rate. It also reduces the force of the contractions. Blood pressure then returns to normal.

These controls are both examples of **reflex actions** (see Chapter 6).

ARTERIES, VEINS AND CAPILLARIES

Arteries carry blood from the heart to the organs of the body. This arterial blood is pumped out by the ventricles at a high pressure. Elastic tissue in the walls of the arteries allows them to stretch and recoil (spring back into shape), maintaining the high blood pressure. A thick muscular wall helps control the flow of blood by dilating (widening) or constricting (narrowing) the vessels.

Veins carry blood from organs back towards the heart. The pressure of this venous blood is much lower than that in the arteries. It puts very little pressure on the walls of the veins, so they can be thinner than those of arteries, and contain less elastic tissue and muscle. Figure 5.9 shows the structure of a typical artery and a typical vein with the same diameter.

EXTENSION WORK

Arterioles are small arteries. They carry blood into organs from arteries. Their structure is similar to the larger arteries, but they have a larger proportion of muscle fibres in their walls. They are also supplied with nerve endings in their walls and so can be made to dilate (become wider) or constrict (become narrower) to allow more or less blood into the organ.

If *all* the arterioles constrict, it is harder for blood to pass through them – there is more resistance. This increases blood pressure. Prolonged stress can cause arterioles to constrict and so increase blood pressure.

KEY POINT

All arteries carry oxygenated blood (blood containing a lot of oxygen) except the pulmonary artery and the umbilical artery of an unborn baby. All veins carry deoxygenated blood (blood containing less oxygen) except the pulmonary vein and umbilical vein.

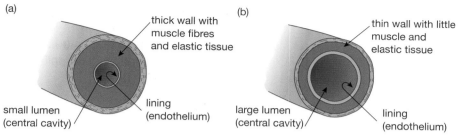

▲ Figure 5.9 The structure of (a) an artery and (b) a vein as seen in cross section.

Veins also have **semilunar** (half-moon shaped) **valves**, which prevent the backflow of blood. The action of these valves is explained in Figure 5.10.

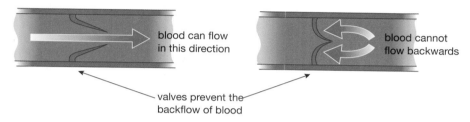

▲ Figure 5.10 The action of semilunar valves in veins

Capillaries carry blood through organs, bringing the blood close to every cell in the organ. Substances are transferred between the blood in the capillary and the cells. To do this, capillaries must be small enough to 'fit' between cells, and allow materials to pass through their walls easily. Figure 5.11 shows the structure of a capillary and how exchange of substances takes place between the capillary and nearby cells. The walls of capillaries are one cell thick, providing a short distance for diffusion of materials into and out of the blood. Red blood cells just fit through the tiny diameter of capillaries, so they are close to the capillary wall. This means that there is a short distance for oxygen to diffuse. Figure 5.12 shows a photograph of a cross-section through an artery and a vein.

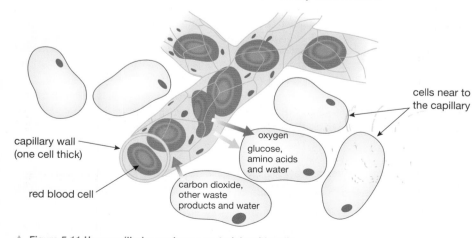

▲ Figure 5.11 How capillaries exchange materials with cells

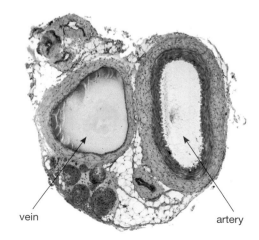

▲ Figure 5.12 The lumen of the artery is the same size as the lumen of the vein – but note the difference in the thickness of the walls of these two vessels.

THE COMPOSITION OF BLOOD

Blood is a lot more than just a red liquid flowing through your arteries and veins! In fact, blood is a complex tissue. Figure 5.13 illustrates the main types of cells found in blood.

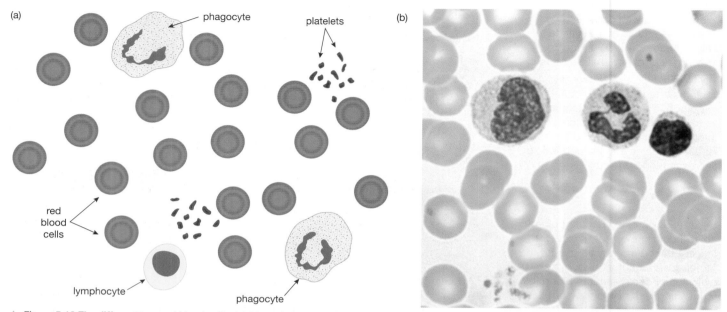

▲ Figure 5.13 The different types of blood cells. (a) Diagram of the different cells. (b) A blood smear seen through a light microscope. The smear shows many red blood cells and three different kinds of white blood cell.

The different parts of blood have different functions. These are described in Table 5.1.

Table 5.1 Functions of the different components of blood.

COMPONENT OF BLOOD	DESCRIPTION OF COMPONENT	FUNCTION OF COMPONENT
plasma	liquid part of blood: mainly water	carries the blood cells around the body; carries dissolved nutrients, hormones, carbon dioxide and urea; also distributes heat around the body
red blood cells (erythrocytes)	biconcave, disc-like cells with no nucleus; millions in each mm³ of blood	transport of oxygen – contain mainly haemoglobin, which loads oxygen in the lungs and unloads it in other regions of the body
WHITE BLOOD CELLS:		
lymphocytes	about the same size as red cells with a large spherical nucleus	produce antibodies to destroy microorganisms – some lymphocytes persist in our blood after infection and give us immunity to specific diseases
phagocytes	much larger than red cells, with a large spherical or lobed nucleus	digest and destroy bacteria and other microorganisms that have infected our bodies
platelets	the smallest cells – are really fragments of other cells	release chemicals to make blood clot when we cut ourselves

RED BLOOD CELLS

The red blood cells or **erythrocytes** are highly specialised cells made in the bone marrow. They have a limited life span of about 100 days after which time they are destroyed in the spleen. They have only one function – to transport oxygen. Several features enable them to carry out this function very efficiently.

Red blood cells contain **haemoglobin**. This is an iron-containing protein that associates (combines) with oxygen to form **oxyhaemoglobin** when there is a high concentration of oxygen in the surroundings. We say that the red blood cell is loading oxygen. When the concentration of oxygen is low, oxyhaemoglobin turns back into haemoglobin and the red blood cell unloads its oxygen.

As red blood cells pass through the lungs, they load oxygen. As they pass through active tissues they unload oxygen.

$$\text{haemoglobin} + \text{oxygen} \underset{\text{in the tissues}}{\overset{\text{in the lungs}}{\rightleftharpoons}} \text{oxyhaemoglobin}$$

Red blood cells do not contain a nucleus. It is lost during their development in the bone marrow. This means that more haemoglobin can be packed into each red blood cell so more oxygen can be transported. Their biconcave shape allows efficient exchange of oxygen in and out of the cell. Each red blood cell has a high surface area to volume ratio, giving a large area for diffusion. The thin shape of the cell results in a short diffusion distance to the centre of the cell.

WHITE BLOOD CELLS

There are several types of white blood cell. Their main role is to protect the body against invasion by disease-causing microorganisms (pathogens), such as bacteria and viruses. They do this in two main ways: **phagocytosis** and production of **antibodies**.

About 70% of white blood cells can ingest (take in) microorganisms such as bacteria. This is called phagocytosis, and the cells are **phagocytes**. They do this by changing their shape, producing extensions of their cytoplasm, called pseudopodia. The pseudopodia surround and enclose the microorganism in a vacuole. Once it is inside, the phagocyte secretes enzymes into the vacuole to break the microorganism down (Figure 5.14). Phagocytosis means 'cell eating' – you can see why it is called this.

Approximately 25% of white blood cells are **lymphocytes**. Their function is to make chemicals called antibodies. Antibodies are soluble proteins that pass into the plasma. Pathogens such as bacteria and viruses have chemical 'markers' on their surfaces, which the antibodies recognise. These markers are called **antigens**. The antibodies stick to the surface antigens and destroy the pathogen. They do this in a number of ways, for example by:

- causing bacteria to stick together, so that phagocytes can ingest them more easily
- acting as a 'label' on the pathogen, so that it is more easily recognised by a phagocyte
- causing bacterial cells to burst open
- neutralising poisons (toxins) produced by pathogens.

The production of antibodies following the first exposure to a foreign antigen is called the primary immune response.

BIOLOGY ONLY

(a)

bacterium

pseudopodia surround bacterium

bacterium enclosed in a vacuole

digestive enzymes destroy bacterium

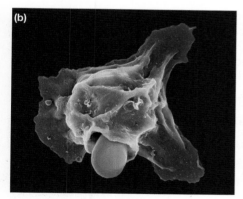
(b)

▲ Figure 5.14 (a) Phagocytosis by a white blood cell. (b) A phagocyte engulfing a pathogenic microorganism (a species of yeast).

IMMUNITY

Some lymphocytes do not get involved in killing microorganisms straight away. Instead, they develop into **memory cells**. These cells remain in the blood for many years, sometimes a lifetime. If the same microorganism re-infects a person, the memory lymphocytes start to reproduce and produce antibodies, so that the pathogen can be quickly dealt with. This is known as immunity.

This secondary **immune response** is much faster and more effective than the primary response. The number of antibodies in the blood quickly rises to a high level, killing the microorganisms before they have time to multiply to a point where they would cause disease. This is shown in Figure 5.15.

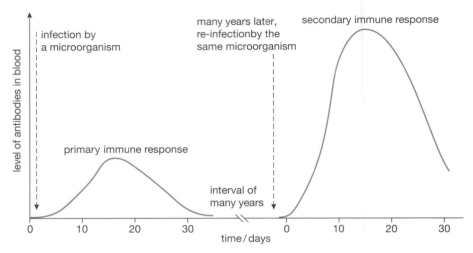

▲ Figure 5.15 The primary and secondary immune responses

A person can be given artificial immunity to a disease-causing organism without ever actually contracting (having) the disease itself. This is done by **vaccination**. A person is injected with an 'agent' that carries the same antigens as a specific pathogen. Lymphocytes recognise the antigens and multiply exactly as if that microorganism had entered the bloodstream. They produce memory cells and make the person immune to the disease. If the person now comes into contact with the 'real' pathogen, they will experience a secondary immune response. Antibody production will happen sooner, faster and in greater quantity than if they had not been vaccinated, and may be enough to prevent the pathogen reproducing in the body and causing the disease.

Some agents used as vaccines are:

■ a weakened strain of the actual microorganism, e.g. vaccines against polio, tuberculosis (TB) and measles

■ dead microorganisms, e.g. typhoid and whooping cough vaccines

■ modified toxins of the bacteria, e.g. tetanus and diphtheria vaccines

■ just the antigens themselves, e.g. influenza vaccine

■ harmless bacteria, genetically engineered to carry the antigens of a different, disease-causing microorganism, e.g. the vaccine against hepatitis B.

PLATELETS

Platelets are not whole cells, but fragments of large cells made in the bone marrow. If the skin is cut, exposure to the air stimulates the platelets and damaged tissue to produce a chemical. This chemical causes the soluble plasma protein **fibrinogen** to change into insoluble fibres of another protein, **fibrin**. The fibrin forms a network across the wound, in which red blood cells become trapped (Figure 5.16) This forms a clot, which prevents further loss of blood and entry of microorganisms that may be pathogens. The clot develops into a scab, which protects the damaged tissue while new skin grows.

▲ Figure 5.16 Red blood cells trapped in fibres of fibrin, forming a blood clot.

END OF BIOLOGY ONLY

CHAPTER QUESTIONS

More questions on blood and circulation can be found at the end of Unit 2 on page 130

SKILLS   CRITICAL THINKING

1   After a period of exercise, which blood vessel will contain the highest concentration of carbon dioxide?

   A   aorta

   B   vena cava

   C   hepatic artery

   D   pulmonary vein

2   When the right ventricle contracts, to which of the following structures does the blood flow next?

   A   aorta

   B   left atrium

   C   pulmonary artery

   D   left ventricle

3   The diagram below shows sections through three blood vessels (not drawn to scale).

Which row in the table shows the correct names of vessels X, Y and Z?

	X	Y	Z
A	vein	capillary	artery
B	artery	capillary	vein
C	vein	artery	capillary
D	capillary	vein	artery

4   Which component of the blood makes antibodies?

   A   red blood cells

   B   white blood cells

   C   plasma

   D   platelets

5   Some animals have a single circulatory system, some have a double circulatory system and some organisms have no circulatory system at all.

   a   Name one type of animal with a single circulatory system and one type of animal with a double circulatory system.

   b   Explain:

     i   the difference between single and double circulatory systems

     ii   why a double circulatory system is more efficient than a single circulatory system.

   c   Explain why single-celled organisms do not need a circulatory system.

SKILLS CRITICAL THINKING

6 Blood transports oxygen and carbon dioxide around the body. Oxygen is transported by the red blood cells.

   a Give three ways in which a red blood cell is adapted to its function of transporting oxygen.

   b Describe how oxygen:

     i   enters a red blood cell from the alveoli in the lungs

     ii passes from a red blood cell to an actively respiring muscle cell.

   c Describe how carbon dioxide is transported around the body.

SKILLS REASONING

7 Blood is carried around the body in arteries, veins and capillaries.

   a Describe two ways in which the structure of an artery is adapted to its function.

   b Describe three differences between arteries and veins.

   c Describe two ways in which the structure of a capillary is adapted to its function.

SKILLS ANALYSIS

8 The diagram shows a section through a human heart.

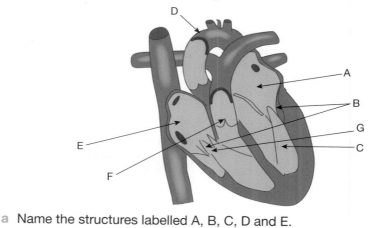

   a Name the structures labelled A, B, C, D and E.

SKILLS CRITICAL THINKING

   b What is the importance of the structures labelled B and F?

SKILLS ANALYSIS

   c Which letters represent the chambers of the heart to which blood returns:

     i   from the lungs

     ii from all the other organs of the body?

9 The diagram shows three types of cells found in human blood.

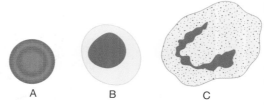

   a Giving a reason for each answer, identify the blood cell which:

     i   transports oxygen around the body

     ii produces antibodies to destroy bacteria

     iii engulfs and digests bacteria.

SKILLS CRITICAL THINKING

   b Name one other component of blood found in the plasma and state its function.

10 The graph shows changes in a person's heart rate over a period of time.

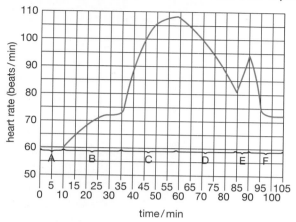

Giving reasons for your answers, give the letter of the time period when the person was:

a running

b frightened by a sudden loud noise

c sleeping

d waking.

11 The graph shows the changes that take place in heart rate before, during and after a period of exercise.

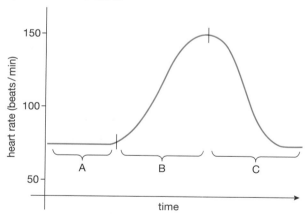

a Describe and explain the heart rates found:
  i at rest, before exercise (period A)
  ii as the person commences the exercise (period B)
  iii as the person recovers from the exercise (period C).

b How can the recovery period (period C) be used to assess a person's fitness?

6 COORDINATION

In the body 'coordination' means making things happen at the right time by linking up different body activities. Humans and other animals have two organ systems which do this. The first is the nervous system, which is the subject of this chapter. The second is the hormone or endocrine system, which is dealt with in Chapter 7.

LEARNING OBJECTIVES

- Understand how organisms are able to respond to changes in their environment

- Understand that a coordinated response requires a stimulus, a receptor and an effector

- Understand that the central nervous system consists of the brain and spinal cord and is linked to sense organs by nerves

- Describe how the nervous system controls responses

- Understand how stimulation of receptors in the sense organs sends electrical impulses along nerves into and out of the central nervous system, resulting in rapid responses

- Describe the structure and function of the eye as a receptor

- Understand the function of the eye in focusing on near and distant objects, and in responding to changes in light intensity

- Describe the structure and functioning of a simple reflex arc, illustrated by the withdrawal of a finger from a hot object

- Understand the role of neurotransmitters at synapses

STIMULUS AND RESPONSE

Suppose you are walking along when you see a football coming at high speed towards your head. If your nerves are working properly, you will probably move or duck quickly to avoid contact. Imagine another situation where you are very hungry, and you smell food cooking. Your mouth might begin to 'water', in other words secrete saliva.

KEY POINT

The surroundings outside the body are called the *external* environment. The inside of the body is known as the *internal* environment. The body also responds to changes in its internal environment, such as temperature and blood glucose levels. You will read about these responses in Chapters 7 and 8.

Each of these situations is an example of a **stimulus** and a **response**. A *stimulus* is a change in an animal's surroundings, and a *response* is a reaction to that change. In the first example, the approaching ball was the stimulus, and your movement to avoid it hitting you was the response. The change in your environment was detected by your eyes, which are an example of a **receptor** organ. The response was brought about by contraction of muscles, which are a type of **effector** organ (they produce an effect). The nervous system links the two, and is an example of a coordination system. A summary of the sequence of events is:

stimulus → receptor → coordination → effector → response

In the second example, the receptor for the smell of food was the nose, and the response was the secretion of saliva from glands. Glands secrete (release) chemical substances, and they are the second type of effector organ. Again, the link between the stimulus and the response is the nervous system. The information in the nerve cells is transmitted in the form of tiny electrical signals called **nerve impulses**.

▲ Figure 6.1 This yellow flower (a) looks very different to a bee, which sees patterns on the petals reflecting UV light (b).

RECEPTORS

The role of any receptor is to detect the stimulus by changing its energy into the electrical energy of the nerve impulses. For example, the eye converts light energy into nerve impulses, and the ear converts sound energy into nerve impulses (Table 6.1).

Table 6.1 Human receptors and the energy they receive.

Receptor	Type of energy received
eye (retina)	light
ear (organ of hearing)	sound
ear (organ of balance)	mechanical (kinetic)
tongue (taste buds)	chemical
nose (organ of smell)	chemical
skin (touch/pressure/pain receptors)	mechanical (kinetic)
skin (temperature receptors)	heat
muscle (stretch receptors)	mechanical (kinetic)

Notice how a 'sense' like touch is made up of several components. When we touch a warm surface we will be stimulating several types of receptor, including touch and temperature receptors, as well as stretch receptors in the muscles (see the section on skin in Chapter 8). As well as this, each sense detects different aspects of the energy it receives. For example, the ears don't just detect sounds, but different loudness and frequencies of sound, while the eye not only forms an image, but also detects brightness of light and in humans can tell the difference between different light wavelengths (colours). Senses tell us a great deal about changes in our environment.

THE CENTRAL NERVOUS SYSTEM

The biological name for a nerve cell is a **neurone**. The impulses that travel along a neurone are not an electric current, as in a wire. They are caused by movements of charged particles (ions) in and out of the neurone. Impulses travel at speeds between about 10 and 100 metres per second, which is much slower than an electric current, but fast enough to produce a rapid response (see the 'Looking ahead' feature at the end of this chapter).

Impulses from receptors pass along nerves containing **sensory neurones**, until they reach the brain and spinal cord. These two organs are together known as the **central nervous system**, or **CNS** (Figure 6.2).

Other nerves contain **motor neurones**, transmitting impulses to the muscles and glands. Some nerves contain only sensory or motor cells, while other nerves contain both – they are 'mixed'. A typical nerve contains thousands of individual neurones.

DID YOU KNOW?

The CNS is well protected by the skeleton. The brain is inside the skull or cranium (nerves connected to the brain are *cranial* nerves) and the spinal cord runs down the middle of the spinal column, passing through a hole in each vertebra. Nerves connected to the spinal cord are called *spinal* nerves.

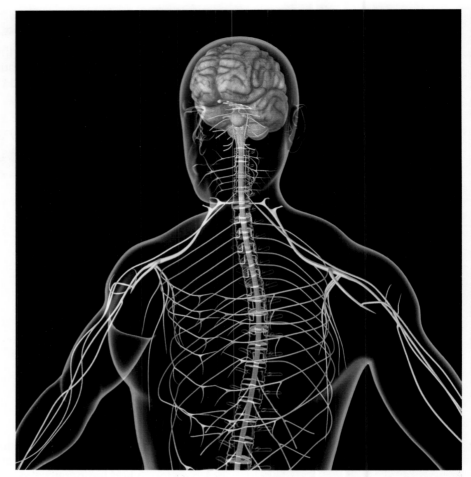

▲ Figure 6.2 The brain and spinal cord form the central nervous system. Cranial and spinal nerves lead to and from the CNS. The CNS sorts out information from the senses and sends messages to muscles.

THE STRUCTURE OF NEURONES

Both sensory and motor neurones can be very long. For example, a motor neurone leading from the CNS to the muscles in the finger has a fibre about a metre in length, which is 100 000 times the length of the cell body (Figure 6.3).

The cell body of a motor neurone is at one end of the fibre, in the CNS. The cell body has fine cytoplasmic extensions, called **dendrons**. These in turn form finer extensions, called **dendrites**. There can be junctions with other neurones on any part of the cell body, dendrons or dendrites. These junctions are called **synapses**. Later in this chapter we will deal with the importance of synapses in nerve pathways. One of the extensions from the motor neurone cell body is much longer than the other dendrons. This is the fibre that carries impulses to the effector organ, and is called the **axon**. At the end of the axon furthest from the cell body, it divides into many nerve endings. These fine branches of the axon connect with a muscle at a special sort of synapse called a **neuromuscular junction**. In this way impulses are carried from the CNS out to the muscle. The signals from nerve impulses are transmitted across the neuromuscular junction, causing the muscle fibres to contract. The axon is covered by a sheath made of a fatty material called myelin. The **myelin sheath** insulates the axon, preventing 'short circuits' with other axons, and also speeds up the conduction of the impulses. The sheath is formed by the membranes of special cells that wrap themselves around the axon as it develops.

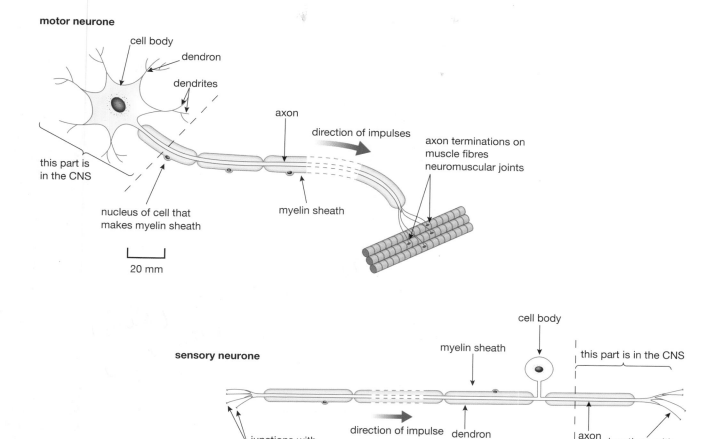

▲ Figure 6.3 The structure of motor and sensory neurones. The cell fibres (axon/dendron) are very long, which is indicated by the dashed sections.

A **sensory neurone** has a similar structure to the motor neurone, but the cell body is located on a side branch of the fibre, just outside the CNS. The fibre from the sensory receptor to the cell body is actually a dendron, while the fibre from the cell body to the CNS is a short axon. As with motor neurones, fibres of sensory neurones are often myelinated.

THE EYE

Many animals have eyes, but few show the complexity of the human eye. Simpler animals, such as snails, use their eyes to detect light but cannot form a proper image. Other animals, such as dogs, can form images but cannot distinguish colours. The human eye does all three. Of course it is not really the eye that 'sees' anything at all, but the brain that interprets the impulses from the eye. To find out how light from an object is converted into impulses representing an image, we need to look at the structure of this complex organ (Figure 6.4).

The tough outer coat of the eye is called the **sclera**, which is the visible, white part of the eye. At the front of the eye the sclera becomes a transparent 'window' called the **cornea**, which lets light into the eye. Behind the cornea is the coloured ring of tissue called the **iris**. In the middle of the iris is a hole called the **pupil**, which lets the light through. It is black because there is no light escaping from the inside of the eye.

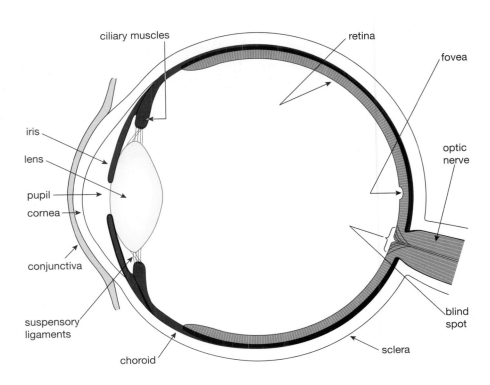

▲ Figure 6.4 A horizontal section through the human eye

Underneath the sclera is a dark layer called the **choroid**. It is dark because it contains many pigment cells, as well as blood vessels. The pigment stops light being reflected around inside the eye.

The innermost layer of the back of the eye is the **retina**. This is the light-sensitive layer, the place where light energy is converted into the electrical energy of nerve impulses. The retina contains receptor cells called **rods** and **cones**. These cells react to light, producing impulses in sensory neurones. The sensory neurones then pass the impulses to the brain through the **optic nerve**. Rod cells work well in dim light, but they cannot distinguish between different colours, so the brain 'sees' an image produced by the rods in black and white. This is why we can't see colours very well in dim light: only our rods are working properly. The cones, on the other hand, will only work in bright light, and there are three types which respond to different wavelengths or colours of light – red, green and blue. We can see all the colours of visible light as a result of these three types of cones being stimulated to different degrees. For example, if red, green and blue are stimulated equally, we see white. Both rods and cones are found throughout the retina, but cones are particularly concentrated at the centre of the retina, in an area called the **fovea**. Cones give a sharper image than rods, which is why we can only see objects clearly if we are looking directly at them, so that the image falls on the fovea.

FORMING AN IMAGE

To form an image on the retina, light needs to be bent or *refracted*. Refraction takes place when light passes from one medium to another of a different density. In the eye, this happens first at the air/cornea boundary, and again at the lens (Figure 6.5). In fact the cornea acts as the first lens of the eye.

DID YOU KNOW?

The fact that the inverted image is seen the right way up by the brain makes the point that it is the brain which 'sees' things, not the eye. An interesting experiment was carried out to test this. Volunteers were made to wear special inverting goggles for long periods. These turned the view of their surroundings upside down. At first this completely disorientated them, and they found it difficult to make even simple coordinated movements. However, after a while their brains adapted, until the view through the goggles looked normal. In fact, when the volunteers removed the goggles, the world then looked upside down!

As a result of refraction at the cornea and lens, the image on the retina is inverted (upside down). The brain interprets the image the right way up.

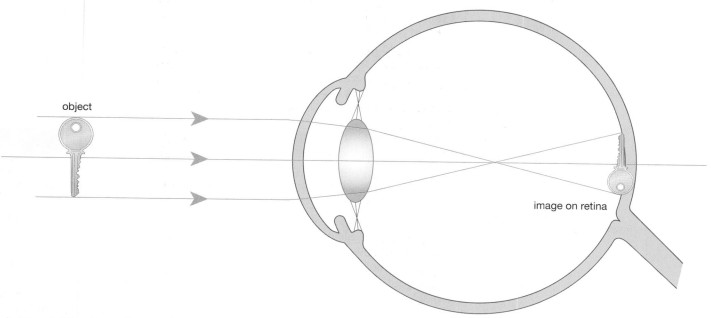

▲ Figure 6.5 How the eye forms an image. Refraction of light occurs at the cornea and lens, producing an inverted image on the retina.

THE IRIS REFLEX

The role of the iris is to control the amount of light entering the eye, by changing the size of the pupil. The iris contains two types of muscles. Circular muscles form a ring shape in the iris, and radial muscles lie like the spokes of a wheel. In bright light, the pupil is constricted (made smaller).

This happens because the circular muscles contract and the radial muscles relax. In dim light, the opposite happens. The radial muscles contract and the circular muscles relax, dilating (widening) the pupil (Figure 6.6).

circular muscles contract

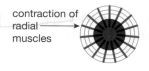

bright light
- circular muscles contract
- radial muscles relax
- pupil constricts

contraction of radial muscles

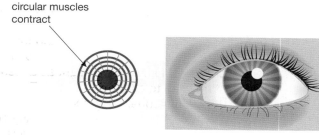

dim light
- circular muscles relax
- radial muscles contract
- pupil dilates

▲ Figure 6.6 The amount of light entering the eye is controlled by the iris, which alters the diameter of the pupil.

In the iris reflex, the route from stimulus to response is this:

stimulus (light intensity)

↓

retina (receptor)

↓

sensory neurones in optic nerve

↓

unconscious part of brain

↓

motor neurones in nerve to iris

↓

iris muscles (effector)

↓

response (change in size of pupil)

Whenever our eyes look from a dim light to a bright one, the iris rapidly and automatically adjusts the pupil size. This is an example of a **reflex action**. You will find out more about reflexes later in this chapter. The purpose of the iris reflex is to allow the right intensity (brightness) of light to fall on the retina. Light that is too bright could damage the rods and cones, and light that is too dim would not form an image. The intensity of light hitting the retina is the stimulus for this reflex. Impulses pass to the brain through the optic nerve, and straight back to the iris muscles, adjusting the diameter of the pupil. It all happens without the need for conscious thought – in fact we are not even aware of it happening.

THE BLIND SPOT

There is one area of the retina where an image cannot be formed; this is where the optic nerve leaves the eye. At this position there are no rods or cones, so it is called the **blind spot**. The retina of each eye has a blind spot, but they are not a problem, because the brain puts the images from each eye together, cancelling out the blind spots of both eyes. As well as this, the optic nerve leaves the eye towards the edge of the retina, where vision is not very sharp anyway. To 'see' your own blind spot you can do a simple experiment. Cover or close your right eye. Hold this page about 30 cm from your eyes and look at the black dot below. Now, without moving the book or turning your head, read the numbers from left to right by moving your left eye slowly towards the right.

● **1 2 3 4 5 6 7 8 9 10 11 12 13 14 15**

You should find that when the image of the dot falls on the blind spot it disappears. If you try doing this with both eyes open, the image of the dot will not disappear.

> **DID YOU KNOW?**
> A way to prove to yourself that the eyes form two overlapping images is to try the 'sausage test'. Focus your eyes on a distant object. Place your two index fingers tip to tip, and bring them up in front of your eyes, about 30 cm from your face, while still focusing at a distance. You should see a finger 'sausage' between the two fingers. Now try this with one eye closed. What is the difference?

ACCOMMODATION

The changes that take place in the eye which allow us to see objects at different distances are called **accommodation**.

You have probably seen the results of a camera or projector not being in focus – a blurred picture. In a camera, we can focus light from objects that are different distances away by moving the lens backwards or forwards, until the picture is sharp. In the eye, a different method is used. Rather than altering its position, the shape of the lens can be changed. A lens that is fatter in the middle (more convex) will refract light rays more than a thinner (less convex) lens. The lens in the eye can change shape because it is made of cells containing an elastic crystalline protein.

Figure 6.4 shows that the lens is held in place by a series of fibres called the **suspensory ligaments**. These are attached like the spokes of a wheel to a ring of muscle, called the **ciliary muscle**. The inside of the eye is filled with a transparent watery fluid which pushes outwards on the eye. In other words, there is a slight positive pressure within the eye. The changes to the eye that take place during accommodation are shown in Figure 6.7.

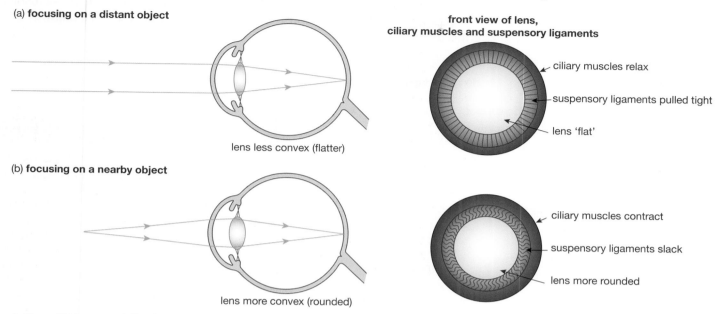

(a) focusing on a distant object

lens less convex (flatter)

front view of lens, ciliary muscles and suspensory ligaments

ciliary muscles relax

suspensory ligaments pulled tight

lens 'flat'

(b) focusing on a nearby object

lens more convex (rounded)

ciliary muscles contract

suspensory ligaments slack

lens more rounded

▲ Figure 6.7 Accommodation: how the eye focuses on objects at different distances

When the eye is focused on a distant object, the rays of light from the object are almost parallel when they reach the cornea (Figure 6.7(a)). The cornea refracts the rays, but the lens does not need to refract them much more to focus the light on the retina, so it does not need to be very convex. The ciliary muscles relax and the pressure in the eye pushes outwards on the lens, flattening it and stretching the suspensory ligaments. This is the condition when the eye is at rest – our eyes are focused for long distances.

When we focus on a nearby object, for example when reading a book, the light rays from the object are spreading out (diverging) when they enter the eye (Figure 6.7(b)). In this situation, the lens has to be more convex in order to refract the rays enough to focus them on the retina. The ciliary muscles now contract, the suspensory ligaments become slack and the elastic lens bulges outwards into a more convex shape.

REFLEX ACTIONS

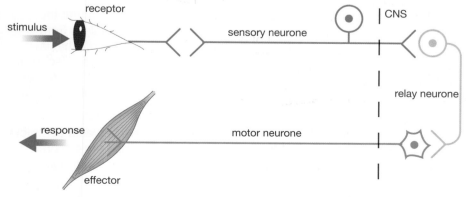

stimulus

receptor

sensory neurone

CNS

relay neurone

response

effector

motor neurone

You saw on page 89 that the dilation and constriction of the pupil by the iris is an example of a reflex action. You now need to understand a little more about the nerves involved in a reflex. The nerve pathway of a reflex is called the **reflex arc**. The 'arc' part means that the pathway goes into the CNS and then straight back out again, in a sort of curve or arc (Figure 6.8).

▲ Figure 6.8 Simplified diagram of a reflex arc

The iris–pupil reflex protects the eye against damage by bright light. Other reflexes are protective too, preventing serious harm to the body. Take, for example, the reflex response to a painful stimulus. This happens when part of your body, such as your hand, touches a sharp or hot object. The reflex results in your hand being quickly withdrawn. Figure 6.9 shows the nerve pathway of this reflex in more detail.

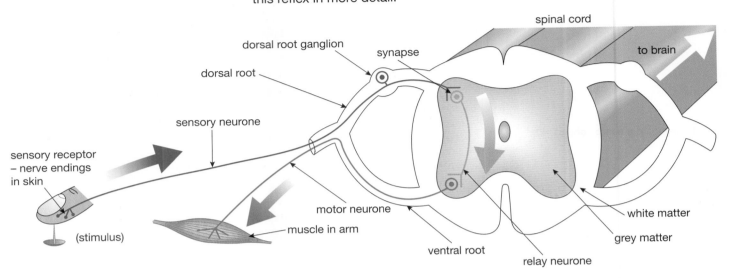

▲ Figure 6.9 A reflex arc in more detail

DID YOU KNOW?

'Dorsal' and 'ventral' are words describing the back and front of the body. The dorsal roots of spinal nerves emerge from the spinal cord towards the back of the person, while the ventral roots emerge towards the front. Notice that the cell bodies of the sensory neurones are all located in a swelling in the dorsal root, called the **dorsal root ganglion**.

The stimulus is detected by temperature or pain receptors in the skin. These generate impulses in sensory neurones. The impulses enter the CNS through a part of the spinal nerve called the **dorsal root**. In the spinal cord the sensory neurones connect by synapses with short **relay neurones**, which in turn connect with motor neurones. The motor neurones emerge from the spinal cord through the **ventral root**, and send impulses back out to the muscles of the arm. These muscles then contract, pulling the arm (and thus finger) away from the harmful stimulus.

The middle part of the spinal cord consists mainly of nerve cell bodies, which gives it a grey colour. This is why it is known as **grey matter**. The outer part of the spinal cord is called **white matter**, and has a whiter appearance because it contains many axons with their fatty myelin sheaths. (In the brain this is reversed – the grey matter is on the outside and the white matter in the middle of the brain.)

Impulses travel through the reflex arc in a fraction of a second, so that the reflex action is very fast, and doesn't need to be started by impulses from the brain. However, this doesn't mean that the brain is unaware of what is going on. This is because in the spinal cord, the reflex arc neurones also form connections called synapses (see below) with nerve cells leading to and from the brain. The brain therefore receives information about the stimulus. This is how we feel the pain.

Movements are sometimes a result of reflex actions, but we can also contract our muscles as a voluntary action, using nerve cell pathways from the brain linked to the same motor neurones. A voluntary action is under *conscious control*.

SYNAPSES

Synapses are critical to the working of the nervous system. The CNS is made of many billions of nerve cells, and these have links with many others, through synapses. In the brain, each neurone may form synapses with thousands of other neurones. It is estimated that there are between 100 and 1000 million *million* synapses in the CNS. Since impulses can take different routes through these, there is an almost infinite number of possible pathways through the system.

A synapse is actually a *gap* between two nerve cells. The gap is not crossed by the electrical impulses passing through the neurones, but by chemicals. Impulses arriving at a synapse cause the ends of the fine branches of the axon to secrete a chemical, called a neurotransmitter. This chemical diffuses across the gap and attaches to the membrane of the second neurone. It then starts off impulses in the second cell (Figure 6.10). After the neurotransmitter has 'passed on the message', it is broken down by an enzyme.

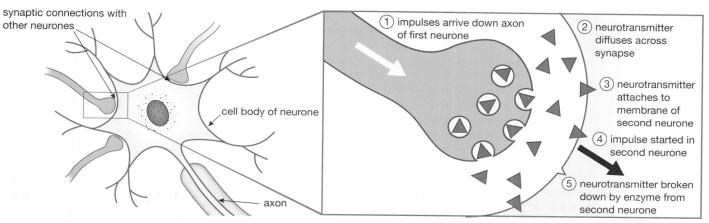

▲ Figure 6.10 The sequence of events happening at a synapse

Remember that many nerve cells, particularly those in the brain, have thousands of synapses with other neurones. The output of one cell may depend on the inputs from many cells adding together. In this way, synapses are important for integrating information in the CNS (Figure 6.11).

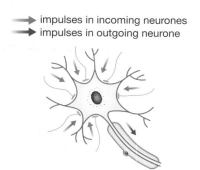

▲ Figure 6.11 Synapses allow the output of one nerve cell to be a result of integration of information from many other cells.

Because synapses are crossed by chemicals, it is easy for other chemicals to interfere with the working of the synapse. They may imitate the neurotransmitter, or block its action. This is the way that many drugs work.

LOOKING AHEAD – WHAT IS A NERVE IMPULSE?

A nerve impulse is an electrical signal that travels along the axon of a nerve cell. When the cell is not transmitting an impulse, there is small potential difference (voltage) across the nerve cell membrane. The potential inside the axon is about –70 mV lower than outside. This 'resting potential' is caused by differences in concentrations of various ions inside and outside the cell.

When the cell is stimulated, sodium ions (Na^+) rush into the axon through the membrane. The inflow of positively charged ions causes the potential to become positive – we say that it is *depolarised*. This sudden switch in voltage is called an 'action potential'. It only lasts for a few milliseconds (Figure 6.12). After a brief 'overshoot' below –70 mV it returns to normal (*repolarised*), as other positive ions (potassium, K^+) pass out through the membrane.

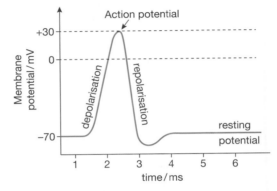

▲ Figure 6.12 Nerve cell action potential.

A nerve impulse is a *propagated* action potential. The action potential stimulates the next part of the cell membrane, so that the depolarisation spreads along the axon. After the action potential has passed, ion exchange pumps in the membrane sort out the imbalance of Na^+ and K^+ ions. The pumps use ATP for active transport – this is one reason why nerve cells need a lot of metabolic energy from respiration.

Nerve cells are called 'excitable cells' because they can change their membrane potential in this way. Other excitable cells include muscle and receptor cells. If you continue to study biology beyond International GCSE you will probably learn more about this interesting topic.

CHAPTER QUESTIONS

SKILLS CRITICAL THINKING

More questions on coordination can be found at the end of Unit 2 on page 130.

1 Which row in the table correctly describes the three types of neurone?

	Sensory neurone	Relay neurone	Motor neurone
A	connects neurones within the CNS	connects impulses to the effector from the CNS	connects impulses from the receptor to the CNS
B	connects impulses from the effector to the CNS	connects neurones within the CNS	connects impulses from the CNS to the receptor
C	connects neurones within the CNS	connects impulses from the receptor to the CNS	connects impulses to the effector from the CNS
D	connects impulses from the receptor to the CNS	connects neurones within the CNS	connects impulses to the effector from the CNS

SKILLS CRITICAL THINKING

2 The diagram below shows a section through the eye.

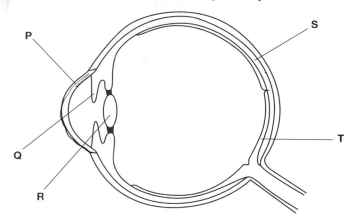

Which row in the table shows the cornea and the choroid?

	Cornea	Choroid
A	R	T
B	P	S
C	R	P
D	Q	T

3 A boy sits in the shade under a tree, reading a book. He looks up into the sunny sky at an aeroplane. Which of the following changes will take place in his eyes?

A The pupils dilate and the lens becomes less convex

B The pupils dilate and the lens becomes more convex

C The pupils constrict and the lens becomes less convex

D The pupils constrict and the lens becomes more convex

4 Below are two statements about how nerve cells work.

1. Neurotransmitters carry a nerve impulse along a neurone

2. An electrical charge carries a nerve impulse across a synapse

Which of these statements is/are true?

A 1   B 2

C 1 and 2   D neither

5 A cataract is an eye problem suffered by some people, especially the elderly. The lens of the eye becomes opaque (cloudy) which blocks the passage of light. It can lead to blindness. Cataracts can be treated by a simple eye operation, where a surgeon removes the lens and replaces it with an artificial lens. After the operation, the patient is able to see again, but the eye is unable to carry out accommodation, and the patient will probably need to wear glasses for close-up work, such as reading.

a What is meant by 'accommodation'?

SKILLS REASONING

b Why is accommodation not possible after a cataract operation?

c Explain how a normal eye accommodates to focus on a nearby object.

6 The diagram shows a section through a human eye.

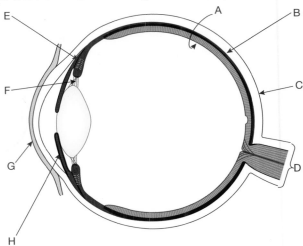

a The table below lists the functions of some of parts A to H. Copy the table and write the letters of the correct parts in the boxes.

Function	Letter
refracts light rays	
converts light into nerve impulses	
contains pigment to stop internal reflection	
contracts to change the shape of the lens	
takes nerve impulses to the brain	

b i Which label shows the iris?

ii Explain how the iris controls the amount of light entering the eye.

iii Why is this important?

7 The diagram shows some parts of the nervous system involved in a simple reflex action that happens when a finger touches a hot object.

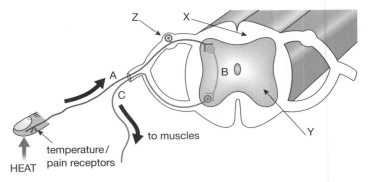

a What type of neurone is:

i neurone A

ii neurone B

iii neurone C?

b Describe the function of each of these types of neurone.

c Which parts of the nervous system are shown by the labels X, Y and Z?

d In what form is information passed along neurones?

e Explain how information passes from one neurone to another.

f Some drugs act at a synapse to prevent a person feeling pain. From your knowledge of synapses, suggest how they might work.

8 The diagram shows a motor neurone.

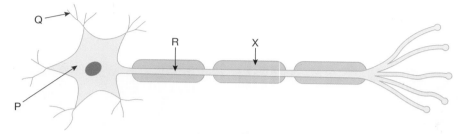

Name the parts of the neurone labelled P, Q and R.

a This motor neurone is 1.2 metres in length. It takes 0.016 seconds for an impulse to pass along the neurone. Calculate the speed of conduction of the impulse.

b Neurones need energy from ATP to conduct impulses. Name the organelle in the cell that provides most of this ATP.

c Structure R is surrounded by a sheath, labelled X in the diagram. What is the function of this sheath?

d Some diseases can cause damage to this sheath. Suggest what would happen to a person's nervous responses if this sheath were to be damaged.

9 a List five examples of stimuli that affect the body and state the response produced by each stimulus.

b For one of your five examples, explain:
  i the nature and role of the receptor
  ii the nature and role of the effector organ.

c For the same example, describe the chain of events from stimulus to response.

7 CHEMICAL COORDINATION

The nervous system (Chapter 6) is a coordination system forming a link between stimulus and response. The body has a second coordination system, which does not involve nerves. This is the endocrine system. It consists of organs called endocrine glands, which make chemical messenger substances called hormones. Hormones are carried in the bloodstream.

LEARNING OBJECTIVES

- Describe how responses can be controlled by hormonal communication

- Understand the differences between nervous and hormonal control

- Understand the sources, roles and effects of the following hormones:
  - adrenaline
  - insulin
  - testosterone*

- progesterone*
- oestrogen*

BIOLOGY ONLY
  - antidiuretic hormone (ADH)*
  - follicle stimulating hormone (FSH)*
  - luteinising hormone (LH)*

*These hormones will be dealt with in more detail in later chapters.

GLANDS AND HORMONES

A gland is an organ that releases or *secretes* a substance. This means that cells in the gland make a chemical which leaves the cells through the cell membrane. The chemical then travels somewhere else in the body, where it carries out its function. There are two types of glands – exocrine and endocrine glands. **Exocrine glands** secrete their products through a tube called a duct. For example, salivary glands in your mouth secrete saliva down salivary ducts, and tear glands secrete tears through ducts that lead to the surface of the eye. **Endocrine glands** have no duct, and so are called ductless glands. Instead, their products, the **hormones**, are secreted into the blood vessels that pass through the gland (Figure 7.1).

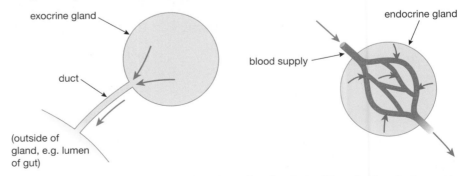

exocrine gland

endocrine gland

blood supply

duct

(outside of gland, e.g. lumen of gut)

▲ Figure 7.1 Exocrine glands secrete their products though a duct, while endocrine glands secrete hormones into the blood.

This chapter looks at some of the main endocrine glands and the functions of the hormones they produce. Because hormones are carried in the blood, they can travel to all areas of the body. They usually only affect certain tissues or organs, called 'target organs', which can be a long distance from the gland that made the hormone. Hormones only affect particular tissues or organs if the cells of that tissue or organ have special chemical receptors for the particular hormone. For example, the hormone **insulin** affects the cells of the liver, which have insulin receptors.

THE DIFFERENCES BETWEEN NERVOUS AND ENDOCRINE CONTROL

Although the nervous and endocrine systems both act to coordinate body functions, there are differences in the way that they do this. These are summarised in Table 7.1.

Table 7.1 The nervous and endocrine systems compared.

Nervous system	Endocrine system
works by nerve impulses transmitted through nerve cells (although chemicals are used at synapses)	works by hormones transmitted through the bloodstream
nerve impulses travel fast and usually have an 'instant' effect	hormones travel more slowly and generally take longer to act
response is usually short-lived	response is usually longer-lasting
impulses act on individual cells such as muscle fibres, so have a very localised effect	hormones can have widespread effects on different organs (although they only act on particular tissues or organs if the cells have the correct receptors)

THE ENDOCRINE GLANDS

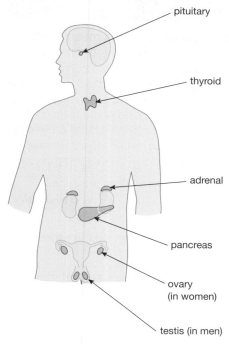

Figure 7.2 The main endocrine glands of the body.

The positions of the main endocrine glands are shown in Figure 7.2. A summary of some of the hormones that they make and their functions is given in Table 7.2.

The **pituitary gland** (often just called 'the pituitary') is found at the base of the brain. It produces a number of hormones, including **antidiuretic hormone (ADH)**, which acts on the kidneys, controlling the amount of water in the blood (see Chapter 8). The pituitary also releases hormones that regulate reproduction (see Chapter 9).

Just above the pituitary is a part of the brain called the **hypothalamus**. The pituitary contains neurones linking it to the hypothalamus, and some of its hormones are produced under the control of the brain.

Table 7.2: Some of the main endocrine glands, the hormones they produce and their functions.

Gland	Hormone	Some functions of the hormones
pituitary	follicle stimulating hormone (FSH)	stimulates egg development and oestrogen secretion in females and sperm production in males
	luteinising hormone (LH)	stimulates egg release (ovulation) in females and testosterone production in males
	antidiuretic hormone (ADH)	controls the water content of the blood
thyroid	thyroxine	controls the body's metabolic rate (how fast chemical reactions take place in cells)
pancreas	insulin	lowers blood glucose
	glucagon	raises blood glucose
adrenals	adrenaline	prepares the body for physical activity
testes	testosterone	controls the development of male secondary sexual characteristics
ovaries	oestrogen	controls the development of female secondary sexual characteristics
	progesterone	regulates the menstrual cycle

The **pancreas** is both an endocrine *and* an exocrine gland. It secretes two hormones involved in the regulation of blood glucose, and is also a gland of the digestive system, secreting enzymes through the pancreatic duct into the small intestine (see Chapter 4). The sex organs of males (**testes**) and females (**ovaries**) are also endocrine organs. In addition to their role in producing sex cells, the testes and ovaries make hormones that are involved in controlling reproduction. This topic is covered more fully in Chapter 9. We will now look at the functions of two hormones in more detail.

ADRENALINE – THE 'FIGHT OR FLIGHT' HORMONE

When you are frightened, excited or angry, your **adrenal glands** secrete the hormone adrenaline.

Adrenaline acts at a number of target organs and tissues, preparing the body for action. In animals other than humans this action usually means dealing with an attack by an enemy, where the animal can stay and fight or run away – hence 'fight or flight'. This is not often a problem with humans, but there are plenty of other times when adrenaline is released (Figure 7.3).

> **DID YOU KNOW?**
> 'Adrenal' means 'next to the kidneys', which describes where the adrenal glands are located – on top of these organs (see Figure 7.2).

▲ Figure 7.3 Many human activities cause adrenaline to be produced, not just a 'fight or flight' situation!

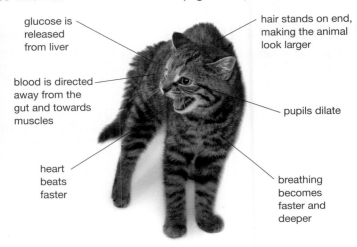

glucose is released from liver

blood is directed away from the gut and towards muscles

heart beats faster

hair stands on end, making the animal look larger

pupils dilate

breathing becomes faster and deeper

▲ Figure 7.4 Adrenaline affects the body of an animal in many ways.

If an animal's body is going to be prepared for action, the muscles need a good supply of oxygen and glucose for respiration. Adrenaline produces several changes in the body that make this happen (Figure 7.4) as well as other changes to prepare for fight or flight.

- The breathing rate increases and breaths become deeper, taking more oxygen into the body.
- The heart beats faster, sending more blood to the muscles, so that they receive more glucose and oxygen for respiration.
- Blood is diverted away from the intestine and into the muscles.
- In the liver, stored carbohydrate (glycogen) is changed into glucose and released into the blood. The muscle cells absorb more glucose and use it for respiration.
- The pupils dilate, increasing visual sensitivity to movement.
- Body hair stands upright, making the animal look larger to an enemy.
- Mental awareness is increased, so reactions are faster.

In humans, adrenaline is not just released in a 'fight or flight' situation, but in many other stressful activities too, such as preparing for a race, going for a job interview or taking an exam.

INSULIN – CONTROL OF BLOOD GLUCOSE

You saw earlier that adrenaline can raise blood glucose from stores in the liver. The liver cells contain carbohydrate in the form of glycogen. Glycogen is made from long chains of glucose sub-units joined together, forming a large insoluble molecule (see Chapter 4). Being insoluble makes glycogen a good storage product. When the body is short of glucose, the glycogen can be broken down into glucose, which then passes into the bloodstream.

Adrenaline raises blood glucose concentration in an emergency, but other hormones act all the time to control the level, keeping it fairly constant at a little less than 1 g of glucose in every dm^3 (cubic decimetre) of blood. The main hormone controlling glucose is insulin. Insulin is made by special cells in the pancreas. It stimulates the liver cells to take up glucose and convert it into glycogen, lowering the level of glucose in the blood.

The concentration of glucose in your blood will start to rise after you have had a meal. Sugars from digested carbohydrate pass into the blood and are carried to the liver in the hepatic portal vein (Chapter 5). In the liver the glucose is converted to glycogen, so the blood leaving the liver in the hepatic vein has a lower concentration of glucose than when it enters the liver.

DIABETES

Some people have a disease where their pancreas cannot make enough insulin to keep their blood glucose level constant – it rises to very high concentrations. The disease is called diabetes. One symptom of diabetes can be detected by a chemical test on urine. Normally, people have no glucose at all in their urine. Someone suffering from diabetes may have such a high concentration of glucose in the blood that it is excreted in their urine. This can be shown up by using coloured test strips (Figure 7.5).

KEY POINT

We should really refer to this disease by its full name, which is 'type 1' diabetes. There is also a 'type 2' diabetes, where the pancreas produces insulin but the body shows *insulin resistance*, where insulin has less effect than it should do. At first the pancreas makes extra insulin, but eventually it can't continue to make enough to maintain blood glucose at a normal level. Type 2 diabetes is common in people who are overweight and eat a poor diet that is high in sugar and other carbohydrates. It can be prevented and controlled by eating a good diet and doing regular exercise. Type 2 diabetes also tends to happen in middle-aged or older people, whereas type 1 can happen at any age, and is common in childhood.

Another symptom of diabetes is a constant thirst. This is because the high blood glucose concentration stimulates receptors in the hypothalamus of the brain. These 'thirst centres' are stimulated, so that by drinking, the person will dilute their blood.

Severe diabetes is very serious. If it is untreated, the sufferer loses weight and becomes weak and eventually falls into a coma and dies.

Carbohydrates in the diet, such as starch and sugars, are the source of glucose in the blood, so a person with diabetes can help to control their blood sugar if they limit the amount of carbohydrate that they eat. However a person

▲ Figure 7.5 Coloured test strips are used to detect glucose in urine.

DID YOU KNOW?

Insulin for the treatment of diabetes has been available since 1921, and has kept millions of people alive. It was originally extracted from the pancreases of animals, and much insulin is still obtained in this way. However, since the 1970s, human insulin has been produced commercially, from genetically modified (GM) bacteria. The bacteria have their DNA 'engineered' to contain the gene for human insulin (see Chapter 22).

with diabetes (type 1) also needs to receive daily injections of insulin to keep the glucose in their blood at the right level.

People with diabetes can check their blood glucose using a special sensor. They prick their finger and place a drop of blood on a test strip. The strip is then put into the sensor, which gives them an accurate reading of how much glucose is in their blood (Figure 7.6). They can then tell when to inject insulin and how much to inject.

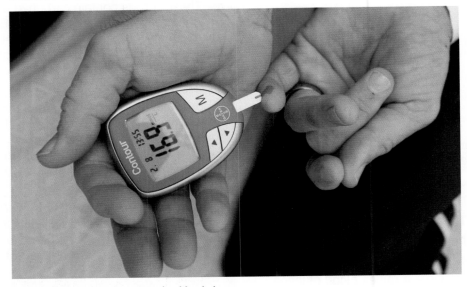

▲ Figure 7.6 Sensor for measuring blood glucose

CHAPTER QUESTIONS

More questions on chemical coordination can be found at the end of Unit 2 on page 130.

SKILLS   CRITICAL THINKING

1 Which of the following statement(s) is/are true?

    1. Insulin converts glucose to glycogen

    2. Insulin causes blood glucose levels to fall

    3. Glucose is stored as glucagon in the liver

    4. Glycogen can be broken down to release glucose into the blood

    A 2 only

    B 2 and 4

    C 3 and 4

    D 1 and 2

BIOLOGY ONLY

2 Which hormone controls the development of the male secondary sexual characteristics?

    A testosterone

    B oestrogen

    C progesterone

    D follicle stimulating hormone (FSH)

END OF BIOLOGY ONLY

SKILLS — CRITICAL THINKING

3 Which of the following will *not* happen when the hormone adrenaline is released?

A an increase in heart rate

B an increase in blood flow to the gut

C dilation of the pupils

D an increase in breathing rate

4 Which of the following is a symptom of type 1 diabetes?

A high insulin level in the blood

B low glucose level in the blood

C glucose present in the urine

D insulin present in the urine

SKILLS — ANALYSIS

5 a *Hormones* are *secreted* by *endocrine glands*. Explain the meaning of the four words in italics

b Identify the hormones A to D in the table.

Hormone	One function of this hormone
A	stimulates the liver to convert glucose to glycogen
B	controls the 'fight or flight' responses
C	controls the breaking of the voice at puberty in boys
D	completes the development of the uterus lining during the menstrual cycle

6 The graph shows the changes in blood glucose in a healthy woman over a 12-hour period.

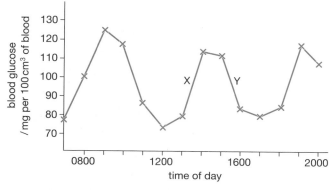

a Explain why there was a rise in blood glucose at X.

b How does the body bring about a decrease in blood glucose at Y? Your answer should include the words insulin, liver and pancreas.

c Diabetes is a disease where the body cannot control the concentration of glucose in the blood.

SKILLS — CRITICAL THINKING

i Why is this dangerous?

ii Describe two ways a person with diabetes can monitor their blood glucose level.

iii Explain two ways that a person with diabetes can help to control their blood glucose level.

8 HOMEOSTASIS AND EXCRETION

The kidneys have major roles to play in both homeostasis and excretion. This chapter is mainly concerned with the structure and function of the kidneys. It also deals with another important aspect of homeostasis – maintaining a steady body temperature.

LEARNING OBJECTIVES

- Understand that homeostasis is the maintenance of a constant internal environment
- Understand that control of body water content and body temperature are examples of homeostasis
- Know the excretory products of the lungs, kidneys and skin
- Understand the origin of carbon dioxide and oxygen as waste products of metabolism and their loss from the stomata of a leaf

BIOLOGY ONLY

- Understand that urine contains water, urea and ions
- Understand how the kidney carries out its roles of excretion and osmoregulation
- Describe the structure of the urinary system, including the kidneys, ureters, bladder and urethra

BIOLOGY ONLY

- Describe the structure of a nephron, including the Bowman's capsule and glomerulus, convoluted tubules, loop of Henlé and collecting duct
- Describe ultrafiltration in the Bowman's capsule and the composition of the glomerular filtrate
- Understand why selective reabsorption of glucose occurs at the proximal convoluted tubule
- Understand how water is reabsorbed into the blood from the collecting duct
- Describe the role of ADH in regulating the water content of the blood

- Describe the role of the skin in temperature regulation, with reference to sweating, vasoconstriction and vasodilation

Inside our bodies, conditions are kept relatively constant. This is called **homeostasis**. The kidneys are organs which have a major role to play in both homeostasis and in the removal of waste products, or **excretion**. They filter the blood, removing substances and controlling the concentration of water and solutes (dissolved substances) in the blood and other body fluids.

HOMEOSTASIS

If you were to drink a litre of water and wait for half an hour, your body would soon respond to this change by producing about the same volume of urine. In other words, it would automatically balance your water input and water loss. Drinking is the main way that our bodies gain water, but there are other sources (Figure 8.1). Some water is present in the food that we eat, and a small amount is formed by cell respiration. The body also loses water, mostly in urine, but also smaller volumes in sweat, faeces and exhaled air. Every day, we gain and lose about the same volume of water, so that the total content of our bodies stays more or less the same. This is an example of homeostasis. The word 'homeostasis' means 'steady state', and refers to keeping conditions inside the body relatively constant.

The inside of the body is known as the **internal environment.** You have probably heard of the 'environment', which means the 'surroundings' of an organism. The *internal* environment is the surroundings of the cells inside the body. It particularly means the blood, together with another liquid called **tissue fluid**.

KEY POINT

The meaning of 'homeostasis' is 'keeping the conditions in the internal environment of the body relatively constant'. One aspect of homeostasis is the maintenance of the water and salt content of the internal environment. This is called **osmoregulation**.

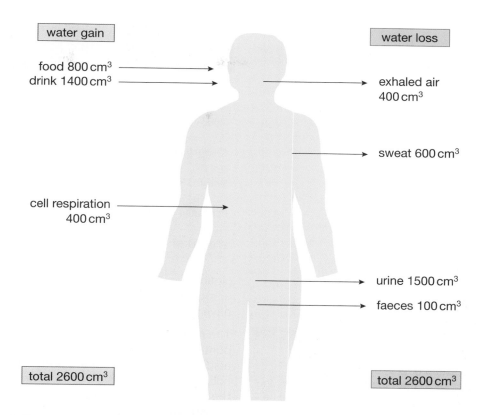

▲ Figure 8.1 The daily water balance of an adult

Tissue fluid is a watery solution of salts, glucose and other solutes. It surrounds all the cells of the body, forming a pathway for the transfer of nutrients between the blood and the cells. Tissue fluid is formed by leakage from blood capillaries. It is similar in composition to blood plasma, but lacks the plasma proteins.

It is not just water and salts that are kept constant in the body. Many other components of the internal environment are maintained. For example, the level of carbon dioxide in the blood is regulated, along with the blood pH, the concentration of dissolved glucose (see Chapter 7) and the body temperature.

Homeostasis is important because cells will only function properly if they are bathed in a tissue fluid which provides them with their optimum conditions. For instance, if the tissue fluid contains too many solutes, the cells will lose water by osmosis, and become dehydrated. If the tissue fluid is too dilute, the cells will swell up with water. Both conditions will prevent them working efficiently and might cause permanent damage. If the pH of the tissue fluid is not correct, it will affect the activity of the cell's enzymes, as will a body temperature much different from 37 °C. It is also important that excretory products are removed. Substances such as urea must be prevented from building up in the blood and tissue fluid, where they would be toxic to cells.

DID YOU KNOW?

'Salts' in urine or in the blood are present as ions. For example, the sodium chloride in Table 8.1 will be in solution as sodium ions (Na^+) and chloride ions (Cl^-). Urine contains many other ions, such as potassium (K^+), phosphate (HPO_4^{2-}) and ammonium (NH_4^+), and removes excess ions from the blood.

URINE

An adult human produces about 1.5 dm^3 of urine every day, although this volume depends very much on the amount of water drunk and the volume lost in other forms, such as sweat. Every litre of urine contains about 40 g of waste products and salts (Table 8.1).

Table 8.1 Some of the main solutes in urine.

Substance	Amount / g per dm³
urea	23.3
ammonia	0.4
other nitrogenous waste	1.6
sodium chloride (salt)	10.0
potassium	1.3
phosphate	2.3

KEY POINT

Excretion is the process by which waste products of metabolism are removed from the body. In humans, the main nitrogenous excretory substance is urea. Another excretory product is carbon dioxide, produced by cell respiration and excreted by the lungs (Chapter 3). The human skin is also an excretory organ, since the sweat that it secretes contains small amounts of urea.

Plants also excrete waste products of metabolism. In light, their leaves produce oxygen from photosynthesis. In the dark they excrete carbon dioxide from respiration. Plants exchange these gases through pores in the leaf called stomata (see Chapter 11).

DID YOU KNOW?

A baby cannot control its voluntary sphincter. When the bladder is full, the baby's involuntary sphincter relaxes, releasing the urine. A toddler learns to control this muscle and hold back the urine.

Notice the words *nitrogenous waste*. Urea and ammonia are two examples of nitrogenous waste. It means that they contain the element nitrogen. All animals have to excrete a nitrogenous waste product.

The reason behind this is quite complicated. Carbohydrates and fats only contain the elements carbon, hydrogen and oxygen. However, proteins also contain nitrogen. If the body has too much carbohydrate or fat, these substances can be stored, for example as glycogen in the liver, or as fat under the skin and around other organs. Excess proteins, or their building blocks (called amino acids) *cannot* be stored. The amino acids are first broken down in the liver. They are converted into carbohydrate (which is stored as glycogen) and the main nitrogen-containing waste product, urea. The urea passes into the blood, to be filtered out by the kidneys during the formation of urine. Notice that the urea is made by chemical reactions in the cells of the body (the body's metabolism). 'Excretion' means getting rid of waste of this kind. When the body gets rid of solid waste from the digestive system (faeces), this is not excretion, since it contains few products of *metabolism*, just the 'remains' of undigested food, along with bacteria and dead cells.

So the kidney is in fact carrying out two functions. It is a *homeostatic* organ, controlling the water and salt (ion) concentration in the body as well as an *excretory* organ, concentrating nitrogenous waste in a form that can be eliminated.

BIOLOGY ONLY

THE URINARY SYSTEM

The human urinary system is shown in Figure 8.2.

Each kidney is supplied with blood through a short **renal artery**. This leads straight from the body's main artery, the aorta, so the blood entering the kidney is at a high pressure. Inside each kidney the blood is filtered, and the 'cleaned' blood passes out through each **renal vein** to the main vein, or vena cava. The urine leaves the kidneys through two tubes, the **ureters**, and is stored in a muscular bag called the **bladder**.

The bladder has a tube leading to the outside, called the **urethra**. The wall of the urethra contains two ring-shaped muscles, called **sphincter muscles**. They can contract to close the urethra and hold back the urine. The lower sphincter muscle is under conscious control, or 'voluntary', while the upper one is involuntary – it automatically relaxes when the bladder is full.

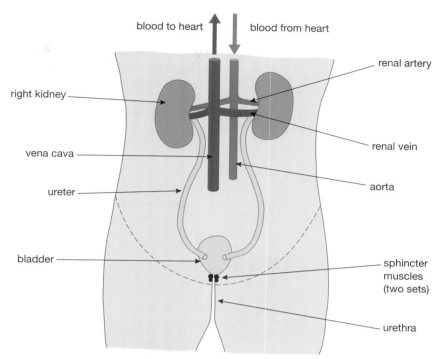

▲ Figure 8.2 The human urinary system

THE KIDNEYS

If you were to cut a kidney lengthwise you would be able to see the structures shown in Figure 8.3.

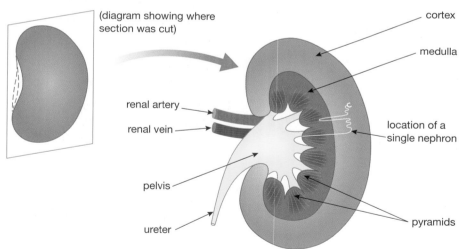

▲ Figure 8.3 Section through a kidney cut along the plane shown.

There is not much that you can make out without the help of a microscope. The darker outer region is called the **cortex**. This contains many tiny blood vessels that branch from the renal artery. It also contains microscopic tubes that are not blood vessels. They are the filtering units, called kidney tubules or **nephrons** (from the Greek word *nephros*, meaning kidney). The tubules then run down through the middle layer of the kidney, called the **medulla**. The medulla has bulges called 'pyramids' pointing inwards towards the concave side of the kidney. The tubules in the medulla eventually join up and lead to the tips of these pyramids, where they empty urine into a funnel-like structure called the **pelvis**. The pelvis connects with the ureter, carrying the urine to the bladder.

STRUCTURE OF THE NEPHRON

By careful dissection, biologists have been able to work out the structure of a single tubule and its blood supply (Figure 8.4). There are about a million of these in each kidney.

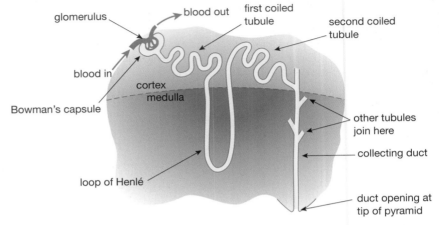

▲ Figure 8.4 A single nephron, showing its position in the kidney. Each kidney contains about a million of these filtering units.

ULTRAFILTRATION IN THE BOWMAN'S CAPSULE

At the start of the nephron is a hollow cup of cells called the **Bowman's capsule**. It surrounds a ball of blood capillaries called a **glomerulus** (plural glomeruli). It is here that the blood is filtered. Blood enters the kidney through the renal artery, which divides into smaller and smaller arteries. The smallest arteries (arterioles) supply the capillaries of the glomerulus (Figure 8.5).

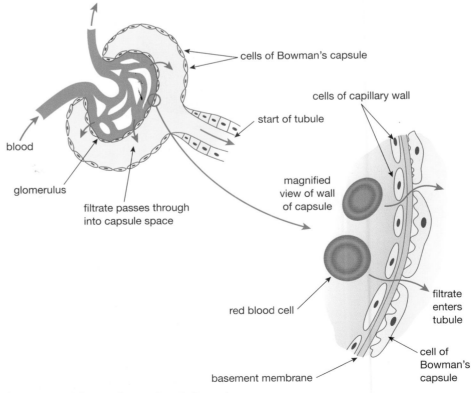

▲ Figure 8.5 A Bowman's capsule and glomerulus

A blood vessel with a smaller diameter carries blood away from the glomerulus, leading to capillary networks which surround the other parts of the nephron. Because of the resistance to flow caused by the glomerulus, the pressure of the blood in the arteriole leading to the glomerulus is very high. This pressure forces fluid from the blood through the walls of the capillaries and the Bowman's capsule, into the space in the middle of the capsule. Blood in the glomerulus and the space in the capsule are separated by two layers of cells, the capillary wall and the wall of the capsule. Between the two cell layers is a third layer called the **basement membrane**, which is not made of cells. These layers act like a filter, allowing water, ions, and small molecules like glucose and urea to pass through, but holding back blood cells and large molecules such as proteins. The fluid that enters the capsule space is called the **glomerular filtrate**. This process, where the filter separates different-sized molecules under pressure, is called **ultrafiltration**.

EXTENSION WORK

The cells of the glomerulus capillaries do not fit together very tightly, there are spaces between them making the capillary walls much more permeable than others in the body. The cells of the Bowman's capsule also have gaps between them, so only act as a coarse filter. It is the basement membrane that acts as the fine molecular filter.

CHANGES TO THE FILTRATE IN THE REST OF THE NEPHRON

The kidneys produce about 125 cm³ (0.125 dm³) of glomerular filtrate per minute. This works out at 180 dm³ per day. Remember though, only 1.5 dm³ of urine is lost from the body every day, which is less than 1% of the volume filtered through the capsules. The other 99% of the glomerular filtrate is *reabsorbed* back into the blood.

We know this because scientists have actually analysed samples of fluid from the space in the middle of the nephron. Despite the diameter of the space being only 20 µm (0.02 mm), it is possible to pierce the tubule with microscopic glass pipettes and extract the fluid for analysis. Figure 8.6 shows the structure of the nephron and the surrounding blood vessels in more detail.

There are two coiled regions of the tubule in the cortex, separated by a U-shaped loop that runs down into the medulla of the kidney, called the **loop of Henlé**. After the second coiled tubule, several nephrons join up to form a **collecting duct**, where the final urine passes out into the pelvis.

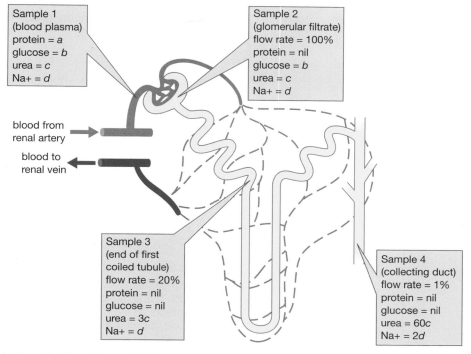

Sample 1
(blood plasma)
protein = a
glucose = b
urea = c
Na+ = d

Sample 2
(glomerular filtrate)
flow rate = 100%
protein = nil
glucose = b
urea = c
Na+ = d

blood from renal artery

blood to renal vein

Sample 3
(end of first coiled tubule)
flow rate = 20%
protein = nil
glucose = nil
urea = $3c$
Na+ = d

Sample 4
(collecting duct)
flow rate = 1%
protein = nil
glucose = nil
urea = $60c$
Na+ = $2d$

▲ Figure 8.6 A nephron and its blood supply. Samples 1–4 show what is happening to the fluid as it travels along the nephron.

Samples 1–4 show the results of analysing the blood before it enters the glomerulus, and the fluid at three points inside the tubule. The flow rate is a measure of how much water is in the tubule. If the flow rate falls from 100% to 50%, this is because 50% of the water in the tubule has passed back into the blood. To make the explanation easier, the concentrations of dissolved protein, glucose, urea and sodium are shown by different letters (*a* to *d*). You can tell the relative concentration of one substance at different points along the tubule from this. For example, urea at a concentration '3*c*' is three times more concentrated than when it is '*c*'.

In the blood (sample 1) the plasma contains many dissolved solutes, including protein, glucose, urea and salts (just sodium ions, Na^+, are shown here). As we saw above, protein molecules are too big to pass through into the tubule, so the protein concentration in sample 2 is zero. The other substances are at the same concentration as in the blood.

Now look at sample 3, taken at the end of the first coiled part of the tubule. The flow rate that was 100% is now 20%. This must mean that 80% of the water in the tubule has been reabsorbed back into the blood. If no solutes were reabsorbed along with the water, their concentrations should be *five times* what they were in sample 2. Since the concentration of sodium hasn't changed, 80% of this substance must have been reabsorbed (and some of the urea too). However, the glucose concentration is now zero – *all* of the glucose is taken back into the blood in the first coiled tubule. This is necessary because glucose is a useful substance that is needed by the body.

Finally, look at sample 4. By the time the fluid passes through the collecting duct, its flow rate is only 1%. This is because 99% of the water has been reabsorbed. Protein and glucose are still zero, but most of the urea is still in the fluid. The level of sodium is only 2*d*, so not all of it has been reabsorbed, but it is still twice as concentrated as in the blood.

This description has only looked at a few of the more important substances. Other solutes are concentrated in the urine by different amounts. Some, like ammonium ions, are secreted *into* the fluid as it passes along the tubule. The concentration of ammonium ions in the urine is about 150 times what it is in the blood.

THE LOOP OF HENLÉ

You might be wondering what the role of the loop of Henlé is. The full answer to this is rather complicated. You may meet it again if you study biology beyond International GCSE, but for now a brief explanation will be enough. It is involved with concentrating the fluid in the tubule by causing more water to be reabsorbed into the blood. Mammals with long loops of Henlé can make a more concentrated urine than ones with short loops. Desert animals have many long loops of Henlé, so they are able to produce very concentrated urine, conserving water in their bodies. Animals which have easy access to water, such as otters or beavers, have short loops of Henlé. Humans have a mixture of long and short loops.

KEY POINT

Here is a summary of what happens in the kidney nephron.

Part of the plasma leaves the blood in the Bowman's capsule and enters the nephron. The filtrate consists of water and small molecules. As the fluid passes along the nephron, all the glucose is absorbed back into the blood in the first coiled part of the tubule, along with most of the sodium and chloride ions. In the rest of the tubule, more water and ions are reabsorbed, and some solutes like ammonium ions are secreted into the tubule. The final urine contains urea at a much higher concentration than in the blood. It also contains controlled quantities of water and ions.

CONTROL OF THE BODY'S WATER CONTENT

Not only can the kidney produce urine that is more concentrated than the blood, it can also *control* the concentration of the urine, and so *regulate* the water content of the blood. This chapter began by asking you to think what would happen if you drank a litre of water. The kidneys respond to this 'upset' to the body's water balance by making a larger volume of more dilute urine. Conversely, if the blood becomes too concentrated, the kidneys produce a smaller volume of urine. These changes are controlled by a hormone produced by the pituitary gland, at the base of the brain. The hormone is called **antidiuretic hormone**, or **ADH**.

'Diuresis' means the flow of urine from the body, so 'antidiuresis' means producing less urine. ADH starts to work when your body loses too much water, for example if you are sweating heavily and not replacing lost water by drinking.

The loss of water means that the concentration of the blood starts to increase. This is detected by receptor cells in a region of the brain called the hypothalamus, situated above the pituitary gland. These cells are sensitive to the solute concentration of the blood, and cause the pituitary gland to release more ADH. The ADH travels in the bloodstream to the kidney. At the kidney tubules it causes the collecting ducts to become more permeable to water, so that more water is reabsorbed back into the blood. This makes the urine more concentrated, so that the body loses less water and the blood becomes more dilute.

The action of ADH illustrates the principle of **negative feedback**. A change in conditions in the body is detected, and starts a process that works to return conditions to normal. When the conditions are returned to normal, the corrective process is switched off (Figure 8.7).

In the situation described in the text, where the body loses too much water, the blood becomes too concentrated. This switches on ADH release, which acts at the kidneys to correct the problem. The word 'negative' means that the process works to eliminate the change. When the blood returns to normal, ADH release is switched off. The feedback pathway forms a 'closed loop'. Many conditions in the body are regulated by negative feedback loops like this.

When the water content of the blood returns to normal, this acts as a signal to 'switch off' the release of ADH. The kidney tubules then reabsorb less water. Similarly, if someone drinks a large volume of water, the blood will become too dilute. This leads to lower levels of ADH secretion, the kidney tubules become less permeable to water, and more water passes out of the body in the urine. In this way, through the action of ADH, the level of water in the internal environment is kept constant.

END OF BIOLOGY ONLY

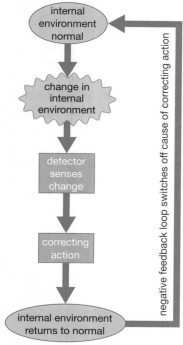

▲ Figure 8.7 In homeostasis, the extent of a correction is monitored by negative feedback.

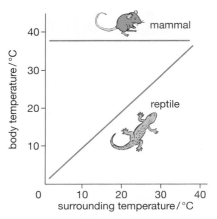

▲ Figure 8.8 The temperature of a homeotherm, such as a mammal, is kept constant at different external temperatures, whereas the lizard's body temperature changes.

DID YOU KNOW?

Physiology is the branch of biology that deals with how the bodies of animals or plants work, for example how muscles contract, how nerves send impulses, or how xylem carries water through plants. In this chapter you have read about kidney physiology.

CONTROL OF BODY TEMPERATURE

You may have heard mammals and birds described as 'warm blooded'. A better word for this is **homeothermic**. It means that they keep their body temperature constant, despite changes in the temperature of their surroundings. For example, the body temperature of humans is kept steady at about 37 °C, give or take a few tenths of a degree. This is another example of homeostasis. All other animals are 'cold blooded'. For example, if a lizard is kept in an aquarium at 20 °C, its body temperature will be 20 °C too. If the temperature of the aquarium is raised to 25 °C, the lizard's body temperature will rise to 25 °C as well. We can show this difference between **homeotherms** and other animals as a graph (Figure 8.8).

In the wild, lizards keep their temperature more constant than in Figure 8.8, by adapting their behaviour. For example, in the morning they may stay in the sun to warm their bodies, or at midday, if the sun is too hot, retreat to holes in the ground to cool down.

The real difference between homeotherms and all other animals is that homeotherms can keep their temperatures constant by using *physiological* changes for generating or losing heat. For this reason, mammals and birds are also called 'endotherms', meaning 'heat from inside'.

An endotherm uses heat from the chemical reactions in its cells to warm its body. It then controls its heat loss by regulating processes like sweating and blood flow through the skin. Endotherms use behavioural ways to control their temperature too. For example, penguins 'huddle' together in groups to keep warm, and humans put on extra clothes in winter.

What is the advantage of a human maintaining a body temperature of 37 °C? It means that all the chemical reactions taking place in the cells of the body can go on at a steady, predictable rate. The metabolism doesn't slow down in cold environments. If you watch goldfish in a garden pond, you will notice that in summer, when the pond water is warm, they are very active, swimming about quickly. In winter, when the temperature drops, the fish slow down and become very sluggish in their actions. This would happen to a mammal too, if its body temperature varied.

It is also important that the body does not become *too* hot. The cells' enzymes work best at 37 °C. At higher temperatures enzymes, like all proteins, are destroyed by **denaturing** (see Chapter 1). Endotherms have all evolved a body temperature of around 40 °C (Table 8.2) and enzymes that work best at this temperature.

Table 8.2 The body temperatures of a range of mammals and birds

Species	Average and normal range of body temperature / °C	Species	Average and normal range of body temperature / °C
brown bear	38.0 ± 1.0	shrew	35.7 ± 1.2
camel	37.5 ± 0.5	whale	35.7 ± 0.1
elephant	36.2 ± 0.5	duck	43.1 ± 0.3
fox	38.8 ± 1.3	ostrich	39.2 ± 0.7
human	36.9 ± 0.7	penguin	39.0 ± 0.2
mouse	39.3 ± 1.3	thrush	40.0 ± 1.7
polar bear	37.5 ± 0.4	wren	41.0 ± 1.0

MONITORING BODY TEMPERATURE

In humans and other mammals the core body temperature is monitored by a part of the brain called the **thermoregulatory centre**. This is located in the hypothalamus of the brain. It acts as the body's 'thermostat'.

If a person goes into a warm or cold environment, the first thing that happens is that temperature receptors in the skin send electrical impulses to the hypothalamus, which stimulates the brain to alter our behaviour. We start to feel hot or cold, and usually do something about it, such as finding shade, having a cold drink, or putting on more clothes.

If changes to our behaviour are not enough to keep our body temperature constant, the thermoregulatory centre in the hypothalamus detects a change in the temperature of the blood flowing through it. It then sends signals via nerves to other organs of the body, which regulate the temperature by physiological means.

> **DID YOU KNOW?**
> A thermostat is a switch that is turned on or off by a change in temperature. It is used in electrical appliances to keep their temperature steady. For example, a thermostat in an iron can be set to 'hot' or 'cool' to keep the temperature of the iron set for ironing different materials.

THE SKIN AND TEMPERATURE CONTROL

The human skin has a number of functions related to the fact that it forms the outer surface of the body. These include:

- forming a tough outer layer able to resist mechanical damage
- acting as a barrier to the entry of pathogens
- forming an impermeable surface, preventing loss of water
- acting as a sense organ for touch and temperature changes
- controlling the loss of heat through the body surface.

Figure 8.9 shows the structure of human skin. It is made up of three layers: the epidermis, dermis and hypodermis.

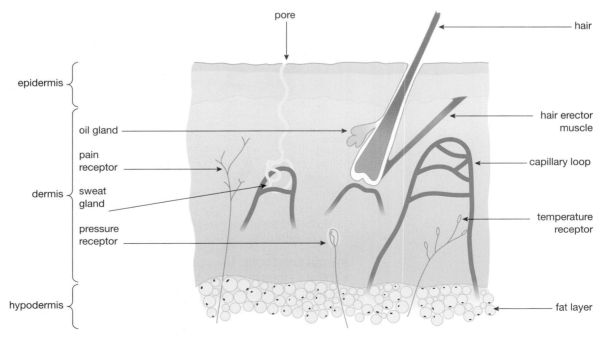

▲ Figure 8.9 A section through human skin

The outer **epidermis** consists of dead cells that stop water loss and protect the body against invasion by microorganisms such as bacteria. The **hypodermis** contains fatty tissue, which insulates the body against heat loss and is a store of energy. The middle layer, the **dermis**, contains many sensory receptors. It is also the location of sweat glands and many small blood vessels, as well as hair follicles. These last three structures are involved in temperature control.

Imagine that the hypothalamus detects a rise in the central (core) body temperature. Immediately it sends nerve impulses to the skin. These bring about changes to correct the rise in temperature.

First of all, the **sweat glands** produce greater amounts of sweat. This liquid is secreted onto the surface of the skin. When a liquid evaporates, it turns into a gas. This change needs energy, called the latent heat of vaporisation. When sweat evaporates, the energy is supplied by the body's heat, cooling the body down. It is not that the sweat is cool – it is secreted at body temperature. It only has a cooling action when it evaporates. In very humid atmospheres (e.g. a tropical rainforest) the sweat stays on the skin and doesn't evaporate. It then has very little cooling effect.

Secondly, hairs on the surface of the skin lie flat against the skin's surface. This happens because of the relaxation of tiny muscles called **hair erector muscles** attached to the base of each hair. In cold conditions, these contract and the hairs are pulled upright. The hairs trap a layer of air next to the skin, and since air is a poor conductor of heat, this acts as insulation. In warm conditions, the thinner layer of trapped air means that more heat will be lost. This is not very effective in humans, because the hairs over most of our body do not grow very large. It is very effective in hairy mammals like cats or dogs. The same principle is used by birds, which 'fluff out' their feathers in cold weather.

Lastly, there are tiny blood vessels called capillary loops in the dermis. Blood flows through these loops, radiating heat to the outside, and cooling the body down. If the body is too hot, arterioles (small arteries) leading to the capillary loops dilate (widen). This increases the blood flow to the skin's surface (Figure 8.10) and is called **vasodilation**.

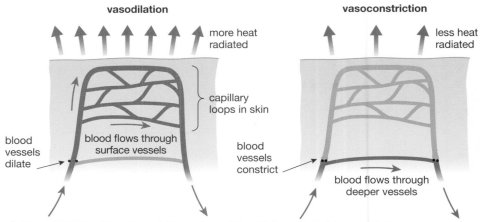

▲ Figure 8.10 Blood flow through the surface of the skin is controlled by vasodilation or vasoconstriction.

In cold conditions, the opposite happens. The arterioles leading to the surface capillary loops constrict (become narrower) and blood flow to the surface of the skin is reduced, so that less heat is lost. This is called **vasoconstriction**. Vasoconstriction and vasodilation are brought about by tiny rings of muscles in the walls of the arterioles, called sphincter muscles, like the sphincters you read about earlier in this chapter, at the outlet of the bladder.

There are other ways that the body can control heat loss and heat gain. In cold conditions, the body's metabolism speeds up, generating more heat. The liver, a large organ, can produce a lot of metabolic heat in this way. The hormone adrenaline stimulates the increase in metabolism (see Chapter 7). **Shivering** also takes place, where the muscles contract and relax rapidly. This also generates a large amount of heat.

Sweating, vasodilation and vasoconstriction, hair erection, shivering and changes to the metabolism, along with behavioural actions, work together to keep the body temperature to within a few tenths of a degree of the 'normal' 37 °C. If the difference is any bigger than this it shows that something is wrong. For instance, a temperature of 39 °C might be due to an illness.

CHAPTER QUESTIONS

More questions on homeostasis and excretion can be found at the end of Unit 2 on page 130.

SKILLS   CRITICAL THINKING

1 Which of the following is *not* an example of excretion?

    A loss of carbon dioxide by a plant in the dark

    B removal of carbon dioxide from the lungs

    C removal of urea in urine

    D elimination of faeces from the alimentary canal

BIOLOGY ONLY

2 The following structures are parts of the human urinary system.

    1. ureter

    2. kidney

    3. urethra

    4. bladder

    In which order does a molecule of urea pass through these structures?

    A $2 \rightarrow 1 \rightarrow 4 \rightarrow 3$

    B $2 \rightarrow 3 \rightarrow 4 \rightarrow 1$

    C $1 \rightarrow 3 \rightarrow 2 \rightarrow 4$

    D $3 \rightarrow 4 \rightarrow 1 \rightarrow 2$

3 Below are three statements about the action of antidiuretic hormone (ADH).

    1. More ADH is released when the water content of the blood rises

    2. ADH increases the permeability of the collecting duct

    3. When ADH is released, more water is reabsorbed.

    Which of these statements are true?

    A 1 and 2

    B 1 and 3

    C 2 and 3

    D 1, 2 and 3

END OF BIOLOGY ONLY

4 If the human body temperature starts to rise, which of the following happens?

A vasoconstriction of arterioles in the skin

B contraction of hair erector muscles

C decrease in the rate of metabolism

D decrease in production of sweat by glands in the skin

BIOLOGY ONLY

5 Explain the meaning of the following terms:

a homeostasis

b excretion

c ultrafiltration

d selective reabsorption

e endotherm.

6 The diagram below shows a simple diagram of a nephron (kidney tubule).

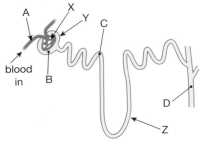

a What are the names of the parts labelled X, Y and Z?

b Four places in the nephron and its blood supply are labelled A, B, C and D. Which of the following substances are found at each of these four places?

water   urea   protein   glucose   salt

7 The hormone ADH controls the amount of water removed from the blood by the kidneys. Write a short description of the action of ADH in a person who has lost a lot of water by sweating, but has been unable to replace this water by drinking. Explain how this is an example of negative feedback. (You will need to write about 250 words to answer this question fully.)

8 The bar chart shows the volume of urine collected from a person before and after drinking $1000\,cm^3$ ($1\,dm^3$) of distilled water. The person's urine was collected immediately before the water was drunk and then at 30-minute intervals for four hours.

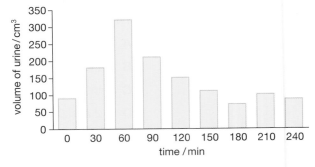

SKILLS ANALYSIS

5

9

a   Describe how the output of urine changed during the course of the experiment.

b   Explain the difference in urine produced at 60 minutes and at 90 minutes.

SKILLS REASONING

c   The same experiment was repeated with the person sitting in a very hot room. How would you expect the volume of urine collected to differ from the first experiment? Explain your answer.

SKILLS PROBLEM SOLVING

d   Between 90 and 120 minutes, the person produced $150 \, cm^3$ of urine. If the rate of filtration at the glomeruli during this time was $125 \, cm^3$ per minute, calculate the percentage of filtrate reabsorbed by the kidney tubules.

END OF BIOLOGY ONLY

SKILLS INTERPRETATION

7

9   Construct a table like the one below to show the changes that take place when a person is put in a hot or cold environment. Your table should have three columns.

Changes taking place	Hot environment	Cold environment
sweating		
blood flow through capillary loops		vasoconstriction decreases blood flow through surface capillaries so that less heat is radiated from the skin
hairs in skin		
shivering		
metabolism		

SKILLS ANALYSIS, REASONING

10   Look at the body temperatures of mammals and birds shown in Table 8.2 on page 112. Use the information in the table to answer these questions:

a   How does the average temperature of birds differ from the average temperature of mammals? Can you suggest why this is an advantage for birds?

b   Is there a relationship between the body temperature of a mammal and the temperature of its habitat? Give an example to support your answer.

6

c   Polar bears have thick white fur covering their bodies. Explain two ways in which this is an adaptation to their habitat (the place where the animal lives).

9 REPRODUCTION IN HUMANS

One of the characteristics of living organisms that makes them different from non-living things is their ability to produce offspring, or **reproduce**. Reproduction is all about an organism passing on its genes. This can be through special sex cells, or gametes. It can also be asexually, without the production of gametes. In this chapter we look at the differences between sexual and asexual reproduction, and study in detail the process of human reproduction.

LEARNING OBJECTIVES

- Understand the differences between sexual and asexual reproduction

- Understand that fertilisation involves the fusion of a male and female gamete to produce a zygote

- Understand that a zygote undergoes cell division and develops into an embryo

- Understand how the structure and function of the human male and female reproductive systems are adapted for their functions

- Describe the role of the placenta in the nutrition of the developing embryo

- Understand how the developing embryo is protected by amniotic fluid

- Understand the roles of oestrogen and testosterone in the development of secondary sexual characteristics

- Understand the roles of oestrogen and progesterone in the menstrual cycle

BIOLOGY ONLY

- Understand the roles of FSH and LH in the menstrual cycle

SEXUAL AND ASEXUAL REPRODUCTION COMPARED

In any method of reproduction, the end result is the production of more organisms of the same species. Humans produce more humans, pea plants produce more pea plants and salmonella bacteria produce more salmonella bacteria. However, the way in which they reproduce differs. There are two types of reproduction: **sexual reproduction** and **asexual reproduction**.

In sexual reproduction, specialised sex cells called **gametes** are produced. There are usually two types, a mobile male gamete called a **sperm** and a stationary female gamete called an egg cell or **ovum** (plural **ova**).

The sperm must move to the egg and fuse (join) with it. This is called **fertilisation** (Figure 9.1). The single cell formed by fertilisation is called a **zygote**. This cell will divide many times by mitosis to form all the cells of the new animal.

In asexual reproduction, there are no specialised gametes and there is no fertilisation. Instead, cells in one part of the body divide by mitosis to form a structure that breaks away from the parent body and grows into a new organism. Not many animals reproduce in this way. Figure 9.2 shows *Hydra* (a small animal similar to jellyfish) reproducing by budding. Cells in the body wall divide to form a small version of the adult. This eventually breaks off and becomes a free-living *Hydra*. One animal may produce several 'buds' in a short space of time.

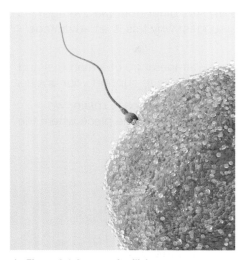

▲ Figure 9.1 A sperm fertilising an egg

KEY POINT

Individuals produced asexually from the same adult organism are called **clones**.

▲ Figure 9.2 *Hydra* reproducing asexually by budding

KEY POINT

A **gene** is a section of DNA that determines a particular characteristic or feature. Genes are found in the nucleus of a cell on the chromosomes (see Chapter 18).

All the offspring produced from *Hydra* buds are genetically identical – they have exactly the same genes. This is because all the cells of the new individual are produced by **mitosis** from just one cell in the body of the adult. When cells divide by mitosis, the new cells that are produced are exact copies of the original cell (see Chapter 17 for a description of mitosis). As a result, all the cells of an organism that are produced asexually have the same genes as the cell that produced them – the original adult cell. So *all* asexually produced offspring from one adult will have the same genes as the cells of the adult. They will *all* be genetic copies of that adult and so will be identical to each other.

Asexual reproduction is useful to a species when the environment in which it lives is relatively stable. If an organism is well adapted to this stable environment, asexual reproduction will produce offspring that are also well adapted. However, if the environment changes significantly, then *all* the individuals will be affected equally by the change. It may be such a dramatic change that none of the individuals are adapted well enough to survive. The species will die out in that area.

SEXUAL REPRODUCTION

There are four stages in any method of sexual reproduction.

- Gametes (sperm and egg cells) are produced.
- The male gamete (sperm) is transferred to the female gamete (egg cell).
- Fertilisation must occur – the sperm fuses with the egg.
- The zygote formed develops into a new individual.

The offspring produced by sexual reproduction show a great deal of genetic variation as a result of both gamete production and fertilisation.

PRODUCTION OF GAMETES

Sperm are produced in the male sex organs – the testes. Eggs are produced in the female sex organs – the ovaries. Both are produced when cells inside these organs divide. These cells do not divide by mitosis but by **meiosis** (see Chapter 17). Meiosis produces cells that are not genetically identical and have only half the number of chromosomes as the original cell.

KEY POINT

Cells that have the full number of chromosomes are called **diploid** cells. Cells that only have half the normal number of chromosomes are called **haploid** cells.

TRANSFER OF THE SPERM TO THE EGG

Sperm are specialised for swimming. They have a tail-like **flagellum** that moves them through a fluid. Figure 9.3 shows the structure of a sperm.

Some male animals, such as those of most fish, release their sperm into the water in which they live. The female animals release their eggs into the water and the sperm then swim through the water to fertilise the eggs. This is called *external* fertilisation as it takes place *outside* the body. Before the release takes place, there is usually some mating behaviour to ensure that male and female are in the same place at the same time. This gives the best chance of fertilisation occurring before water currents sweep the sex cells away.

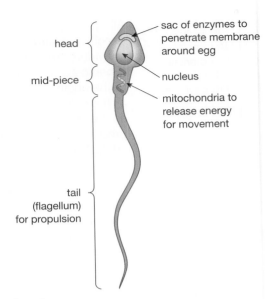

▲ Figure 9.3 The structure of a sperm

Other male animals, such as those of birds and mammals, **ejaculate** their sperm in a special fluid into the bodies of the females. *Internal* fertilisation then takes place *inside* the female's body. Fertilisation is much more likely as there are no external factors to prevent the sperm from reaching the eggs. Some form of sexual intercourse precedes ejaculation.

FERTILISATION

REMINDER

Red blood cells are exceptions. They have no nucleus, so have no chromosomes.

Once the sperm has reached the egg, its nucleus must enter the egg and fuse with the egg nucleus. As each gamete has only half the normal number of chromosomes, the zygote formed by fertilisation will have the full number of chromosomes. In humans, the sperm and egg each have only 23 chromosomes. The zygote has 46 chromosomes, like all other cells in the body. Figure 9.4 shows the main stages in fertilisation.

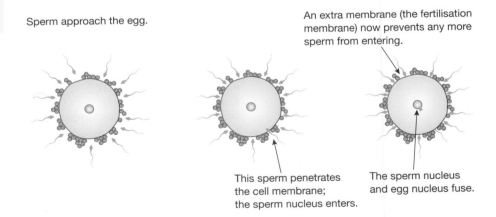

▲ Figure 9.4 The main stages in fertilisation

Fertilisation does more than just restore the diploid chromosome number; it provides an additional source of genetic variation. The sperm and eggs are all genetically different because they are formed by meiosis. Therefore, each time fertilisation takes place, it brings together a different combination of genes.

SUMMARY OF THE HUMAN LIFE CYCLE

Each zygote that is formed must divide to produce all the cells that will make up the adult. All these cells must have the full number of chromosomes, so the zygote divides repeatedly by mitosis. Figure 9.5 shows the importance of meiosis, mitosis and fertilisation in the human life cycle.

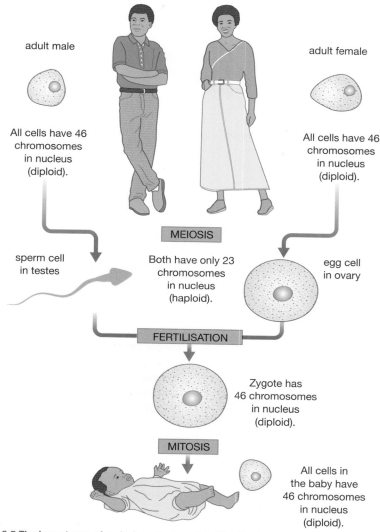

adult male

All cells have 46 chromosomes in nucleus (diploid).

adult female

All cells have 46 chromosomes in nucleus (diploid).

MEIOSIS

sperm cell in testes

Both have only 23 chromosomes in nucleus (haploid).

egg cell in ovary

FERTILISATION

Zygote has 46 chromosomes in nucleus (diploid).

MITOSIS

All cells in the baby have 46 chromosomes in nucleus (diploid).

▲ Figure 9.5 The importance of meiosis, mitosis and fertilisation in the human life cycle.

Mitosis is not the only process involved in development, otherwise all that would be produced would be a ball of cells. During the process, cells move around and different shaped structures are formed. Also, different cells specialise to become bone cells, nerve cells, muscle cells, and so on (the process called **differentiation** – see Chapter 1).

REPRODUCTION IN HUMANS

Humans reproduce sexually and fertilisation is internal. Figures 9.6 and 9.7 show the structure of the human female and male reproductive systems.

The sperm are produced in the testes by meiosis. During sexual intercourse, they pass along the sperm duct and are mixed with a fluid from the seminal vesicles. This mixture, called **semen**, is ejaculated through the urethra into the vagina of the female. The sperm then begin to swim towards the oviducts.

Side view

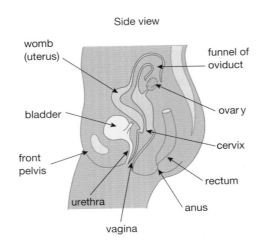

Front view

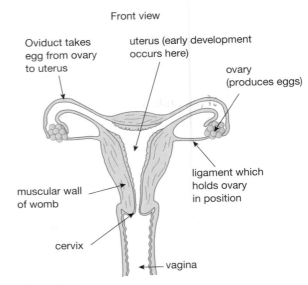

▲ Figure 9.6 The human female reproductive system

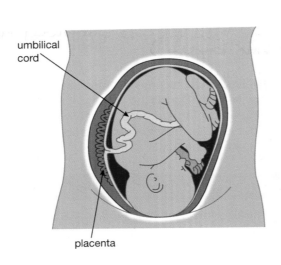

▲ Figure 9.7 The human male reproductive system.

Each month, an egg is released into an **oviduct** from one of the ovaries. (The oviduct is also known as the Fallopian tube.) This is called **ovulation**. If an egg is present in the oviduct, then it may be fertilised by sperm introduced during intercourse. The zygote formed will begin to develop into an **embryo**, which will implant in the lining of the uterus. Here, the embryo will develop a **placenta**, which will allow the embryo to obtain materials such as oxygen and nutrients from the mother's blood. It also allows the embryo to get rid of waste products such as urea and carbon dioxide, as well as anchoring the embryo in the uterus. The placenta secretes female hormones, in particular **progesterone**, which maintain the pregnancy and prevent the embryo from aborting (being rejected by the mother's body). Figure 9.8 shows the structure and position of the placenta.

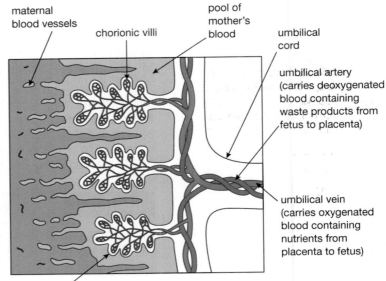

▲ Figure 9.8 The position of the fetus just before birth, and the structure of the placenta.

During pregnancy, a membrane called the **amnion** encloses the developing embryo. The amnion secretes a fluid called **amniotic fluid**, which protects the developing embryo against sudden movements and bumps. As the embryo develops, it becomes more and more complex. When it becomes recognisably human, we no longer call it an embryo but a **fetus**. At the end of nine months of development, there just isn't any room left for the fetus to grow and it sends a hormonal 'signal' to the mother to begin the birth process. Figure 9.8 also shows the position of a human fetus just before birth.

There are three stages to the birth of a child.

1. **Dilation of the cervix.** The **cervix** is the 'neck' of the uterus. It gets wider to allow the baby to pass through. The muscles of the uterus contract quite strongly and tears the amnion, allowing the amniotic fluid to escape. (In some countries the woman describes this as 'her waters have broken'.)

2. **Delivery of the baby.** Strong contractions of the muscles of the uterus push the baby's head through the cervix and then through the vagina to the outside world.

3. **Delivery of the afterbirth.** After the baby has been born, the uterus continues to contract and pushes the placenta out, together with the membranes that surrounded the baby. These are known as the afterbirth.

Figure 9.9 shows the stages of birth.

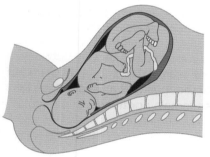
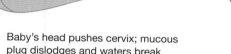

1 Baby's head pushes cervix; mucous plug dislodges and waters break.

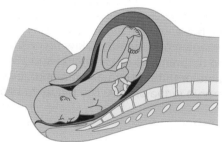

2 Uterus contracts to push baby out through the vagina.

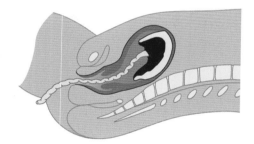

3 The placenta becomes detached from the wall of the uterus and is expelled through the vagina as the afterbirth.

▲ Figure 9.9 The stages of birth

HORMONES CONTROLLING REPRODUCTION

Most animals are unable to reproduce when they are young. We say that they are sexually immature. When a baby is born, it is recognisable as a boy or girl by its sex organs.

The presence of male or female sex organs is known as the primary sex characteristics. During their teens, changes happen to boys and girls that lead to sexual maturity. These changes are controlled by hormones, and the time when they happen is called **puberty**. Puberty involves two developments. The first is that the gametes (eggs and sperm) start to mature and be released. The second is that the bodies of both sexes adapt to allow reproduction to take place. These events are started by hormones released by the pituitary gland (see Chapter 7, Table 7.2) called **follicle stimulating hormone** (**FSH**) and **luteinising hormone** (**LH**).

In boys, FSH stimulates sperm production, while LH instructs the testes to secrete the male sex hormone, **testosterone**. Testosterone controls the development of the male **secondary sexual characteristics**. These include growth of the penis and testes, growth of facial and body hair, muscle development and breaking of the voice (Table 9.1).

In girls, the pituitary hormones control the release of a female sex hormone called **oestrogen**, from the ovaries. Oestrogen produces the female secondary sexual characteristics, such as breast development and the beginning of menstruation ('periods').

Table 9.1 Changes at puberty.

In boys	In girls
sperm production starts	the menstrual cycle begins, and eggs are released by the ovaries every month
growth and development of male sexual organs	growth and development of female sexual organs
growth of armpit and pubic hair, and chest and facial hair (beard)	growth of armpit and pubic hair
increase in body mass; growth of muscles, e.g. chest	increase in body mass; development of 'rounded' shape to hips
voice breaks	voice deepens without sudden 'breaking'
sexual 'drive' develops	sexual 'drive' develops
	breasts develop

The age when puberty takes place can vary a lot, but it is usually between about 11 and 14 years in girls and 13 and 16 years in boys. It takes several years for puberty to be completed. Some of the most complex changes take place in girls, with the start of menstruation.

HORMONES AND THE MENSTRUAL CYCLE

'Menstrual' means 'monthly', and in most women the **menstrual cycle** takes about a month, although it can vary from as little as two weeks to as long as six weeks (Figures 9.10 and 9.11). In the middle of the cycle is an event called **ovulation**, which is the release of a mature egg cell, or egg.

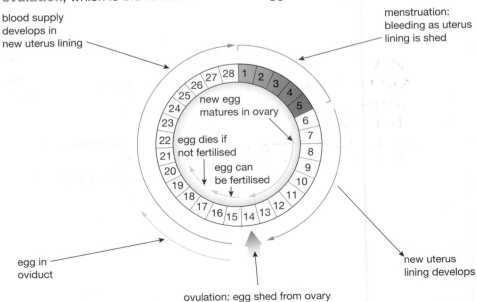

Figure 9.10 The menstrual cycle

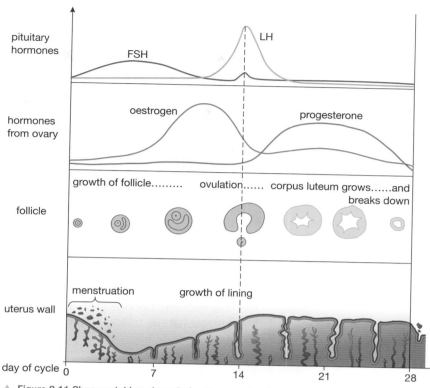

Figure 9.11 Changes taking place during the menstrual cycle

One function of the cycle is to control the development of the lining of the uterus (womb), so that if the egg is fertilised, the lining will be ready to receive the fertilised egg. If the egg is not fertilised, the lining of the uterus is lost from the woman's body as the flow of menstrual blood and cells of the lining, called a **period**.

A cycle is a continuous process, so it doesn't really have a beginning, but the first day of menstruation is usually called day 1.

Inside a woman's ovaries are hundreds of thousands of cells that could develop into mature eggs. Every month, one of these grows inside a ball of cells called a **follicle** (Figure 9.12). This is why the pituitary hormone which switches on the growth of the follicle is called 'follicle stimulating hormone'. At the middle of the cycle (about day 14) the follicle moves towards the edge of the ovary and the egg is released as the follicle bursts open. This is the moment of ovulation.

DID YOU KNOW?
A small percentage of women are able to sense the exact moment that ovulation happens, as the egg bursts out of an ovary.

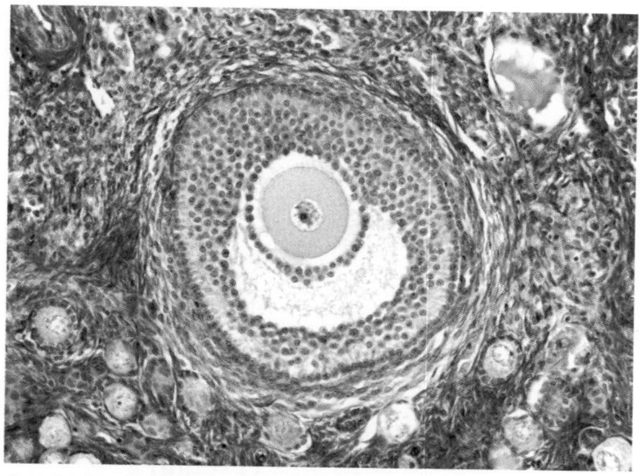

Figure 9.12 Eggs developing inside the follicles of an ovary. The large follicle contains a fully developed egg ready for ovulation.

While this is going on, the lining of the uterus has been repaired after menstruation, and has thickened. This change is brought about by the hormone oestrogen, which is secreted by the ovaries in response to FSH. Oestrogen also has another job. It slows down production of FSH, while stimulating secretion of LH. It is a peak of LH that causes ovulation.

After the egg has been released, it travels down the oviduct to the uterus. It is here in the oviduct that fertilisation may happen, if sexual intercourse has taken place. What's left of the follicle now forms a structure in the ovary called the **corpus luteum**. The corpus luteum makes another hormone called **progesterone**. Progesterone completes the development of the uterus lining, by thickening and maintaining it, ready for the fertilised egg to sink into it and develop into an embryo. Progesterone also inhibits (prevents) the release of FSH and LH by the pituitary, stopping ovulation.

If the egg is not fertilised, the corpus luteum breaks down and stops making progesterone. The lining of the uterus then passes out through the woman's vagina during menstruation. If, however, the egg is fertilised, the corpus luteum carries on making progesterone, the lining is not shed, and menstruation doesn't happen. The first sign that tells a woman she is pregnant is when her monthly periods stop. Later on in pregnancy, the **placenta** secretes progesterone, taking over the role of the corpus luteum.

EXTENSION WORK

'Corpus luteum' is Latin for 'yellow body'. A corpus luteum appears as a large yellow swelling in an ovary after the egg has been released. The growth of the corpus luteum is under the control of luteinising hormone (LH) from the pituitary.

CHAPTER QUESTIONS

More questions on human reproduction can be found at the end of Unit 2 on page 130.

SKILLS   CRITICAL THINKING

1 Which of the following describes sex cells (gametes)?

    A diploid cells produced by meiosis

    B diploid cells produced by mitosis

    C haploid cells produced by meiosis

    D haploid cells produced by mitosis

2 Which of the following is *not* a function of the ovaries?

    A the secretion of progesterone

    B the production of eggs

    C the secretion of oestrogen

    D the site of fertilisation

3 Which of the following organs produce(s) the hormone progesterone?

    1. pituitary gland

    2. ovary

    3. uterus

    4. placenta

    A 1 and 2

    B 2 and 3

    C 3 and 4

    D 2 and 4

4 A woman's first day of menstruation was on 1st June. Assuming she has a 28-day menstrual cycle, when was she most likely to ovulate?

A 7 June

B 10 June

C 14 June

D 21 June

5 The diagram shows a baby about to be born.

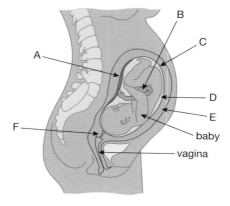

a Name parts A to F on the diagram.

b What is the function of A during pregnancy?

c What must happen to D and E just before birth?

d What must E and F do during birth?

6 The diagram shows *Hydra* (a small water animal) reproducing in two ways.

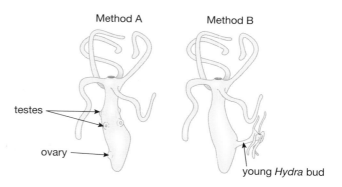

a Which of the two methods shows asexual reproduction? Give a reason for your answer.

b Explain why organisms produced asexually are genetically identical to each other and to the organism that produced them.

c When the surroundings do not change for long periods, *Hydra* reproduces mainly asexually. When the conditions change, *Hydra* begins to reproduce sexually. How does this pattern of sexual and asexual reproduction help *Hydra* to survive?

7 a The diagram shows the female reproductive system.

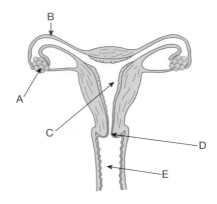

Which letter represents:

i the site of production of oestrogen and progesterone

ii the structure where fertilisation usually occurs

iii the structure that must dilate when birth commences

iv the structure that releases ova?

b The graph shows the changes in the thickness of the lining of a woman's uterus over 100 days.

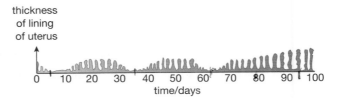

i Name the hormone that causes the thickening of the uterine lining.

ii Use the graph to determine the duration of *this* woman's menstrual cycle. Explain how you arrived at your answer.

iii From the graph, deduce the approximate day on which fertilisation leading to pregnancy took place. Explain how you arrived at your answer.

iv Why must the uterus lining remain thickened throughout pregnancy?

8 The number of sperm cells per cm^3 of semen (the fluid containing sperm) is called the 'sperm count'. Some scientists believe that over the last 50 years, the sperm counts of adult male humans have decreased. They think that this is caused by a number of factors, including drinking water polluted with oestrogens and other chemicals. Carry out an Internet search to find out the evidence for this. Summarise your findings in about two sides of A4 paper.

9 The graph shows some of the changes taking place during the menstrual cycle.

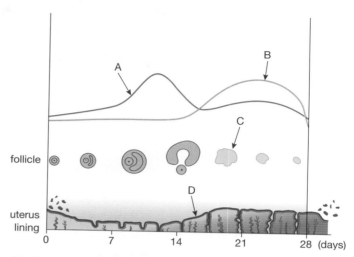

a Identify the two hormones produced by the ovary, which are shown by the lines A and B on the graph.

b Name the structure C.

c What is the purpose of the thickening of the uterus lining at D?

d When is sexual intercourse most likely to result in pregnancy: at day 6, 10, 13, 20 or 23?

e Why is it important that the level of progesterone remains high in the blood of a woman during pregnancy? How does her body achieve this:

i just after she becomes pregnant

ii later on in pregnancy?

BIOLOGY ONLY

10 Construct a table of the hormones involved in the menstrual cycle. Use these headings:

Name of hormone	Place where the hormone is made	Function(s) of the hormone

END OF BIOLOGY ONLY

UNIT QUESTIONS

SKILLS CRITICAL THINKING **1** The table shows the percentage of gases in inhaled and exhaled air.

Gas	Inhaled air / %	Exhaled air / %
nitrogen	78	79
oxygen		
carbon dioxide		
other gases (mainly argon)	1	1

a Copy the table and fill in the gaps by choosing from the following numbers:

   21   4   0.04   16 **(2)**

SKILLS REASONING **6** b Explain why the percentage of carbon dioxide is different in inhaled and exhaled air. **(2)**

c Explain why exhaling is a form of excretion. **(2)**

d The following features can be seen in the lungs:

   i thin membranes between the alveoli and the blood supply

7   ii a good blood supply

   iii a large surface area.

In each case explain how the feature helps gas exchange to happen quickly. **(6)**

Total 12 marks

SKILLS ANALYSIS **7** **2** Digestion is brought about by enzymes converting large insoluble molecules into smaller soluble molecules that can be more easily absorbed.

a The activity of enzymes is influenced by pH and temperature. The graph shows the activity of two human enzymes from different regions of the gut at different pH values.

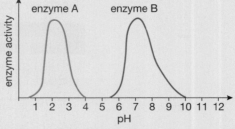

   i Suggest which regions of the gut the two enzymes come from. Explain your answer. **(4)**

6   ii Which nutrient does enzyme A digest? **(1)**

b Farmers sometimes include urea in cattle food. The microorganisms in the rumen can use urea to make protein.

SKILLS CRITICAL THINKING **4**   i In mammals, where in the body is urea made? **(1)**

5   ii What is urea made from? **(1)**

8   iii Suggest how feeding urea to cattle can result in an increased growth rate. **(1)**

BIOLOGY ONLY

SKILLS   CRITICAL THINKING   7

iv The Bowman's capsule and the loop of Henlé are both parts of a nephron. Explain how each of them helps to remove urea from the bloodstream.

(4)

END OF BIOLOGY ONLY

Total 12 marks

SKILLS   ANALYSIS   6

3   The diagram shows the human circulatory system. The system carries nutrients, oxygen and carbon dioxide around the body.

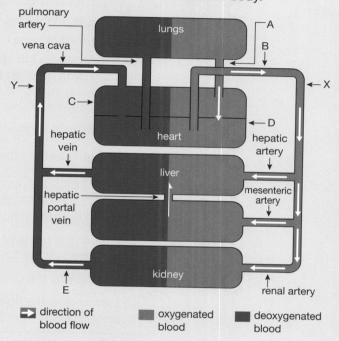

a Write down the names of the structures labelled A to E.   (5)

7

b Give two differences between the blood vessels at point X and point Y.   (2)

c During exercise, the adrenal gland releases the hormone adrenaline. Reflexes involving the medulla of the brain influence the heartbeat and breathing.

SKILLS   CRITICAL THINKING   6

    i Describe two effects of adrenaline on the heartbeat.   (2)

    ii Describe two other effects of adrenaline on the body.   (2)

SKILLS   REASONING   7

d How is a reflex action different from a voluntary action?   (2)

6

e Following exercise there is a recovery period in which breathing rate and heart rate gradually return to pre-exercise levels. Explain why they do not return immediately to these levels.   (3)

Total 16 marks

SKILLS INTERPRETATION

4 The immune system responds to infections using white blood cells. A phagocyte is one type of white blood cell.

a Draw and label a phagocyte. **(3)**

SKILLS CRITICAL THINKING

b State one way that the structure of a phagocyte differs from that of a red blood cell. **(1)**

c Phagocytes carry out phagocytosis. Describe this process. **(2)**

d Describe how other white blood cells are involved in the immune response. **(3)**

Total 9 marks

SKILLS REASONING

5 Humans and other mammals are able to maintain a constant body temperature which is usually higher than that of their surroundings.

a Explain the advantage in maintaining a constant, high body temperature. **(1)**

b The temperature of the blood is constantly monitored by the brain. If it detects a drop in blood temperature, the following things happen: the arterioles leading to the skin capillaries constrict, less sweat is formed and shivering begins.

   i Explain how each response helps the body to keep warm. **(3)**

   ii Explain how the structure of arterioles allows them to constrict. **(2)**

BIOLOGY ONLY

c When the weather is hot we produce less urine.

   i What is the name of the hormone that controls the amount of urine produced by the body? **(1)**

   ii Explain why the body produces less urine in hot weather. **(1)**

   iii Explain how the hormone in i) works in the kidney to produce less urine. **(3)**

END OF BIOLOGY ONLY

Total 11 marks

SKILLS ANALYSIS

6 The diagram represents a typical menstrual cycle.

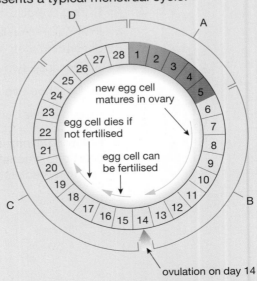

new egg cell matures in ovary

egg cell dies if not fertilised

egg cell can be fertilised

ovulation on day 14

a Using evidence from the diagram, answer the following questions.

During which of the stages A, B, C or D does:

i the level of the hormone oestrogen increase in the blood

ii the level of the hormone progesterone increase in the blood

iii the uterine lining become more vascular

iv the levels of oestrogen and progesterone in the blood fall

v the uterine lining begin to break down? (5)

b Explain how knowledge of the menstrual cycle can be used to avoid pregnancy. (3)

Total 8 marks

7 Cells can divide by mitosis or by meiosis. Human cells contain 46 chromosomes. The graphs show the changes in the number of chromosomes per cell as two different human cells undergo cell division.

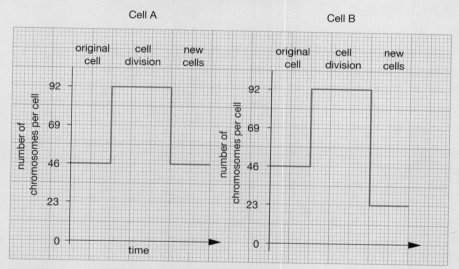

a Which of the two cells, A or B, is dividing by meiosis? Explain how you arrived at your answer. (3)

b Explain the importance of meiosis, mitosis and fertilisation in maintaining the human chromosome number constant at 46 chromosomes per cell, generation after generation. (6)

c Give *three* differences between mitosis and meiosis. (3)

Total 12 marks

8 Protein supplements are foods that some body-builders use to increase the growth of their muscles. Describe an investigation to find out if adding a protein supplement to the diet of rats will increase their growth.

Your answer should include experimental details and be written in full sentences. (6)

UNIT 3
PLANT PHYSIOLOGY

Plant physiology is the branch of biology that deals with the internal activities of plants. It covers many aspects of plant life, including nutrition (photosynthesis), movement of materials around the plant, and reproduction. There are many reasons why we need to study how plants work, not least because they provide the world's food supply and oxygen. Animals, including humans, could not exist without plants. Modern agriculture depends on the scientific study of plant physiology, as does our understanding of eccosystems.

10 PLANTS AND FOOD

This chapter looks at photosynthesis, the process by which plants make starch, and the structure of leaves in relation to photosynthesis. It also deals with how plants obtain other nutrients, and their uses in the plant.

LEARNING OBJECTIVES

- Investigate photosynthesis by testing for the production of starch in a leaf

- Investigate the need for light, carbon dioxide and chlorophyll for photosynthesis

- Understand the process of photosynthesis and its importance in the conversion of light energy to chemical energy

- Know the word equation and the balanced symbol equation for photosynthesis

- Describe the structure of the leaf and explain how it is adapted for photosynthesis

BIOLOGY ONLY

- Understand how the structure of the leaf is adapted for gas exchange

- Understand the role of diffusion and stomata in gas exchange by a leaf

BIOLOGY ONLY

- Understand how the exchange of carbon dioxide and oxygen in plants depends on the balance between respiration and photosynthesis

- Understand how respiration in plants continues during the day and night, and that the net exchange of carbon dioxide and oxygen depends on the intensity of light

- Understand how light intensity, carbon dioxide concentration and temperature affect the rate of photosynthesis

BIOLOGY ONLY

- Investigate the effect of light on net gas exchange from a leaf, using hydrogen carbonate indicator

- Investigate photosynthesis using the evolution of oxygen from a water plant

- Understand that plants require mineral ions for growth, and that magnesium ions are needed to make chlorophyll and nitrate ions are needed to make amino acids

PLANTS MAKE STARCH

All the foods shown in Figure 10.1 are products of plants. Some, such as potatoes, rice and bread (made from cereals such as wheat or rye), form the staple diet of humans. They all contain starch, which is the main storage carbohydrate made by plants. Starch is a good way of storing carbohydrate because it insoluble, compact and can be broken down easily.

Figure 10.1 All these foods are made by plants or made from products of plants. They all contain starch. ▶

ACTIVITY 1

▼ PRACTICAL: TESTING LEAVES FOR STARCH

You can test for starch in food by adding a few drops of yellow-brown iodine solution (see Chapter 4). If the food contains starch, a blue-black colour is produced.

Leaves that have been in sunlight also contain starch, but you can't test for it by adding iodine solution to a fresh leaf. The outer waxy surface of the leaf will not absorb the solution, and besides, the green colour of the leaf would hide the colour change. To test for starch in a leaf, the outer waxy layer needs to be removed and the leaf decolourised. This is done by placing the leaf in boiling ethanol (see Figure 10.2).

A beaker of water is set up on a tripod and gauze and the water heated until it boils. A leaf is removed from a plant and killed by placing it in boiling water for 30 seconds (this stops all chemical reactions in the leaf).

The Bunsen burner is turned off (this is important because ethanol is highly flammable), the leaf is placed in a boiling tube containing ethanol, and the tube is placed in the beaker of hot water. The boiling point of ethanol (about 78 °C) is lower than that of water (100 °C) so the ethanol will boil for a few minutes, until the water cools down. This is long enough to remove most of the chlorophyll from the leaf.

When the leaf has turned colourless or pale yellow, it is removed and washed in cold water to soften it, then spread out on a tile and covered with a few drops of iodine solution. After a few minutes, any parts of the leaf that contain starch will turn a dark blue-black colour. This only works if the plant has had plenty of light for some hours before the test.

> **!** Safety Note: Wash your hands after handling the leaves. Take care not to splash boiling water. Do not heat the ethanol directly with any flame: instead use a beaker of hot water. Iodine solution will badly stain everything – including skin.

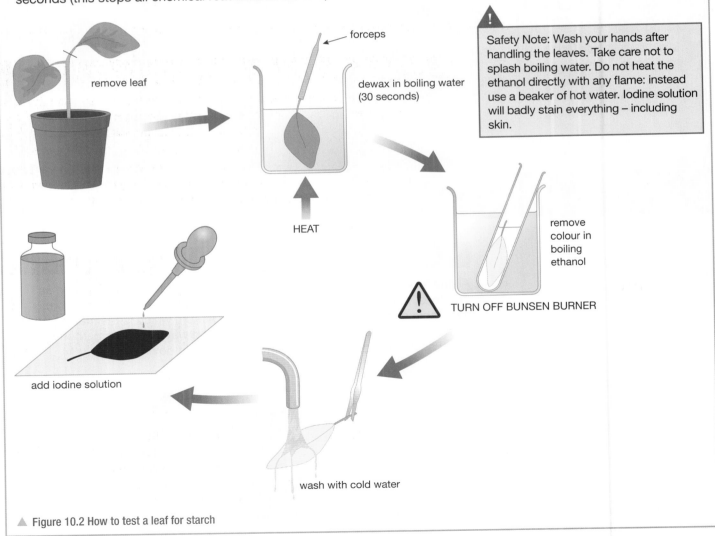

remove leaf

forceps

dewax in boiling water (30 seconds)

HEAT

remove colour in boiling ethanol

TURN OFF BUNSEN BURNER

add iodine solution

wash with cold water

▲ Figure 10.2 How to test a leaf for starch

KEY POINT

You can 'de-starch' a plant by placing it in the dark for 2 or 3 days. The plant uses up the starch stores in its leaves. De-starched plants are used to find out the conditions needed for the plant to make more starch by photosynthesis.

Starch is only made in the parts of plants that contain chlorophyll. You can show this by testing a leaf from a variegated plant, which has green and white areas to its leaves. The white parts of the leaf will give a negative starch test, staining yellow-brown with iodine solution. Only the green areas will stain blue-black. Figure 10.3 shows the results of a starch test on a leaf. The leaf was taken from a plant that had been under a bright light for 24 hours.

(a)   (b)

(c)

▲ Figure 10.3 Testing a leaf for starch. (a) Leaf before test (b) Decolourised leaf (c) Leaf after test, stained blue-black with iodine solution.

HINT

You might think that the results of the test on the variegated leaf prove that chlorophyll is needed for photosynthesis. But is this conclusion fully justified? The leaf could have photosynthesised in the white areas and the sugars been transported elsewhere in the plant. Similarly, the green areas may not be photosynthesising at all, but simply laying down starch from glucose made somewhere else. All it really shows is that starch is present in the green areas and not in the white areas of the leaf. We *assume* this is because chlorophyll is needed for photosynthesis.

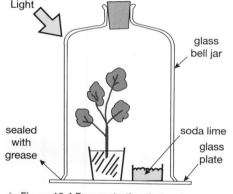

▲ Figure 10.4 Demonstration that carbon dioxide is needed for photosynthesis. The soda lime absorbs carbon dioxide from the air in the bell jar. A Control experiment should be set up, using exactly the same apparatus but without the soda lime.

Taking away the source of light is not the only way you can prevent a plant making starch in its leaves. You can also place it in a closed container containing a chemical called soda lime (Figure 10.4). This substance absorbs carbon dioxide from the air around the plant. If the plant is kept under a bright light but with no carbon dioxide, it again won't be able to make starch.

WHERE DOES THE STARCH COME FROM?

You have now found out three important facts about starch production by leaves:

- it uses carbon dioxide from the air
- it needs light
- it needs chlorophyll in the leaves.

As well as starch, there is another product of this process which is essential to the existence of most living things on the Earth – oxygen. When a plant is in the light, it makes oxygen gas. You can show this using an aquatic plant such as *Elodea* (Canadian pondweed). When a piece of this plant is placed in a test tube of water under a bright light, it produces a stream of small bubbles. If the bubbles are collected and their contents analysed, they are found to contain a high concentration of oxygen (Figure 10.5).

▲ Figure 10.5 The bubbles of gas released from this pondweed contain a higher concentration of oxygen than in atmospheric air.

Starch is composed of long chains of glucose (see Chapter 4). A plant does not make starch directly, but first produces glucose, which is then joined together in chains to form starch molecules. A carbohydrate made of many sugar sub-units is called a **polysaccharide**. Glucose has the formula $C_6H_{12}O_6$. The carbon and oxygen atoms of the glucose molecule come from the carbon dioxide gas in the air around the plant. The hydrogen atoms come from another molecule essential to the living plant – water.

It would be very difficult in a school laboratory to show that a plant uses water to make starch. If you deprived a plant of water in the same way as you deprived it of carbon dioxide, it would soon wilt and eventually die. However, scientists have proved that water is used in photosynthesis. They have done this by supplying the plant with water with 'labelled' atoms, for example using the 'heavy' isotope of oxygen (^{18}O). This isotope ends up in the oxygen gas produced by the plant. A summary of the sources of the atoms in the glucose and oxygen looks like this:

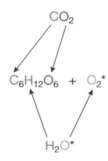

(*oxygen labelled with ^{18}O)

EXTENSION WORK

Isotopes are forms of the same element with the same atomic number but different mass numbers (due to extra neutrons in the nucleus). Isotopes of some elements are radioactive and can be used as 'labels' to follow chemical pathways. Others like ^{18}O are identified by their mass.

PHOTOSYNTHESIS

Plants use the simple inorganic molecules carbon dioxide and water, in the presence of chlorophyll and light, to make glucose and oxygen. This process is called **photosynthesis**.

It is summarised by the equation:

$$\text{carbon dioxide} + \text{water} \xrightarrow{\text{light}} \text{glucose} + \text{oxygen}$$

$$\text{or:} \quad 6CO_2 + 6H_2O \xrightarrow{\text{chlorophyll}} C_6H_{12}O_6 + 6O_2$$

The role of the green pigment, chlorophyll, is to absorb the light energy needed for the reaction to take place. The products of the reaction (glucose and oxygen) contain more energy than the carbon dioxide and water. In other words, photosynthesis converts light energy into chemical energy.

You will probably have noticed that the overall equation for photosynthesis is the reverse of the one for aerobic respiration (see Chapter 1):

$$C_6H_{12}O_6 + 6O_2 \rightarrow 6CO_2 + 6H_2O \quad \text{(plus energy)}$$

Respiration, which is carried out by both animals and plants, *releases* energy (but not as light) from the breakdown of glucose. The chemical energy in the glucose came originally from light 'trapped' by the process of photosynthesis.

DID YOU KNOW?

The 'photo' in photosynthesis comes from the Greek word *photos*, meaning light, and a 'synthesis' reaction is one where small molecules are built up into larger ones.

THE STRUCTURE OF LEAVES

Most green parts of a plant can photosynthesise, but the leaves are the plant organs which are best adapted for this function. To be able to photosynthesise efficiently, leaves need to have a large surface area to absorb light, many chloroplasts containing the chlorophyll, a supply of water and carbon dioxide, and a system for carrying away the products of photosynthesis to other parts of the plant. They also need to release oxygen (and water vapour) from the leaf cells. Most leaves are thin, flat structures supported by a leaf stalk which can grow to allow the blade of the leaf to be angled to receive the maximum amount of sunlight (Figure 10.6).

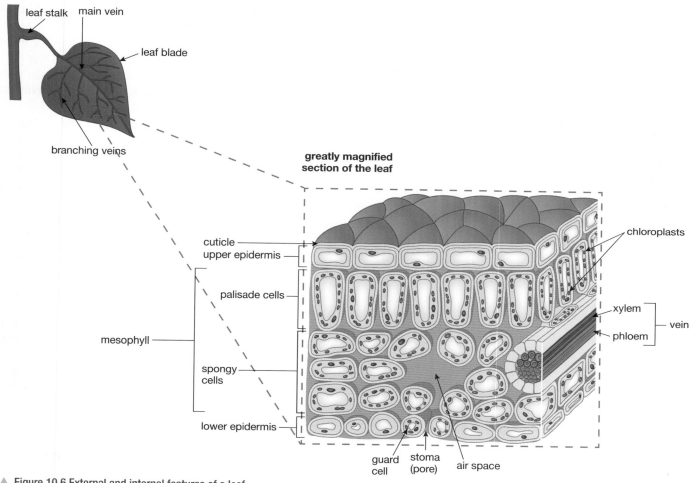

▲ Figure 10.6 External and internal features of a leaf

Inside the leaf are layers of cells with different functions.

KEY POINT

You will only need to know about adaptations for a gas exchange in a leaf if you are taking the Biology course; but not if you are taking the Double Award course.

- The two outer layers of cells (the upper and lower **epidermis**) have few chloroplasts and are covered by a thin layer of a waxy material called the **cuticle**. This reduces water loss by evaporation, and acts as a barrier to the entry of disease-causing microorganisms such as bacteria and fungi.

- The lower epidermis has many holes or pores called **stomata** (a single pore is a **stoma**). Usually the upper epidermis contains fewer or no stomata. The stomata allow carbon dioxide to diffuse into the leaf, to reach the photosynthetic tissues. They also allow oxygen and water vapour to diffuse out. Each stoma is formed as a gap between two highly specialised cells called **guard cells**, which can change their shape to open or close the stoma (see Chapter 11).

PHOTOSYNTHESIS AND RESPIRATION

- In the middle of the leaf are two layers of photosynthetic cells called the mesophyll ('mesophyll' just means 'middle of the leaf'). Just below the upper epidermis is the **palisade mesophyll** layer. This is a tissue made of long, narrow cells, each containing hundreds of chloroplasts, and is the main site of photosynthesis. The palisade cells are close to the source of light, and the upper epidermis is relatively transparent, allowing light to pass through to the enormous numbers of chloroplasts which lie below.

- Below the palisade layer is a tissue made of more rounded, loosely packed cells, with air spaces between them, called the **spongy mesophyll** layer. These cells also photosynthesise, but have fewer chloroplasts than the palisade cells. They form the main gas exchange surface of the leaf, absorbing carbon dioxide and releasing oxygen and water vapour. The air spaces allow these gases to diffuse in and out of the mesophyll.

- Water and mineral ions are supplied to the leaf by vessels in a tissue called the **xylem**. This forms a continuous transport system throughout the plant. Water is absorbed by the roots and passes up through the stem and through veins in the leaves in the **transpiration stream**. In the leaves, the water leaves the xylem and supplies the mesophyll cells.

- The products of photosynthesis, such as sugars, are carried away from the mesophyll cells by another transport system, the **phloem**. The phloem supplies all other parts of the plant, so that tissues and organs that can't make their own food receive products of photosynthesis. The veins in the leaf contain both xylem and phloem tissue, and branch again and again to supply all parts of the leaf.

You will find out more about both plant transport systems in Chapter 11.

BIOLOGY ONLY

Through photosynthesis, plants supply animals with two of their essential needs – food and oxygen – as well as removing carbon dioxide from the air. But remember that living cells, including plant cells, respire *all the time*, and they need oxygen for this. When the light intensity is high, a plant carries out photosynthesis at a much higher rate than it respires. So in bright light, there is an overall uptake of carbon dioxide from the air around a plant's leaves, and a surplus production of oxygen that animals can use. A plant only produces more carbon dioxide than it uses up in dim light. We can show this as a graph of carbon dioxide exchanged at different light intensities (Figure 10.7).

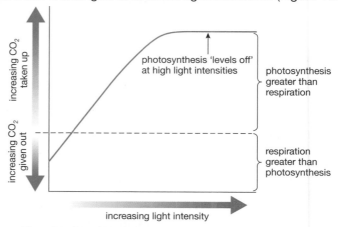

The point where the curve crosses the dashed line shows where photosynthesis is equal to respiration – there is no net gain or loss of CO_2.

▲ Figure 10.7 As the light intensity gets higher, photosynthesis speeds up, but eventually levels off in very bright light.

The concentration of carbon dioxide in the air around plants actually changes throughout the day. Scientists have measured the level of carbon dioxide in the air in the middle of a field of long grass in summer. They found that the air contained least carbon dioxide in the afternoon, when photosynthesis was happening at its highest rate (Figure 10.8). At night, when there was no photosynthesis, the level of carbon dioxide rose. This rise is due to less carbon dioxide being absorbed by the plants, while carbon dioxide was added to the air from the respiration of all organisms in the field.

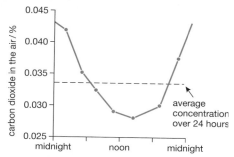

▲ Figure 10.8 Photosynthesis affects the concentration of carbon dioxide in the air around plants. Over a 24-hour period, the concentration rises and falls, as a result of the relative levels of photosynthesis and respiration.

INVESTIGATING THE EFFECT OF LIGHT ON GAS EXCHANGE BY A LEAF

KEY POINT

'Equilibration' means allowing something to reach a 'balanced state'. In this example the solution is left exposed to atmospheric air until the movement of CO_2 into the solution is equal to the movement of CO_2 out. (We also talk about 'equilibrating a liquid to temperature', which means leaving the liquid until it reaches the surrounding temperature – when the movement of heat in is equal to the movement of heat out).

Safety Note: Wash hands after collecting and inserting leaves.

Hydrogen carbonate indicator solution is very sensitive to changes in carbon dioxide concentration. When the solution is made, it is equilibrated with atmospheric air, which has a CO_2 concentration of 0.04%. If extra CO_2 is added to the solution, or if CO_2 is taken away from the solution, it changes colour (Table 10.1).

Table 10.1 Changes to hydrogen carbonate indicator solution with different concentrations of carbon dioxide.

Condition	Indicator colour
high concentrations of CO_2 (more than 0.04%)	yellow
CO_2 in normal air (0.04%)	orange
low concentrations of CO_2 (less than 0.04%)	purple

ACTIVITY 2

▼ PRACTICAL: INVESTIGATING THE EFFECT OF LIGHT ON GAS EXCHANGE BY A LEAF

Hydrogen carbonate indicator solution can be used to show the effect of light on the exchange of carbon dioxide by a leaf.

A volume of 10 cm³ of the indicator solution is placed in the bottom of each of four boiling tubes. Three large leaves are detached from a suitable plant and placed in the tops of three of the tubes. Each tube is sealed with a rubber bung.

One tube is placed in bright light, one in the dark (covered with silver foil) and the third covered with some translucent material, to restrict the intensity of light reaching the leaf. The fourth tube without a leaf acts as a Control, to show that it is the presence of the leaf which affects the colour of the indicator. (Figure 10.9).

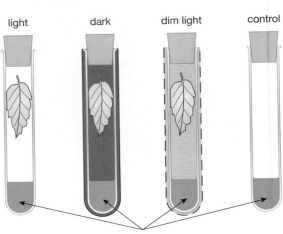

▲ Figure 10.9 Using hydrogen carbonate indicator to test the effect of light on gas exchange.

The four tubes are left set up under a bright light for a few hours, and the colour of the indicator recorded. The results from one experiment like this are shown below.

Tube	Colour of indicator at end of experiment
leaf in light	purple
leaf in dark	Yellow
leaf in dim light	orange
control (no leaf)	orange

Use your knowledge of a plant's rate of respiration and photosynthesis in the light and dark to explain these results. You may be able to carry out a similar investigation.

END OF BIOLOGY ONLY

FACTORS AFFECTING THE RATE OF PHOTOSYNTHESIS

In Figure 10.7, you can see that when the light intensity rises, the rate of photosynthesis begins by rising too, but eventually it reaches a maximum rate. What makes the rate 'level off' (flatten) like this? It is because some other factor needed for photosynthesis is in short supply, so that increasing the light intensity does not affect the rate any more. Normally, the factor which 'holds back' the rate of photosynthesis is the concentration of carbon dioxide in the air. This is only about 0.03 to 0.04%, and the plant can only take up the carbon dioxide and fix it into carbohydrate at a certain rate. If the plant is put in a closed container with a higher than normal concentration of carbon dioxide, it will photosynthesise at a faster rate. If there is both a high light intensity and a high level of carbon dioxide, the temperature may limit the rate of photosynthesis, by limiting the rate of the chemical reactions in the leaf. A rise in temperature will then increase the rate. With normal levels of carbon dioxide, very low temperatures (close to 0 °C) slow the reactions, but high temperatures (above about 35 °C) also reduce photosynthesis by denaturing enzymes in the plant cells (see Chapter 1).

KEY POINT

A limiting factor is the component of a reaction that is in 'shortest supply' so that it prevents the rate of the reaction increasing, in other words sets a 'limit' to it.

KEY POINT

Knowledge of limiting factors is used in some glasshouses (greenhouses) to speed up the growth of crop plants such as tomatoes and lettuces (see Chapter 15). Extra carbon dioxide is added to the air around the plants, by using gas burners. The higher concentration of carbon dioxide, along with the high temperature in the glasshouse, increases the rate of photosynthesis and the growth of the leaves and fruits.

Light intensity, carbon dioxide concentration and temperature can all act as what are called **limiting factors** in this way. This is easier to see as a graph (Figure 10.10).

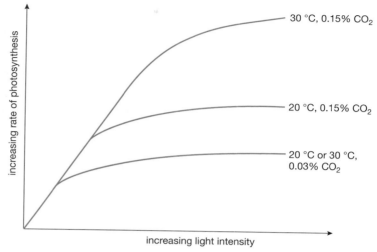

▲ Figure 10.10 Light intensity, carbon dioxide concentration and temperature can all act as limiting factors on the rate of photosynthesis.

ACTIVITY 3

▼ PRACTICAL: MEASURING THE RATE OF PHOTOSYNTHESIS USING PONDWEED

You can measure the rate of photosynthesis of a plant by measuring how quickly it produces oxygen. With a land plant this is difficult, because the oxygen is released into the air, but with an aquatic plant such as the pondweed *Elodea*, bubbles of oxygen are released into the water around the plant (Figure 10.5).

If the bubbles formed per minute are counted, this is a measure of the rate of photosynthesis of the plant. It is easiest to count the bubbles if the cut piece of weed is placed upside down in a test tube, as shown in Figure 10.11. A small paperclip attached to the bottom of the piece of weed makes it sink.

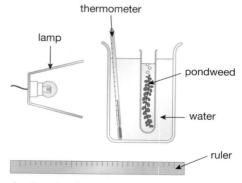

▲ Figure 10.11 Measuring the rate of photosynthesis in an aquatic plant.

The light intensity is changed by moving the lamp, altering the distance between the lamp and the pondweed. The beaker of water keeps the temperature of the plant constant.

! Safety Note: Wash hands after collecting and preparing pondweed. Do not handle the lamp, plug or switch with wet hands.

Design an experiment using this apparatus to find out if the rate of photosynthesis is affected by the light intensity. In your plan you should include:

■ a hypothesis – state what you think will happen when you change the light intensity, and why

■ a systematic way of changing the light intensity

■ how the experiment will be controlled so that nothing else is changed apart from the light intensity (e.g. what will you do about the background light in the laboratory?)

■ a Control that you could use, to show that it is the effect of light on the pondweed that is producing the bubbles

■ how you will ensure that your results are reliable.

When you have completed your plan, you may be allowed to use similar apparatus to carry out the experiment. How could you modify your plan to find the effect of changing the *temperature* on the rate of photosynthesis? What factors would you need to keep constant this time? What would be a suitable range of temperatures to use?

THE PLANT'S USES FOR GLUCOSE

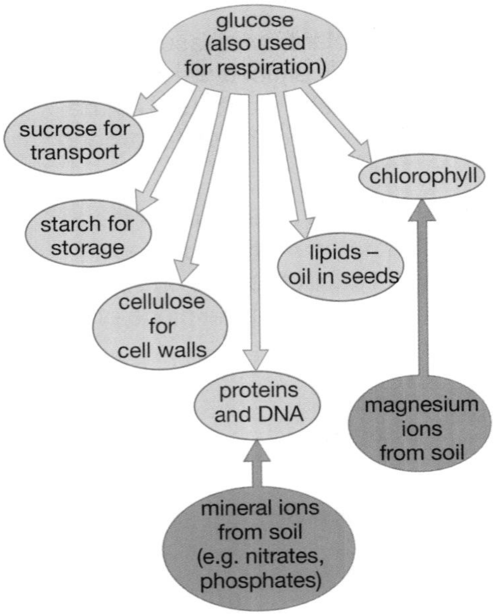

▲ Figure 10.12 Compounds that plant cells can make from glucose.

As you have seen, some glucose that the plant makes is used in respiration to provide the plant's cells with energy. Some glucose is quickly converted into starch for storage. However, a plant is not made up of just glucose and starch, and must make all of its organic molecules, starting from glucose.

Glucose is a single sugar unit (a **monosaccharide**). Plant cells can convert it into other sugars, such as a monosaccharide called **fructose** (found in fruits) and the **disaccharide sucrose**, which is the main sugar carried in the phloem. It can also be changed into another polymer, the polysaccharide called **cellulose**, which forms plant cell walls.

All these compounds are carbohydrates. Plant cells can also convert glucose into lipids (fats and oils). Lipids are needed for the membranes of all cells, and are also an energy store in many seeds and fruits, such as peanuts, sunflower seeds and olives.

Carbohydrates and lipids both contain only three elements – carbon, hydrogen and oxygen – and so they can be inter-converted without the need for a supply of other elements. Proteins contain these elements too, but all amino acids (the building blocks of proteins) also contain nitrogen. This is obtained as nitrate ions from the soil, through the plant's roots. Other compounds in plants contain other elements. For example, chlorophyll contains magnesium ions, which are also absorbed from water in the soil. Some of the products that a plant makes from glucose are summarised in Figure 10.12.

KEY POINT

Glucose from photosynthesis is not just used as the *raw material* for the production of molecules such as starch, cellulose, lipids and proteins. Reactions like these, which synthesise large molecules from smaller ones, also need a source of energy. This energy is provided by the plant's *respiration* of glucose.

MINERAL NUTRITION

Nitrate ions are absorbed from the soil water, along with other minerals such as phosphate, potassium and magnesium ions. The element phosphorus is needed for the plant cells to make many important compounds, including DNA. Potassium ions are required for enzymes in respiration and photosynthesis to work, and magnesium forms a part of the chlorophyll molecule.

WATER CULTURE EXPERIMENTS

A plant takes only water and mineral ions from the soil for growth. Plants can be grown in soil-free cultures (water cultures) if the correct balance of minerals is added to the water. In the nineteenth century, the German biologist Wilhelm Knop invented one example of a culture solution. Knop's solution contains the following chemicals (per dm^3 of water):

0.8 g	calcium nitrate
0.2 g	magnesium sulfate
0.2 g	potassium nitrate
0.2 g	potassium dihydrogenphosphate
(trace)	iron(III) phosphate

DID YOU KNOW?

In fact, in addition to the ions listed in Knop's solution, plants need very small amounts of other mineral ions for healthy growth. Knop's culture solution only worked because the chemicals he used to make his solutions weren't very pure, and supplied enough of these additional ions by mistake!

Notice that these chemicals provide all of the main elements that the plant needs to make proteins, DNA and chlorophyll, as well as other compounds, from glucose. It is called a *complete* culture solution. If you were to make a similar solution, but to replace, for example, magnesium sulfate with more calcium sulfate, this would produce a culture solution which was *deficient* (lacking) in magnesium. You could then grow plants in the complete and deficient solutions, and compare the results. There are several ways to grow the plants, such as using the apparatus shown in Figure 10.13, which is useful for plant cuttings. Seedlings can be grown by packing cotton wool around the seed, instead of using a rubber bung.

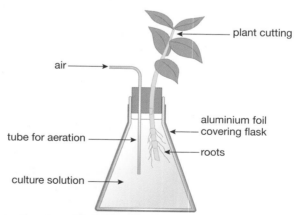

▲ Figure 10.13 A simple water culture method

The plant is kept in bright light, so that it can photosynthesise. The covering around the flask prevents algae from growing in the culture solution. The aeration tube is used for short periods to supply the roots with oxygen for respiration of the root cells, since some ions are taken up by active transport. Using methods like this, it soon becomes clear that mineral deficiencies result in poor plant growth. A shortage of a particular mineral results in particular symptoms in the plant, called a *mineral deficiency disease*. For example, lack of magnesium means that the plant won't be able to make chlorophyll, and the leaves will turn yellow. Lack of nitrate limits a plant's growth, because it is unable to make enough protein. Some of the mineral ions that a plant needs, their uses, and the deficiency symptoms are shown in Table 10.2. Compare the photographs of the mineral deficient plants to those of the control plant in Figure 10.14(a).

▲ Figure 10.14 (a) A healthy bean plant

Table 10.2: Mineral ions needed by plants

Mineral ion	Use	Deficiency symptoms
nitrate	making amino acids, proteins, chlorophyll, DNA and many other compounds	limited growth of plant; older leaves turn yellow ▲ Figure 10.14 (b) A plant showing symptoms of nitrate deficiency.
phosphate	making DNA and many other compounds; part of cell membranes	poor root growth; younger leaves turn purple ▲ Figure 10.14 (c) A plant showing symptoms of phosphate deficiency.
potassium	needed for enzymes of respiration and photosynthesis to work	leaves turn yellow with dead spots ▲ Figure 10.14 (d) A plant showing symptoms of potassium deficiency.
magnesium	part of chlorophyll molecule	leaves turn yellow ▲ Figure 10.14 (e) A plant showing symptoms of magnesium deficiency.

DID YOU KNOW?

Some commercial crops such as lettuces can be grown without soil in culture solutions. This is called **hydroponics**. The plants' roots grow in a long plastic tube that has culture solution passing through it (Figure 10.15). The composition of the solution can be carefully adjusted to ensure the plants grow well. Pests such as insects, which might live in soil, are also less of a problem.

▲ Figure 10.15 Lettuce plants grown by hydroponics

LOOKING AHEAD – MORE ABOUT PHOTOSYNTHESIS

The role of chlorophyll is to absorb the light energy needed to drive photosynthesis. The products of the reaction (glucose and oxygen) contain more chemical energy than the reactants (carbon dioxide and water). The simple equation for photosynthesis that you saw earlier in this chapter is actually a summary of a series of reactions. They take place in two stages (Figure 10.16).

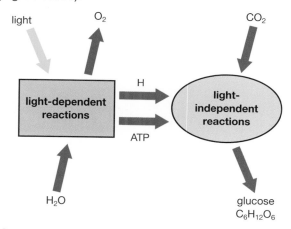

▲ Figure 10.16 Photosynthesis takes place in two stages called light-dependent and light-independent reactions.

The light energy absorbed by the chlorophyll is used to split water molecules into hydrogen and oxygen. The waste product, oxygen, is given off as a gas. At the same time, the light energy is used to convert ADP and phosphate into ATP. Because this stage needs light, the steps in the process are known as the *light-dependent reactions*.

Next, the hydrogen atoms from the water and energy from the ATP are used to reduce carbon dioxide to glucose. This takes place in another series of reactions in the chloroplast. Because these do not need light, they are called the *light-independent reactions*.

If you continue with biology beyond International GCSE, you will find out more about the biochemistry of photosynthesis.

CHAPTER QUESTIONS

More questions on plants and food can be found at the end of Unit 3 on page 182.

SKILLS REASONING

1 A variegated leaf attached to a plant was placed in a test tube as shown below. The test tube contained soda lime to absorb carbon dioxide.

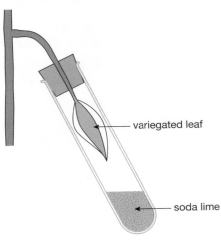

variegated leaf

soda lime

The plant was left in bright light for 24 hours and tested for starch using iodine solution. Which of the following diagrams shows the expected results?

A   B   C   D

BIOLOGY ONLY

SKILLS CRITICAL THINKING

2 At night, the concentration of carbon dioxide in the air in a forest is high. What is the main reason for this?

  A There is increased respiration by animals that are active at night

  B Plants are unable to absorb carbon dioxide because their stomata are shut

  C Plants begin to respire in the dark

  D Plants can only use carbon dioxide for photosynthesis in the light

END OF BIOLOGY ONLY

3 Which of the following is not normally a factor that limits the rate of photosynthesis?

  A temperature

  B oxygen concentration

  C carbon dioxide concentration

  D light intensity

SKILLS   CRITICAL THINKING

4 Which of the following compounds can only be made by a plant if it has access to magnesium ions?

   A chlorophyll

   B cellulose

   C starch

   D protein

5 A plant with variegated leaves had a piece of black paper attached to one leaf as shown in the diagram.

black paper (on both sides of leaf)

edge of leaf lacks chlorophyll

The plant was kept under a bright light for 24 hours. The leaf was then removed, the paper taken off and the leaf was tested for starch.

   a Name the chemical used to test for starch, and describe the colour change if the test is positive.

   b Copy the leaf outline and shade in the areas which would contain starch.

   c Explain how you arrived at your answer to (b).

   d What is starch used for in a plant? How do the properties of starch make it suitable for this function?

6 Copy and complete the following table to show the functions of different parts of a leaf, and how that part is adapted for its function. One row has been done for you.

Part of leaf	Function	How the part is adapted for its function
palisade mesophyll layer	main site of photosynthesis	cells contain many chloroplasts for photosynthesis
spongy mesophyll layer		
stomata		
xylem		
phloem		

BIOLOGY ONLY

SKILLS ANALYSIS

7 The graph shows the changes in the concentration of carbon dioxide in a field of long grass throughout a 24-hour period in summer.

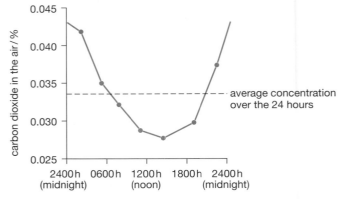

a Explain why the levels of carbon dioxide are high at 0200 hours and low at 1200 hours.

SKILLS REASONING

b What factor will limit the rate of photosynthesis at 0400 hours and at 1400 hours?

END OF BIOLOGY ONLY

SKILLS INTERPRETATION

8 The table below shows some of the substances that can be made by plants. Give one use in the plant for each. The first has been done for you.

Substance	Use
glucose	oxidised in respiration to give energy
sucrose	
starch	
cellulose	
protein	
lipid	

SKILLS REASONING

9 The apparatus shown in the diagram was used to grow a pea seedling in a water culture experiment.

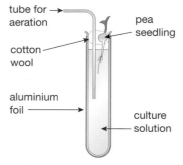

a Explain the purpose of the aeration tube and the aluminium foil around the test tube.

SKILLS REASONING

b After two weeks, the roots of the pea seedling had grown less than normal, although the leaves were well developed. What mineral ion is likely to be deficient in the culture solution?

SKILLS INTERPRETATION

10 A piece of Canadian pondweed was placed upside down in a test tube of water, as shown in the diagram.

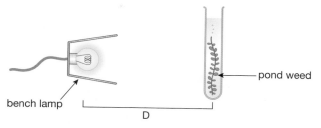

Light from a bench lamp was shone onto the weed, and bubbles of gas appeared at the cut end of the stem. The distance of the lamp from the weed was changed, and the number of bubbles produced per minute was recorded. The results are shown in the table below.

Distance of lamp (D) / cm	Number of bubbles per minute
5	126
10	89
15	64
20	42
25	31
30	17
35	14
40	10

a Plot a graph of the number of bubbles per minute against the distance of the weed from the lamp.

SKILLS REASONING

b Using your graph, predict the number of bubbles per minute that would be produced if the lamp was placed 17 cm from the weed.

c The student who carried out this experiment arrived at the following conclusion:

'The gas made by the weed is oxygen from photosynthesis, so the faster production of bubbles shows that the rate of photosynthesis is greater at higher light intensities.'

Write down three reasons why his conclusion could be criticised. (Hint: is counting the bubbles a reliable method of measuring the rate of photosynthesis?)

SKILLS INTERPRETATION

11 Write a summary account of photosynthesis. You should include a description of the process, a summary equation, an account of how a leaf is adapted for photosynthesis and a note of how photosynthesis is important to other organisms, such as animals. You must keep your summary to less than two sides (about 500 words).

11 TRANSPORT IN PLANTS

Chapter 10 described how plants make food by photosynthesis. Photosynthesis requires water from the roots to be transported to the leaves, and sugars and other products of photosynthesis to be transported away from the leaves to the rest of the plant. This chapter explains how these materials are moved through the plant.

LEARNING OBJECTIVES

- Understand how water can move in or out of plant cells by osmosis

- Investigate osmosis in living and non-living systems (plant cells and Visking tubing)

BIOLOGY ONLY

- Understand how water is absorbed by root hair cells

- Understand that transpiration is the evaporation of water from the surface of a plant

- Describe the role of the xylem in the transport of water and mineral ions from the roots to other parts of a plant

- Describe the role of phloem in the transport of sucrose and amino acids from the leaves to other parts of a plant

- Understand the origin of carbon dioxide and oxygen as waste products of metabolism and their loss from the stomata of a leaf

BIOLOGY ONLY

- Understand how the rate of transpiration is affected by changes in temperature, wind speed, humidity and light intensity

- Investigate the role of environmental factors in determining the rate of transpiration from a leafy shoot.

OSMOSIS

Osmosis is the name of a process by which water moves into and out of cells. In order to understand how water moves through a plant, you need to understand the mechanism of osmosis. Osmosis happens when a material called a **partially permeable membrane** separates two solutions. One artificial partially permeable membrane is called Visking tubing. This is used in kidney dialysis machines. Visking tubing has microscopic holes in it, which let small molecules like water pass through (it is *permeable* to them) but is not permeable to some larger molecules, such as the sugar sucrose. This is why it is called 'partially' permeable. You can show the effects of osmosis by filling a Visking tubing 'sausage' with concentrated sucrose solution, attaching it to a capillary tube and placing the Visking tubing in a beaker of water (Figure 11.1).

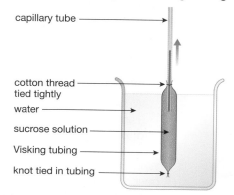

capillary tube

cotton thread tied tightly

water

sucrose solution

Visking tubing

knot tied in tubing

▲ Figure 11.1 Water enters the Visking tubing 'sausage' by osmosis. This causes the level of liquid in the capillary tube to rise. In the photograph, the contents of the Visking tubing have had a red dye added to make it easier to see the movement of the liquid.

The level in the capillary tube rises as water moves from the beaker to the inside of the Visking tubing. This movement is due to osmosis. You can understand what's happening if you imagine a highly magnified view of the Visking tubing separating the two liquids (Figure 11.2).

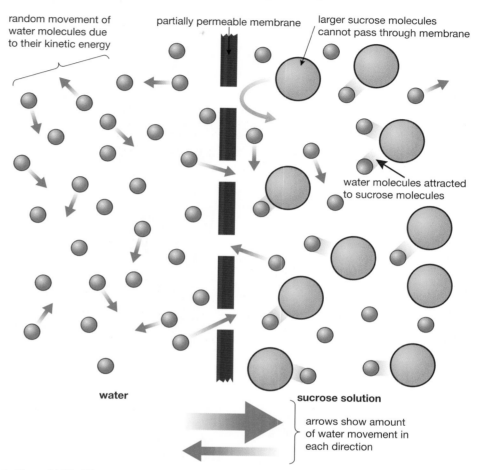

random movement of water molecules due to their kinetic energy

partially permeable membrane

larger sucrose molecules cannot pass through membrane

water molecules attracted to sucrose molecules

water

sucrose solution

arrows show amount of water movement in each direction

▲ Figure 11.2 In this model of osmosis, more water molecules diffuse from left to right than from right to left.

The sucrose molecules are too big to pass through the holes in the partially permeable membrane. The water molecules can pass through the membrane in either direction, but those on the right are attracted to the sugar molecules. This slows them down and means that they are less free to move – they have less kinetic energy. As a result of this, more water molecules diffuse from left to right than from right to left. In other words, there is a greater diffusion of water molecules from the more dilute solution (in this case pure water) to the more concentrated solution.

How 'free' the water molecules are to move is called the **water potential**. The molecules in pure water can move most freely, so pure water has the highest water potential. The more concentrated a solution is, the lower is its water potential. In the model in Figure 11.2, water moves from a high to a low water potential. This is a law which applies whenever water moves by osmosis. We can bring these ideas together in a definition of osmosis.

OSMOSIS IN PLANT CELLS

So far we have only been dealing with osmosis through Visking tubing. However, there are partially permeable membranes in cells too. The cell surface membranes of both animal and plant cells are partially permeable, and so is the inner membrane around the plant cell's sap vacuole (Figure 11.3).

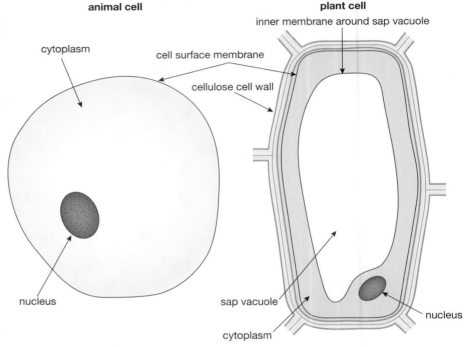

▲ Figure 11.3 Membranes in animal and plant cells

Around the plant cell is the tough cellulose cell wall. This outer structure keeps the shape of the cell, and can resist changes in pressure inside the cell. This is very important, and critical in explaining the way that plants are supported. The cell contents, including the sap vacuole, contain many dissolved solutes, such as sugars and ions.

If a plant cell is put into pure water or a dilute solution, the contents of the cell have a lower water potential than the external solution, so the cell will absorb water by osmosis (Figure 11.4). The cell then swells up and the cytoplasm pushes against the cell wall. A plant cell that has developed an internal pressure like this is called **turgid**.

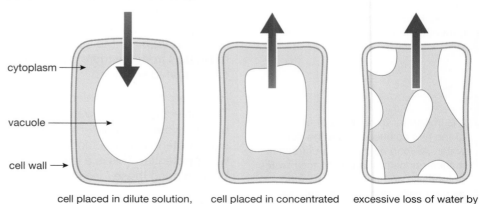

cell placed in dilute solution, or water, absorbs water by osmosis and becomes turgid

cell placed in concentrated solution loses water by osmosis and becomes flaccid

excessive loss of water by osmosis causes the cell to become plasmolysed

▲ Figure 11.4 The effects of osmosis on plant cells

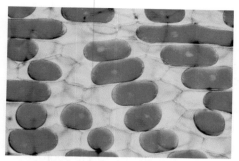

▲ Figure 11.5 Cells of onion epidermis, showing plasmolysis. The cell membranes and cytoplasm can be seen to have pulled away from the cell wall.

On the other hand, if the cell is placed in a concentrated sucrose solution that has a lower water potential than the cell contents, it will *lose* water by osmosis. The cell decreases in volume and the cytoplasm no longer pushes against the cell wall. In this state, the cell is called **flaccid**. Eventually the cell contents shrink so much that the membrane and cytoplasm split away from the cell wall and gaps appear between the wall and the membrane. A cell like this is called **plasmolysed**. You can see plasmolysis happening in the plant cells shown in Figure 11.5. The space between the cell wall and the cell surface membrane will now be filled with the sucrose solution.

KEY POINT

This chapter only covers plant cells. Osmosis also happens in animal cells, but there is much less water movement. This is because animal cells do not have a strong cell wall around them, and can't resist the changes in internal pressure resulting from large movements of water. For example, if red blood cells are put into water, they will swell up and burst. If the same cells are put into a concentrated salt solution, they lose water by osmosis and shrink, producing cells with crinkly edges (Figure 11.6).

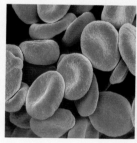
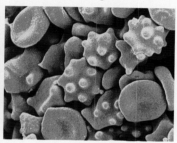

◄ Figure 11.6 Compare the red blood cells on the right, which were placed in a 3% salt solution, with the normal cells on the left. Blood plasma has a concentration equal to a 0.85% salt solution.

Turgor (the state a plant is in when its cells are turgid) is very important to plants. The pressure inside the cells pushes neighbouring cells against each other, like a box full of inflated balloons. This supports the non-woody parts of the plant, such as young stems and leaves, and holds stems upright, so the leaves can carry out photosynthesis properly. Turgor is also important in the functioning of stomata. If a plant loses too much water from its cells so that they become flaccid, this makes the plant **wilt**. You can see wilting in a pot plant which has been left for too long without water. The leaves droop and collapse. In fact this is a protective action. It cuts down water loss by reducing the exposed surface area of the leaves and closing the stomata.

Inside the plant, water moves from cell to cell by osmosis. If a cell has a higher water potential than the cell next to it, water will move from the first cell to the second. In turn, this will dilute the contents of the second cell, so that it has a higher water potential than the next cell. In this way, water can move across a plant tissue, down a gradient of water potential (Figure 11.7).

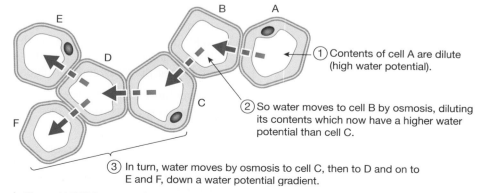

① Contents of cell A are dilute (high water potential).

② So water moves to cell B by osmosis, diluting its contents which now have a higher water potential than cell C.

③ In turn, water moves by osmosis to cell C, then to D and on to E and F, down a water potential gradient.

▲ Figure 11.7 Water moves from cell to cell down a water potential gradient.

ACTIVITY 1

▼ PRACTICAL: INVESTIGATING THE EFFECTS OF OSMOSIS IN ONION EPIDERMIS CELLS

A drop of concentrated (1 mol per dm³) sucrose solution is placed on one microscope slide, and a drop of distilled water on a second slide. Two small squares of inner epidermis are removed from the outer fleshy layers of an onion (Figure 11.8). One square is transferred to the sucrose solution, and the second to the water. This is done as quickly as possible, so that the cells do not dry out.

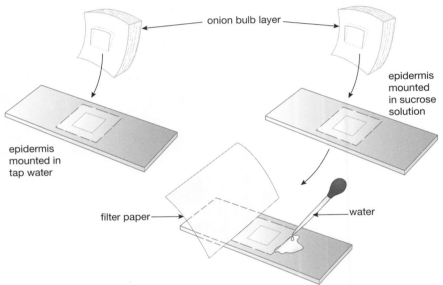

onion bulb layer

epidermis mounted in sucrose solution

epidermis mounted in tap water

filter paper

water

▲ Figure 11.8 Investigating osmosis in onion epidermis cells

A drop of the correct solution is added to the top of each specimen, followed by a cover slip. Any excess liquid is blotted (cleaned) up with filter paper.

Each slide is examined through the microscope for several minutes. The specimen in water will show turgid cells, similar to the first cell in Figure 11.4. The cells in sucrose solution will gradually plasmolyse, as shown in the last diagram of Figure 11.4, and Figure 11.5.

It is interesting to replace the sugar solution on the first slide with distilled water, to see the effect on the plasmolysed cells. This can be done quite easily without removing the cover slip. Some water is placed on one side of the cover slip and drawn across the slide using filter paper, as shown in Figure 11.8. After a while all the sucrose solution is replaced by water. If the cells are now observed for a few minutes, they will gradually recover from their plasmolysed condition until they are fully turgid again.

It may be possible for you to carry out these procedures. If you can, observe and draw one or two turgid cells, and one or two that are plasmolysed.

Can you explain all these observations in terms of osmosis?

If you use a red onion, the contents of the cells are coloured and easier to see.

Safety Note: Take great care when cutting the potato into 'chips' and wash hands afterwards.

ACTIVITY 2

▼ PRACTICAL: INVESTIGATING THE EFFECTS OF OSMOSIS ON POTATO TUBER TISSUE

A potato tuber is a plant storage organ. It is a convenient tissue to use to investigate the effects of osmosis on the mass of the tissue.

A boiling tube is half filled with distilled water and a second with concentrated (1 mol per dm^3) sucrose solution. A third tube is left empty.

A potato is cut into chips 5 cm × 1 cm × 1 cm, making these measurements as accurate as possible so that the three chips are the same size. No skin is left on the potato tissue.

Each chip is gently blotted to remove excess moisture, and the mass of each is found by weighing on a balance. One chip is placed in each of the three boiling tubes (Figure 11.9).

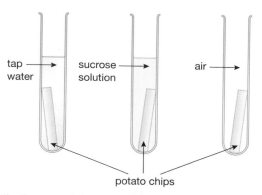

▲ Figure 11.9 Investigating osmosis in potato tissue

After 30 minutes, the chips are removed from the tubes and gently blotted to remove excess liquid, then re-weighed. Each is felt in turn, to compare how flexible or stiff they are.

The change in mass (+ or −) of each chip is calculated, and the percentage change found from the equation:

$$\% \text{ change } = \frac{\text{change in mass}}{\text{starting mass}} \times 100$$

The table below shows a set of results from this experiment.

Tube contents	Starting mass / g	Final mass / g	Change in mass / g	Change in mass / %	Condition (flexible / stiff)
water	6.25	6.76	+0.51	+8.16	stiff
sucrose solution	7.10	6.31	−0.79	−11.1	flexible
air	6.66	6.57	−0.09	−1.35	stiff

What conclusions can you make from these results? Explain them in terms of osmosis.

How large are the percentage changes in mass of the chips in the two liquids, compared with the chip in air?

Can you explain the final 'condition' of the chips, using terms such as 'flaccid' and 'turgid'?

Can you think of any criticisms of this experiment? For example, does using one chip per tube yield reliable evidence?

You may be able to perform an experiment similar to the one above. You don't have to use potato; other tissues such as carrot or sweet potato should work. An alternative to weighing is to measure the length of the chips at the end of the experiment, and compare these measurements with the starting lengths. However, weighing is much more accurate and gives a bigger percentage change.

You could try extending the experiment into a more complete investigation. For example you could make predictions about what would happen to chips placed in solutions that are intermediate in concentration between water and molar sucrose solution.

Plan a method to test your hypothesis. You should include a list of the apparatus and materials to be used, a description of the procedure, and a statement of the expected results.

BIOLOGY ONLY

UPTAKE OF WATER BY ROOTS

The regions just behind the growing tips of the roots of a plant are covered in thousands of tiny **root hairs** (Figure 11.10). These areas are the main sites of water absorption by the roots, where the hairs greatly increase the surface area of the root epidermis.

Each hair is actually a single, specialised cell of the root epidermis. The long, thin outer projection of the root hair cell penetrates between the soil particles, reaching the soil water. The water in the soil has some solutes dissolved in it, such as mineral ions, but their concentrations are much lower than the concentrations of solutes inside the root hair cell.

The soil water therefore has a higher water potential than the inside of the cell. This allows water to enter the root hair cell by osmosis. In turn, this water movement dilutes the contents of the cell, increasing its water potential. Water then moves out of the root hair cell into the outer tissue of the root (called the root cortex). Continuing in this way, a gradient of water potential is set up across the root cortex, kept going by water being taken up by the xylem in the middle of the root (Figure 11.11).

▲ Figure 11.10 Root hairs on the growing root of a radish seedling. The hairs increase the surface area for absorption of water.

KEY POINT

If a plant needs a mineral ion that is at a low concentration in the soil, the root hair cells use active transport to absorb it.

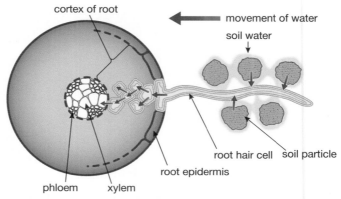

cortex of root

movement of water

soil water

root hair cell   soil particle

root epidermis

phloem   xylem

▲ Figure 11.11 Water is taken up by root hairs of the plant epidermis and carried across the root cortex by a water potential gradient. It then enters the xylem and is transported to all parts of the plant.

LOSS OF WATER BY THE LEAVES – TRANSPIRATION

Osmosis is also involved in the movement of water through leaves. The epidermis of leaves is covered by a waxy cuticle (see Chapter 10), which is impermeable to water. Most water passes out of the leaves as water vapour through pores called stomata (stoma is the singular of stomata).

Water leaves the cells of the leaf mesophyll and evaporates into the air spaces between the spongy mesophyll cells. The water vapour then diffuses out through the stomatal pores (Figure 11.12).

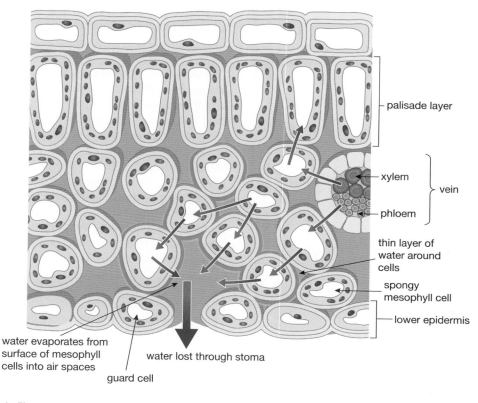

▲ Figure 11.12 Passage of water from the xylem to the stomatal pores of a leaf

Loss of water from the mesophyll cells sets up a water potential gradient which 'draws' water by osmosis from surrounding mesophyll cells. In turn, the xylem vessels supply the leaf mesophyll tissues with water.

This loss of water vapour from the leaves is called **transpiration**. Transpiration causes water to be 'pulled up' the xylem in the stem and roots in a continuous flow known as the **transpiration stream** (Figure 11.13). The transpiration stream has more than one function. It:

- supplies water for the leaf cells to carry out photosynthesis
- carries mineral ions dissolved in the water
- provides water to keep the plant cells turgid
- allows evaporation from the leaf surface, which cools the leaf, in a similar way to sweat cooling the human skin.

DID YOU KNOW?

Plants that live in dry habitats, such as cacti and succulents (plants with fleshy leaves), usually have a very thick layer of waxy cuticle on their leaves. Cacti have leaves reduced to spines. which have a low surface area. These adaptations reduce water loss.

water is lost from leaves (transpiration)

water passes up through the stem and leaves in the xylem

water is absorbed by the roots

▲ Figure 11.13 The transpiration stream

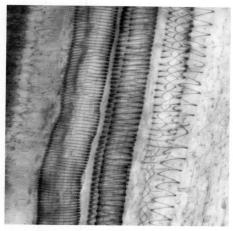

▲ Figure 11.14 Xylem vessels in a stem. The spirals and rings (stained red) are made of lignin.

TRANSPORT IN THE XYLEM

Xylem transports water and minerals throughout the plant. Xylem contains dead cells arranged end-to-end, forming continuous vessels. When they are mature, the vessels contain no cytoplasm. Instead, they have a hollow central space or **lumen** through which the water passes. The walls of the xylem vessels contain a woody material called **lignin** (Figure 11.14).

The xylem vessels begin life as living cells with normal cytoplasm and cellulose cell walls. As they develop they become elongated, and gradually their original cellulose cell walls become impregnated with lignin, made by the cytoplasm. As this happens, the cells die, forming hollow tubes. Lignification makes them very strong, and enables them to carry water up tall plants without collapsing. Lignin is also impermeable to water.

TRANSPORT IN THE PHLOEM

The other plant transport tissue, the phloem, consists of living cells at all stages in its development. Tubes in the phloem are also formed by cells arranged end-to-end, but they have cell walls made of cellulose, and retain their cytoplasm. The end of each cell is formed by a cross-wall of cellulose with holes, called a **sieve plate**. The living cytoplasm extends through the holes in the sieve plates, linking each cell with the next, forming a long **sieve tube** (Figure 11.15). The sieve tubes transport the products of photosynthesis from the leaves to other parts of the plant. Sugars for energy, or amino acids for building proteins, are carried to young leaves and other growing points in the plant. Sugar may also be taken to the roots and converted into starch for storage. Despite being living cells, the phloem sieve tubes have no nucleus. They seem to be controlled by other cells that lie alongside the sieve tubes, called **companion cells** (Figure 11.15).

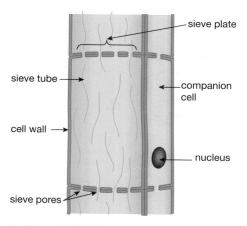

▲ Figure 11.15 Xylem vessels in a stem

REMINDER

Remember: xylem carries water and minerals up from the roots. Phloem carries products of photosynthesis away from the leaves. The contents of the phloem can travel up or down the plant.

STRUCTURE OF A STEM

In a young stem, xylem and phloem are grouped together in areas called **vascular bundles**. Unlike in the root, where the vascular tissue is in the central core, the vascular bundles are arranged in a circle around the outer part of the stem (Figure 11.16).

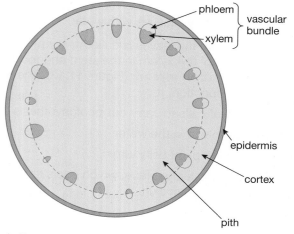

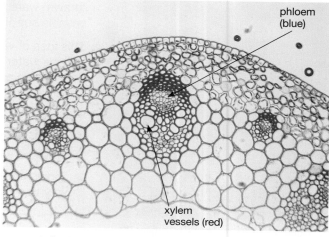

▲ Figure 11.16 (a) This cross-section of a stem shows the arrangement of xylem and phloem tissue in vascular bundles. (b) Three vascular bundles in part of a stem section. The outer red cells are lignified fibres for extra support.

In older stems, the vascular tissue grows to form complete rings around the stem. The inner xylem forms the woody central core of a stem, with the living layer of phloem outside this.

BIOLOGY ONLY

CONTROL OF TRANSPIRATION BY STOMATA

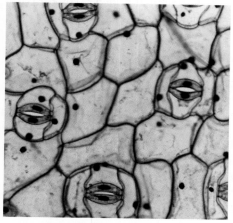

As you saw in Chapter 10, there are usually more stomata on the lower surface of the leaves than the upper surface in most plant species (Figure 11.17). If they were mainly on the upper leaf surface, the leaf would lose too much water. This is because the stomata would be exposed to direct sunlight, which would produce a high rate of evaporation from them. There is also less air movement on the underside of leaves. The evolution of this arrangement of stomata is an adaptation that reduces water loss.

The stomata can open and close. The guard cells that surround each stoma have an unusual 'banana' shape, and the part of their cell wall nearest the stoma is especially thick. In the light, water enters the guard cells by osmosis from the surrounding epidermis cells. This causes the guard cells to become turgid, and, as they swell up, their shape changes. They bend outwards, opening up the stoma. In the dark, the guard cells lose water again, they become flaccid and the stoma closes. No one knows for sure how this change is brought about, but it seems to be linked to the fact that the guard cells are the only cells in the lower epidermis that contain chloroplasts. In the light, the guard cells use energy to accumulate solutes in their vacuoles, causing water to be drawn in by osmosis (Figure 11.18).

▲ Figure 11.17 Stomata in the surface of the lower epidermis of a leaf

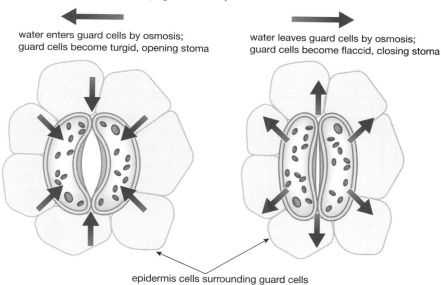

water enters guard cells by osmosis; guard cells become turgid, opening stoma

water leaves guard cells by osmosis; guard cells become flaccid, closing stoma

epidermis cells surrounding guard cells

▲ Figure 11.18 When the guard cells become turgid, the stoma opens. When they become flaccid, it closes.

Closure of stomata in the dark is a useful adaptation. Without the sun there is no need for loss of water vapour from the stomata to cool the leaves. In addition, leaves cannot photosynthesise in the dark, so they don't need water for this purpose. Therefore, it doesn't matter if the transpiration stream is shut down by closure of the stomata. As you will see, other physical factors can also affect the rate of transpiration.

FACTORS AFFECTING THE RATE OF TRANSPIRATION

There are four main factors which affect the rate of transpiration:

■ **Light intensity** The rate of transpiration increases in the light, because of the opening of the stomata in the leaves, so that the leaf can photosynthesise.

■ **Temperature** High temperatures increase the rate of transpiration, by increasing the rate of evaporation of water from the mesophyll cells.

■ **Humidity** When the air around the plant is humid, this reduces the diffusion gradient between the air spaces in the leaf and the external air. The rate of transpiration therefore decreases in humid air and speeds up in dry air.

■ **Wind speed** The rate of transpiration increases with faster air movements across the surface of the leaf. The moving air removes any water vapour which might remain near the stomata. This moist air would otherwise reduce the diffusion gradient and slow down diffusion.

MEASURING THE RATE OF TRANSPIRATION: POTOMETERS

A potometer is a simple piece of apparatus which measures the rate of transpiration or the rate of uptake of water by a plant. (These are not the same thing – some of the water taken up by the plant may stay in the plant cells, or be used for photosynthesis.) There are two types, 'weight' and volume potometers.

A 'weight' potometer measures the rate of loss of mass from a potted plant or leafy shoot over an extended period of time, usually several hours (Figure 11.19).

The polythene bag around the pot prevents loss of moisture by evaporation from the soil. Most of the mass lost by the plant will be due to water evaporating from the leaves during transpiration. (However, there will be small changes in mass due to respiration and photosynthesis, since both of these processes exchange gases with the air).

A volume potometer is used to find the rate of uptake of water by a leafy shoot, by 'magnifying' this uptake in a capillary tube. These potometers come in various shapes and sizes. The simplest is a straight vertical tube joined to the shoot by a piece of rubber tubing. More complex versions have a horizontal capillary tube and a way of refilling the capillary to re-set the water at its starting position (Figure 11.20).

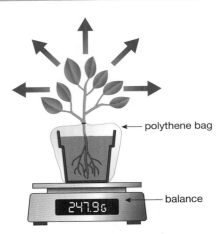

polythene bag

balance

247.9g

▲ Figure 11.19 A 'weight' potometer

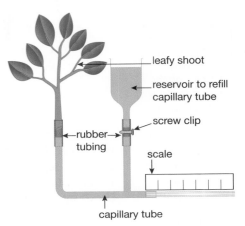

leafy shoot

reservoir to refill capillary tube

screw clip

rubber tubing

scale

capillary tube

▲ Figure 11.20 A commercial potometer with a water reservoir

ACTIVITY 3

▼ PRACTICAL: INVESTIGATING THE EFFECTS OF ENVIRONMENTAL FACTORS ON THE RATE OF TRANSPIRATION USING A SIMPLE POTOMETER

The capillary and attached rubber tubing is placed in a sink of water and any air it contains removed by squeezing the tubing. A shoot is taken from a plant and the end of the shoot is cut at an angle (this is done under water to stop air entering the shoot). The angled cut makes it easier to push the stem into the rubber tubing. The assembled apparatus is allowed to dry and any joints sealed with petroleum jelly. The simple potometer is clamped in a vertical position and left to adjust to its new conditions (Figure 11.21).

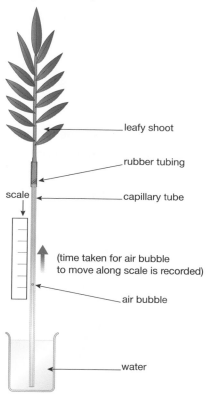

leafy shoot

rubber tubing

scale

capillary tube

(time taken for air bubble to move along scale is recorded)

air bubble

water

◀ Figure 11.21 A simple potometer

The distance moved by the column of water in the capillary tube is measured over a certain time, e.g. 10 minutes, and the rate of movement calculated. The plant is then exposed to different environmental conditions to see how they affect the rate. For example:

■ Using a fan or hair dryer to investigate the effect of moving air

■ Placing the plant under a bright light to investigate the effect of changing light intensity.

For a controlled experiment, only one condition is changed at a time. Each time a condition is changed, the shoot is left to adjust or equilibrate to the new conditions. How can you tell when the shoot has equilibrated to the change? (Hint – what will happen to the rate of transpiration?)

END OF BIOLOGY ONLY

CHAPTER QUESTIONS

More questions on transport in plants can be found at the end of Unit 3 on page 182.

SKILLS CRITICAL THINKING

1 A plant cell was placed in a concentrated sucrose solution. The diagram below shows the appearance of the cell after an hour in the solution.

What is the best description of this cell?

A turgid

B flaccid

C plasmolysed

D wilted

2 What are the functions of xylem and phloem?

A xylem transports water and sugars, phloem transports mineral ions

B xylem transports water and mineral ions, phloem transports sugars

C xylem transports sugars and mineral ions, phloem transports water

D xylem transports sugars, phloem transports water and mineral ions

BIOLOGY ONLY

SKILLS CRITICAL THINKING

3 In the table below, which row shows the correct directions of movement of materials in the xylem and phloem?

	Xylem	Phloem
A	up and down	down only
B	down only	up and down
C	up only	up and down
D	up and down	up only

4 Which of the following conditions result in the highest rate of transpiration?

A warm, windy and dry

B warm, windy and wet

C cold, windy and dry

D warm, still and wet

END OF BIOLOGY ONLY

SKILLS PROBLEM SOLVING

5 Three 'chips' of about the same size and shape were cut from the same potato. Each was blotted, weighed and placed in a different sucrose solution (A, B or C). The chips were left in the solutions for an hour, then removed, blotted and re-weighed. Here are the results:

	Starting mass / g	Final mass / g	Change in mass / %
solution A	7.4	6.5	−12.2
solution B	8.2	8.0	
solution C	7.7	8.5	+10.4

a Calculate the percentage change in mass for the chip in solution B.

SKILLS CRITICAL THINKING

SKILLS ANALYSIS

b Name the process that caused the chips to lose or gain mass.

c Which solution was likely to have been the most concentrated?

d Which solution had the highest water potential?

e Which solution had a water potential most similar to the water potential of the potato cells?

SKILLS CRITICAL THINKING

f The cell membrane is described as 'partially permeable'. Explain the meaning of this.

BIOLOGY ONLY

SKILLS CRITICAL THINKING

6 Explain how each of the following cells is adapted for its function:

a a root hair cell   b a xylem vessel   c a guard cell.

SKILLS REASONING

7 Suggest reasons for each of the following observations.

a When removing a small plant from a pot and re-planting it in a garden, it is important to take the plant out of its pot carefully, leaving the roots in a ball of soil. If this is not done, the plant may wilt after it has been transplanted.

b A plant cutting is more likely to grow successfully if you remove some of its leaves before planting it in compost.

c Plants that live in very dry habitats often have stomata located in sunken pits in their leaves.

d Greenflies feed by sticking their hollow tube-like mouthparts into the phloem of a plant stem.

END OF BIOLOGY ONLY

SKILLS ANALYSIS

8 The diagram shows a cross-section through the stem of a plant.

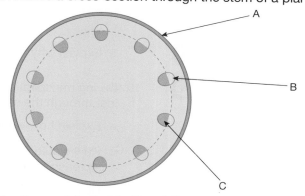

a Identify the tissues labelled A, B and C.

b A young stem was placed in a solution of a red dye for an hour. Which tissue in the diagram would be most likely to contain the dye? Explain your answer.

BIOLOGY ONLY

9 A simple volume potometer was used to measure the uptake of water by a leafy shoot under four different conditions. During the experiment, the temperature, humidity and light intensity were kept constant. The conditions were:

1. Leaves in still air with no petroleum jelly applied to them.

2. Leaves in moving air with no petroleum jelly applied to them.

3. Leaves in still air with the lower leaf surface covered in petroleum jelly.

4. Leaves in moving air with the lower surface covered in petroleum jelly.

The results are shown in the graph.

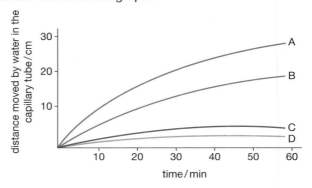

a Copy and complete the table below to show which condition (1, 2, 3 or 4) is most likely to produce curve A, B, C and D. One has been done for you:

Condition	Curve
1	B
2	
3	
4	

b Explain why moving air affects the rate of uptake of water by the shoot.

10 During transpiration, water moves through the cells of a leaf and is finally lost through the stomata by evaporation.

a Explain how water travels from one cell to another in the leaf.

b How would the rate of transpiration be affected if the air temperature increased from 20 °C to 30 °C? Explain your answer.

c Describe an adaptation found in plants living in dry habitats that decreases water loss from the leaves.

SKILLS   INTERPRETATION

11   The diagram shows a section through the root of a plant.

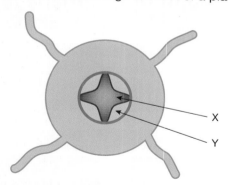

a   Identify the tissues labelled X and Y.

b   Draw and label a root hair cell.

SKILLS   CRITICAL THINKING

c   Describe how a root hair cell absorbs water.

SKILLS   INTERPRETATION

12   Write a short description of how water moves from the soil through the plant. Your description must include these words: xylem, evaporation, root hair cells, water potential, stomata and osmosis.

Underline each of these words in your description.

END OF BIOLOGY ONLY

12 CHEMICAL COORDINATION IN PLANTS

Plants can detect changes in their environment and respond to them, but the responses are much slower than those in animals. This is because movements in a plant are brought about by changes in the plant's growth. This chapter is about these growth responses, and the chemicals that coordinate them.

LEARNING OBJECTIVES

■ Understand that plants respond to stimuli

■ Describe the phototropic response of roots and stems

■ Describe the geotropic responses of roots and stems

■ Understand the role of auxin in the phototropic response of stems

Chapter 6 explains how animals detect and respond to changes in their environment. Animals usually respond very quickly – for example the reflex action resulting from a painful stimulus (page 90) is over in a fraction of a second.

As in animals, some species of plant can respond rapidly to a stimulus, for example the Venus flytrap (Figure 12.1). This plant has modified leaves, which close quickly around their 'prey', trapping it. The plant then secretes enzymes to digest the insect. The movement is brought about by rapid changes in turgor of specialised cells at the base of the leaves.

▲ Figure 12.1 The Venus flytrap catches and digests insects to gain extra nutrients. The plant responds very quickly to a fly landing on one of its leaves.

TROPISMS

Most plants do not respond to stimuli as quickly as the Venus flytrap, because their response normally involves a change in their rate of growth. Different parts of plants may grow at different rates, and a plant may respond to a stimulus by increasing growth near the tip of its shoot or roots.

Imagine a plant growing normally in a pot. Usually, most light will be falling on the plant from above. If you turn the plant on its side and leave it for a day or so, you will see that its shoot starts to grow upwards (Figure 12.2).

▲ Figure 12.2 This bean has responded to being placed horizontally. The growing shoot has started to bend upwards.

▲ Figure 12.3 The shoots of these cress seedlings are showing a positive phototropism.

KEY POINT

Phototropisms are growth responses to light from one direction. Geotropisms are growth responses to the direction of gravity ('geo' refers to the Earth).

KEY POINT

Light from one direction is called 'unidirectional'. If light shines on the plant evenly from all directions, this is called 'uniform' light.

There are two stimuli acting on the plant in Figure 12.2. One is the direction of the light that falls on the plant. The other stimulus is gravity. Both light and gravity are *directional* stimuli (they act in a particular direction). The growth response of a plant to a directional stimulus is called a **tropism**. If the growth response is *towards* the direction of the stimulus, it is a *positive* tropism, and if it is *away* from the direction of the stimulus, it is a *negative* tropism. The stem of the plant in Figure 12.2 is showing a positive **phototropism** and a negative **geotropism**, which both make the stem grow upwards.

The aerial part of a plant (the 'shoot') needs light to carry out photosynthesis. This means that in most species, a positive phototropism is the strongest tropic response of the shoot. If a shoot grows towards the light, it ensures that the leaves, held out at an angle to the stem, will receive the maximum amount of sunlight. This response is easily seen in any plant placed near a window, or another source of 'one-way' or *unidirectional* light (Figure 12.3).

In darkness or uniform light, the shoot shows a negative geotropism. As you might expect, the roots of plants are strongly positively geotropic. This response makes sure that the roots grow down into the soil, where they can reach water and mineral ions. Roots also fix a plant firmly in place – they provide 'anchorage'.

The roots of some species that have been studied are also negatively phototropic, but most roots don't respond to directional light at all. Table 12.1 summarises the main tropisms.

Table 12.1: The main responses of plants to directional stimuli (tropisms).

Stimulus	Name of response	Response of shoots	Response of roots
light	phototropism	grow towards light source (positive phototropism)	most species show no response; some grow away from light (negative phototropism)
gravity	geotropism	grow away from direction of gravity (negative geotropism)	grow towards direction of gravity (positive geotropism)

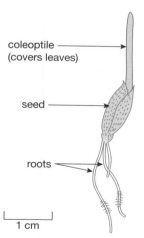

coleoptile (covers leaves)

seed

roots

1 cm

▲ Figure 12.4 Germinating oat seedling

EXTENSION WORK

Charles Darwin is best known for his theory of evolution by natural selection (Chapter 19). However, he carried out research in other areas of biology, including studying plant hormones. You could do some research to find out more about his life and work. Make a list of his achievements.

DETECTING THE LIGHT STIMULUS – PLANT HORMONES

Plants do not have the obvious sense organs and nervous system of animals. However, since they respond to stimuli such as light and gravity, they must have some way of detecting them and coordinating the response. The detection system of phototropism was first investigated by the great English biologist Charles Darwin (see Chapter 19) in the late nineteenth century. Instead of using stems, Darwin (and later scientists) used cereal coleoptiles, which are easier to grow and use in experiments.

A **coleoptile** is a protective sheath that covers the first leaves of a cereal seedling. It protects the delicate leaves as the shoot emerges through the soil (Figure 12.4). Coleoptiles have a simple structure and are easy to grow, so they are often used to investigate tropisms.

Darwin showed that the stimulus of unidirectional light was detected by the tip of the coleoptile, and transmitted to a growth zone, just behind the tip (Figure 12.5).

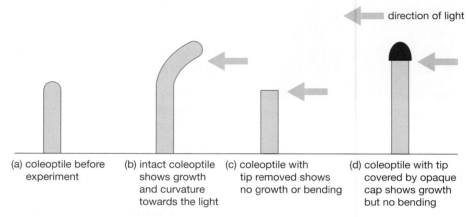

(a) coleoptile before experiment

(b) intact coleoptile shows growth and curvature towards the light

(c) coleoptile with tip removed shows no growth or bending

(d) coleoptile with tip covered by opaque cap shows growth but no bending

▲ Figure 12.5 Darwin's experiments with phototropism (1880)

Since plants don't have a nervous system, biologists began to look for a chemical messenger (or plant **hormone)** that might be the cause of phototropism in coleoptiles. Between 1910 and 1926 several scientists investigated this problem. Some of their results are summarised in Figure 12.6.

Experiment 1

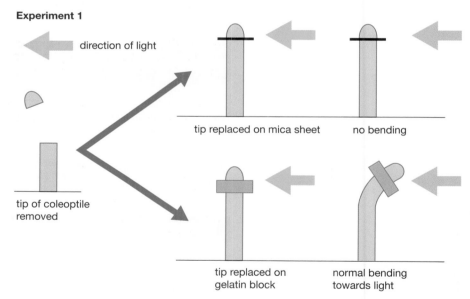

tip of coleoptile removed

tip replaced on mica sheet

no bending

tip replaced on gelatin block

normal bending towards light

Experiment 2

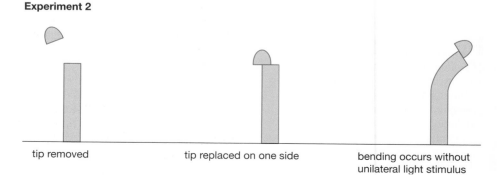

tip removed

tip replaced on one side

bending occurs without unilateral light stimulus

Experiment 3

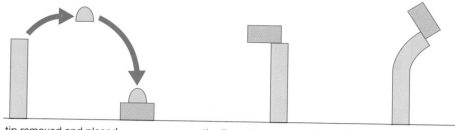

tip removed and placed on agar block for several hours

tip discarded and agar block placed on one side of decapitated coleoptile

bending occurs in absence of unidirectional light

▲ Figure 12.6 Experiments on coleoptiles that helped to explain the mechanism of phototropism.

In experiment 1, the stimulus for growth was found to pass through materials such as gelatine, which absorbs water-soluble chemicals, but not through materials such as mica (a mineral), which is impermeable to water. This made biologists think that the stimulus was a chemical that was soluble in water.

In experiment 2, it was shown that the phototropic response could be brought about, even *without* unidirectional light, by removing a coleoptile tip ('decapitating' the coleoptile) and placing the tip on one side of the decapitated stalk.

In experiment 3, it was found that the hormone could be collected in another water-absorbing material (a block of agar jelly). Placing the agar block on one side of the decapitated coleoptile stalk caused it to bend.

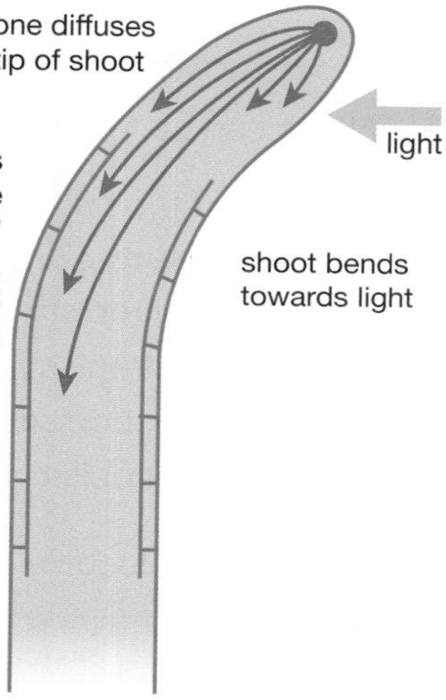

Investigations like experiments 2 and 3 led scientists to believe that the hormone caused bending by stimulating growth on the side of the coleoptile furthest from the light. The theory is that the hormone is produced in the tip of the shoot, and diffuses back down the shoot. If the shoot is in the dark, or if light is all around the shoot, the hormone diffuses at equal rates on each side of the shoot, so it stimulates the shoot equally on all sides. However, if the shoot is receiving light from one direction, the hormone moves away from the light as it diffuses downwards. The higher concentration of hormone on the 'dark' side of the shoot stimulates cells there to grow, making the shoot bend towards the light (Figure 12.7).

Since these experiments were carried out, scientists have identified the hormone responsible. It is called **auxin**. Several other types of plant hormone have been found. Like auxin, they all influence growth and development of plants in one way or another, so that many scientists prefer to call them **plant growth substances** rather than plant hormones.

hormone diffuses from tip of shoot

light

light causes more hormone to reach 'dark' side of shoot, causing cell elongation

shoot bends towards light

▲ Figure 12.7 How movement of a plant hormone causes phototropism.

EXTENSION WORK

'Auxin' should really be 'auxins', since there are a number of chemicals with very similar structures making up a group of closely related plant hormones.

EXTENSION WORK

As well as auxins, there are four other main groups of plant hormones, called gibberellins, cytokinins, abscisic acid and ethene. They control many aspects of plant growth and development, apart from tropisms. These include growth of buds, leaves and fruit, fruit ripening, seed germination, leaf fall and opening of stomata, to name just a few.

USING A CLINOSTAT TO SHOW TROPISMS

A clinostat is a piece of apparatus consisting of an electric motor turning a cork disc. Germinating seeds are attached to the disc. The motor turns the disc and seeds around very slowly, so that the movement eliminates the effect of any directional stimulus that may be acting on the seeds. The clinostat can be turned through 90°, so that the disc rotates either horizontally or vertically.

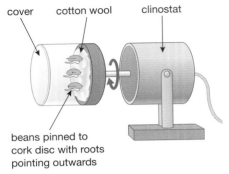

▲ Figure 12.8 A clinostat

The clinostat can be used to demonstrate tropisms, for example geotropism in roots. A few beans are soaked in water overnight and then placed on wet cotton wool for a day or two, until the first root of each seed (called the radicle) grows to a length of about 2 cm.

Wet cotton wool is attached to the cork disc of two clinostats. Three or four of the germinating bean seeds are pinned on to each of the discs, with their radicles pointing outwards. Covers are placed over them to keep the air around the beans moist.

The clinostats are turned on their sides, as shown in Figure 12.8. One clinostat is switched on, with the other left switched off, as a control. Both clinostats are left set up for a few days. The cotton wool is watered regularly to keep it damp.

After this time, the radicles of the beans on the control clinostat will have grown downwards, under the influence of gravity. Those on the moving clinostat will have grown straight out horizontally. The continuously changing direction of the gravitational stimulus acting on the seeds of the moving clinostat cancels out the geotropic response.

CHAPTER QUESTIONS

More questions on chemical coordination in plants can be found at the end of Unit 3 on page 182.

SKILLS CRITICAL THINKING

1 Which of the following statements about tropisms is correct?

A Roots show positive phototropism and negative geotropism

B Roots show negative phototropism and positive geotropism

C Stems show negative phototropism and negative geotropism

D Stems show positive phototropism and positive geotropism

2 Below are three statements about phototropism in a coleoptile.

1. Auxin is made in the tip of the coleoptile

2. Auxin diffuses towards the light

3. A high concentration of auxin stimulates growth of the coleoptile towards the light.

Which of these statements are true?

A 1 and 2

B 1 and 3

C 2 and 3

D 1, 2 and 3

3 The diagram below shows a bean seedling that has been growing in the dark.

What is the name given to the response of the shoot?

A positive phototropism

B negative phototropism

C positive geotropism

D negative geotropism

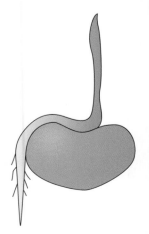

4 Movement of auxin is thought to take place by which of the following methods?

  A diffusion

  B active transport

  C osmosis

  D mass flow in the phloem

5 a What are the main stimuli affecting the growth of

  i the shoot?   ii the root?

  b How does a plant benefit from a positive phototropism in its stem?

6 Draw a labelled diagram to show how auxin brings about phototropism in a coleoptile that has light shining on it from one direction.

7 An experiment was carried out to investigate phototropism in a coleoptile. The diagram shows what was done. Predict the results you would expect to get in each of the experiments a) to c). Explain your answers.

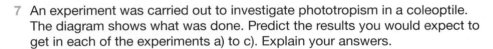

mica sheet

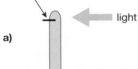
a)

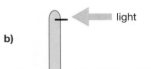
b)

c)

8 An experiment was carried out to investigate which part of a coleoptile is sensitive to a light stimulus. Three dishes containing wheat seedlings were subjected to different treatments:

  ■ Dish 1 – 2 mm was removed from the growing coleoptiles ('decapitated' coleoptiles)

  ■ Dish 2 – the tips of the coleoptiles were covered with caps of aluminium foil

  ■ Dish 3 – the seedlings were left untreated as a control.

The coleoptiles were measured and the mean length of each treatment group was calculated. Each batch of seedlings was then placed in a box with a hole cut in one side to provide a unidirectional source of light.

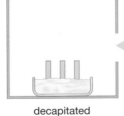
decapitated coleoptiles

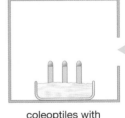
coleoptiles with tips covered

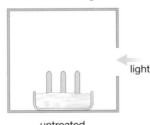
untreated coleoptiles

After 48 hours the new mean length of the coleoptiles was found for each treatment group, and the percentage increase in length calculated. Any change in direction of growth of the coleoptiles was noted.

  a Suggest which treatment would produce the greatest percentage increase in length and which the least. Explain your answer.

  b What differences in direction of growth would you expect to see as a result of the three treatments? Explain your answer.

  c Explain why it is necessary to calculate the *percentage* increase in the length of the coleoptiles, rather than simply the increase in length.

13 REPRODUCTION IN PLANTS

Plants, like animals, can reproduce sexually and asexually. The sexual organs of a flowering plant are its flowers, which produce pollen and ovules containing the flower's gametes. This chapter looks at both types of reproduction in flowering plants.

LEARNING OBJECTIVES

- Describe the differences between sexual and asexual reproduction in plants

- Understand that plants can reproduce asexually by natural methods (illustrated by runners) and by artificial methods (illustrated by cuttings)

- Describe the structures of an insect-pollinated and a wind-pollinated flower and explain how each is adapted for pollination

- Understand that the growth of the pollen tube followed by fertilisation leads to seed and fruit formation

- Understand how germinating seeds utilise food reserves until the seedling can carry out photosynthesis

- Investigate the conditions needed for seed germination

SEXUAL AND ASEXUAL REPRODUCTION

You have seen how animals can reproduce sexually and asexually (Chapter 9). Plants can also carry out both types of reproduction. Table 13.1 summarises some of the differences between the two methods of reproduction.

Table 13.1 Sexual and asexual reproduction compared.

Feature of the process	Sexual reproduction	Asexual reproduction
gametes produced	yes	no
fertilisation takes place	yes	no
genetic variation in offspring	yes	no
has survival value in:	changing environment	stable environment

KEY POINT

As each method of asexual reproduction involves some part of the plant growing, new cells must be produced. These cells are produced by **mitosis** and so are all genetically identical. This means that all the offspring formed by asexual reproduction will also be genetically identical.

ASEXUAL REPRODUCTION IN PLANTS

There are many different methods of asexual reproduction in plants. Most involve some part of the plant growing, and then breaking away from the parent plant before developing into a new plant (Figure 13.1).

Gardeners often take advantage of the ways that plants can reproduce asexually. They use runners, bulbs and tubers to produce more plants.

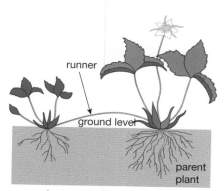

A new plant is produced where the runner touches the ground.

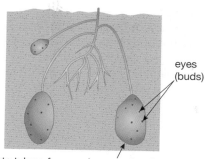

Potato tubers form underground at the ends of branches from the main stem. Each potato can produce several new plants from the 'eyes' which are buds.

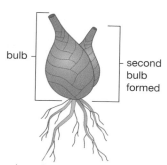

Some plants form bulbs. They are the bases of leaves which have become swollen with food. Buds in them can develop into new plants. Plants can form more than one bulb.

▲ Figure 13.1 Some methods of asexual reproduction in plants

Another type of asexual reproduction is to grow plants from **cuttings**. A piece of a plant's stem, with a few leaves attached, is cut from a healthy plant. This is planted in damp soil or compost, where it will grow roots and develop into a new plant (Figure 13.2).

▲ Figure 13.2 These geranium cuttings have started to grow roots

SEXUAL REPRODUCTION IN PLANTS

Plants produce specialised, haploid gametes in their flowers. The male gametes are contained within the **pollen grains** and the female gametes are egg cells or **ova**. Just as in animals, the male gametes must be transferred to the female gametes. This takes place through **pollination,** which is normally carried out either by wind or insects. Following pollination, fertilisation takes place and the zygote formed develops into a **seed**, which, in turn, becomes enclosed in a **fruit**.

PRODUCTION OF GAMETES AND POLLINATION

The gametes are produced by meiosis in structures in the flowers. Pollen grains are produced in the **anthers** of the **stamens**. The ova are produced in **ovules** in the **ovaries**.

In pollination, pollen grains are transferred from the anthers of a flower to the **stigma**. If this occurs within the same flower it is called **self-pollination**. If the pollen grains are transferred to a different flower, it is called **cross-pollination**. Pollination can occur by wind or by insect in either case.

Plants that are wind-pollinated produce flowers with a different structure to those of insect-pollinated flowers. These differences are related to the different methods of pollination of the flowers. Figure 13.3 shows the structure of a typical insect-pollinated flower and Figure 13.4 shows the structure of a typical wind-pollinated flower. Table 13.2 summarises the main differences between insect-pollinated flowers and wind-pollinated flowers.

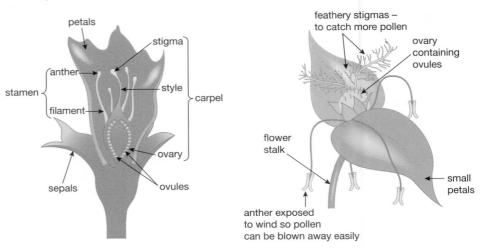

▲ Figure 13.3 The main structures in an insect-pollinated flower.

▲ Figure 13.4 The main structures in an wind-pollinated flower.

Table 13.2 Differences between insect-pollinated and wind-pollinated flowers.

Feature of flower	Type of flower	
	Insect-pollinated	Wind-pollinated
position of stamens	enclosed within flower so that insect must make contact	exposed so that wind can easily blow pollen away
position of stigma	enclosed within flower so that insect must make contact	exposed to catch pollen blowing in the wind
type of stigma	sticky so pollen grains attach from insects	feathery, to catch pollen grains blowing in the wind
size of petals	large to attract insects	small
colour of petals	brightly coloured to attract insects	not brightly coloured, usually green
nectaries	present – they produce nectar, a sweet liquid containing sugars as a 'reward' for insects	absent
pollen grains	larger, sticky grains or grains with hooks, to stick to insects' bodies	smaller, smooth, inflated grains to carry in the wind

FERTILISATION

Pollination transfers the pollen grain to the stigma. However, for fertilisation to take place, the nucleus of the pollen grain (the male gamete) must fuse with the nucleus of the ovum, which is inside an ovule in the ovary. To transfer the nucleus to the ovum, the pollen grain forms a **pollen tube**, which grows down through the tissue of the **style** and into the ovary. Here it curves around to enter the opening in an ovule. The tip of the tube dissolves and allows the pollen grain nucleus to move out of the tube and into the ovule. Here it fertilises the ovum nucleus. These events are summarised in Figure 13.5.

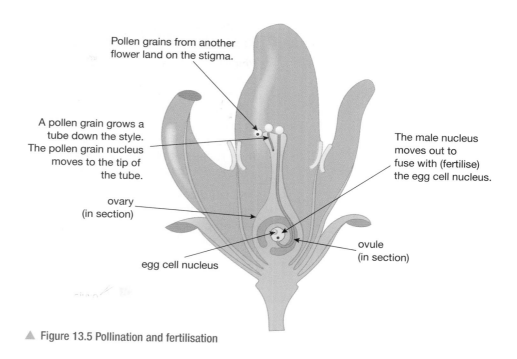

Pollen grains from another flower land on the stigma.

A pollen grain grows a tube down the style. The pollen grain nucleus moves to the tip of the tube.

The male nucleus moves out to fuse with (fertilise) the egg cell nucleus.

ovary (in section)

ovule (in section)

egg cell nucleus

▲ Figure 13.5 Pollination and fertilisation

SEED AND FRUIT FORMATION

Once fertilisation has occurred, a number of changes take place in the ovule and ovary that will lead to the fertilised ovule becoming a seed and the ovary in which it is found becoming a fruit. Different flowers produce different types of fruits, but in all cases the following four changes take place.

1 The zygote develops into an embryonic plant with small root (**radicle**) and shoot (**plumule**).

2 Other contents of the ovule develop into a food store for the young plant when the seed germinates.

3 The ovule wall becomes the seed coat or **testa**.

4 The ovary wall becomes the fruit coat; this can take many forms depending on the type of fruit.

Figure 13.6 summarises these changes as they occur in the plum flower, which forms a fleshy fruit.

> **DID YOU KNOW?**
> Any structure that contains seeds is a fruit. A pea pod is a fruit. The 'pod' is the fruit wall, formed from the ovary wall, and the peas are individual seeds. Each forms from a fertilised ovule.

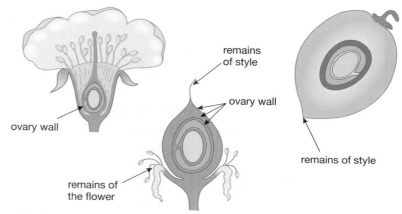

ovary wall

remains of style

ovary wall

remains of style

remains of the flower

▲ Figure 13.6 How a plum fruit forms

GERMINATION

A seed contains a plant embryo, consisting of a root (radicle), shoot (plumule) and one or two seed leaves, called **cotyledons**. It also contains a food store, either in the cotyledons or another part of the seed. During germination, the food store is used up, providing the nutrients to allow the radicle and plumule to grow. The radicle grows down into the soil, where it will absorb water and mineral ions. The plumule grows upwards towards the light, where it can start the process of photosynthesis (Figure 13.7). Once the small plant (seedling) is able to photosynthesise, germination is over.

The seeds of plants such as peas or beans have two cotyledons. They are called dicotyledonous plants, or **dicots**. Seeds of grasses and other narrow-leaved plants, such as irises and orchids have only one cotyledon. They are monocotyledonous plants, or **monocots**.

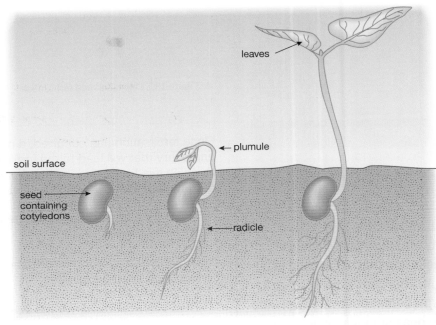

▲ Figure 13.7 Germination in the broad bean. In this species, the cotyledons contain a food store. They remain below the ground when the seed germinates. In other species (e.g. the French bean), they are carried above the ground during germination.

The food store of peas or beans is present in the cotyledons. It consists mainly of starch and protein. In monocots such as maize, there is a separate food store of starch.

THE CONDITIONS NEEDED FOR GERMINATION

The growth of a new plant from a seed is called **germination**. When seeds are dispersed from the parent plant they are usually very dry, containing only about 10% water. This low water content restricts a seed's metabolism, so that it can remain alive but dormant (inactive) for a long time, sometimes for many years.

When a seed germinates, dormancy comes to an end. The seed's food store is broken down by enzymes and respired aerobically. This means that germination needs the following conditions:

■ warm temperatures, so that enzymes can act efficiently (see Chapter 1)

■ water, for chemical reactions to take place in solution

■ oxygen, for respiration.

Safety Note: Wash hands after handling seeds.

ACTIVITY 1

▼ PRACTICAL: INVESTIGATING THE CONDITIONS NEEDED FOR GERMINATION

Small seeds such as peas or mustard will grow on wet cotton wool in a test tube. Figure 13.A shows four tubes set up to investigate the conditions needed for the seeds to germinate.

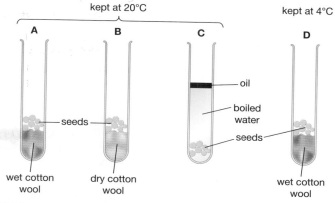

▲ Figure 13.8 Apparatus to investigate the conditions needed for germination

The water in tube C has been boiled to remove dissolved oxygen. The layer of oil (e.g. cooking oil) keeps out oxygen from the air.

■ Tube A: seeds on wet cotton wool, maintained at 20 °C (room temperature)

■ Tube B: seeds on dry cotton wool, maintained at 20 °C

■ Tube C: seeds in boiled water, with a thin surface layer of oil; maintained at 20 °C

■ Tube D: seeds on wet cotton wool, placed in a refrigerator at 4 °C

After a few days the seeds in the control tube (A) will start to germinate. There will be no germination in tubes B or C. The seeds in tube D may eventually start to germinate, but much more slowly than in tube A.

What do the results tell you about the conditions needed for germination? Do you have any criticisms of this method? (Hint – are all the variables fully controlled?)

CHAPTER QUESTIONS

More questions on reproduction in plants can be found at the end of Unit 3 on page 182.

SKILLS ▷ CRITICAL THINKING

1 Which of the following features are shown by wind-pollinated flowers?

 A large petals, a scent and sticky pollen grains

 B small petals, no scent and light pollen grains

 C large petals, no scent and light pollen grains

 D small petals, a scent and sticky pollen grains

2 During pollination, between which of the following structures are pollen grains transferred?

 A from anther to stigma   **B** from anther to ovary

 C from stigma to style   **D** from stigma to ovary

3 Following fertilisation, which of the following takes place?

A The ovule becomes the fruit and the ovary becomes the seed

B The carpel becomes the fruit and the anther becomes the seed

C The anther becomes the fruit and the carpel becomes the seed

D The ovary becomes the fruit and the ovule becomes the seed

4 Read the three statements below.

1. Uptake of water   2. Uptake of oxygen   3. Increase in metabolism

Which of these events are associated with seed germination?

A 1 and 2   **B** 1 and 3   **C** 2 and 3   **D** 1, 2 and 3

5 The diagram below shows a section through an insect-pollinated flower. Pollination happens when pollen grains land on part X.

a Name part X.

b How are insects attracted to a flower like this? Give two ways.

c Copy the diagram and extend the pollen tube to show where it would go when fully grown.

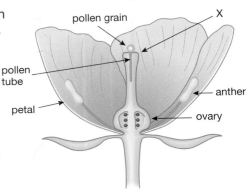

6 A student investigated whether temperature affected the germination of seeds. He set up three boiling tubes, each containing 10 seeds placed on wet cotton wool.

■ Tube A was kept at room temperature (20 °C) in the light.

■ Tube B was placed in a fridge at 4 °C.

■ Tube C was placed in an incubator at 30 °C.

The student monitored the seeds for 3 days, making sure that the cotton wool remained damp. At the end of this time he observed the results and measured the height of any seedlings that had germinated. His results are shown in the table.

Tube	Height of any seedlings that germinated / cm	% seeds germinated	Average height of seeds that germinated / cm
A	2.3, 2.7, 1.9, 2.5, 2.6, 3.1, 1.9, 2.2, 2.6, 3.2	100	2.5
B	0.3, 0.5 (the other seeds did not germinate)	20	0.4
C	3.4, 4.5, 2.5, 3.7, 2.8, 4.4, 4.3, 2.9, 2.1, 3.7	100	

a State the independent variable and the dependent variable in the student's experiment.

b Calculate the average height of the seedlings that germinated in tube C.

c Summarise the conclusions that can be drawn from this experiment.

d Why does temperature affect the germination of seeds?

e Another student suggested that the experiment was not properly controlled. Explain what she meant, and suggest how the experiment could be modified.

7   The diagram shows a potato plant producing new tubers (potatoes). Buds on the parent plant grow into stems that grow downwards, called stolons. The ends of each stolon develop into a new tuber.

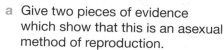
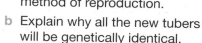
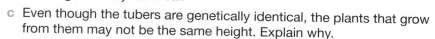

a   Give two pieces of evidence which show that this is an asexual method of reproduction.

b   Explain why all the new tubers will be genetically identical.

c   Even though the tubers are genetically identical, the plants that grow from them may not be the same height. Explain why.

d   Why do wild plants need to reproduce sexually as well as asexually?

8   The drawing shows a wind-pollinated flower.

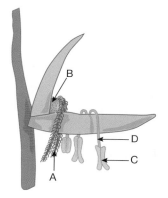

a   Name the structures labelled A, B, C and D.

b   Give *three* pieces of evidence *visible in the diagram*, which show that this flower is wind-pollinated.

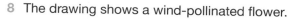

c   Describe how fertilisation takes place once a flower has been pollinated.

d   Describe *four* ways in which you would expect an insect-pollinated flower to be different from the flower shown.

9   The drawing shows a strawberry plant reproducing in two ways.

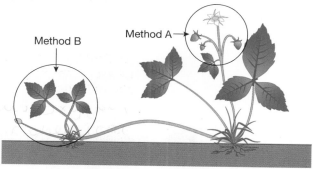

Method B   Method A

a   Which of the two methods of reproduction shown will result in offspring that show genetic variation? Explain your answer.

b   Is the strawberry flower likely to be wind-pollinated or insect-pollinated? Give reasons for your answer.

10 All banana plants are reproduced asexually. Biologists are concerned for their future, as a new strain of fungus has appeared which is killing all the banana plants in some plantations.

a   Explain why the fungus is able to kill all the banana plants in some plantations.

b   Explain why this would be less likely to happen if banana plants reproduced sexually.

c   Describe the benefits of reproducing banana plants asexually.

UNIT QUESTIONS

  1 Light intensity and the concentration of carbon dioxide in the atmosphere influence the rate of photosynthesis.

a The graph shows the effect of changing light intensity on the rate of photosynthesis at two different carbon dioxide concentrations.

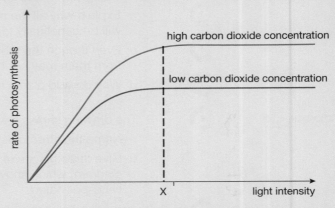

i Describe the effect of light intensity on the rate of photosynthesis at each concentration of carbon dioxide up to light intensity X and beyond light intensity X. **(4)**

ii Which factor limits the rate of photosynthesis up to light intensity X and beyond light intensity X? **(2)**

Explain your answer in each case. **(3)**

b iii Describe two other factors which influence the rate of photosynthesis. **(2)**

ii Explain why each is a limiting factor. **(4)**

c 'Photosynthesis is a means of converting light energy into chemical energy.'

Explain what this statement means. **(2)**

Total 17 marks

2 In an investigation to determine the water potential of potato cells, the following procedure was adopted.

Cylinders of tissue were obtained from a potato. Each was the same diameter and cut to a length of 5 cm.

Each cylinder was dried and then weighed.

Three potato cylinders were placed in each of six different concentrations of sucrose solution and left for two hours.

The cylinders were then removed from the solutions, dried and reweighed. The percentage change in mass for each was calculated, and then an average percentage change in mass calculated for each solution.

The graph summarises the results.

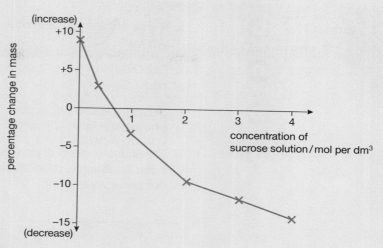

a Explain why:

   i the cylinders were dried before and after being placed in the sucrose solutions **(1)**

   ii three cylinders were used for each solution **(1)**

   iii all the cylinders were the same diameter and were cut to the same length. **(1)**

b i In terms of water potential, explain the result obtained with a 3 mol per dm^3 sucrose solution. **(3)**

   ii What concentration of sucrose has a water potential equivalent to that of the potato cells? Explain your answer. **(3)**

c How could the water potential of the potato tissue be determined more accurately? **(2)**

Total 11 marks

BIOLOGY ONLY

3   Transpiration is the process by which water moves through plants from roots to leaves. Eventually, it is lost through the stomata.

The diagram shows the main stages in the movement of water through a leaf.

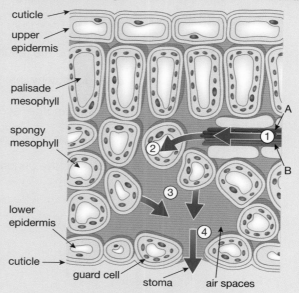

SKILLS ANALYSIS   (6)

a Name the tissues labelled A and B. Explain how you arrived at your answers. **(4)**

(8)

b Describe how water is being moved at each of the stages 1, 2, 3 and 4. **(4)**

(5)

c Describe two ways, visible in the diagram, in which this leaf is adapted to photosynthesise efficiently. **(4)**

(6)

d For plants living in dry areas, explain a possible conflict between the need to obtain carbon dioxide for photosynthesis and the need to conserve water. **(2)**

Total 14 marks

END OF BIOLOGY ONLY

SKILLS CRITICAL THINKING   (5)   **4**

Plants can respond to a range of stimuli.

a Plant shoots detect and grow towards light.

   i What is this process called? **(1)**

(8)

   ii Explain how a plant bends towards the light. **(3)**

(6)

   iii Explain the advantage to the plant of this response. **(1)**

b In an investigation, young plant shoots were exposed to light from one side. The wavelength of the light was varied. The graph summarises the results of the investigation.

SKILLS ANALYSIS   (6)

   i Describe the results shown in the graph. **(2)**

(7)

   ii Suggest why the results show this pattern. **(2)**

SKILLS CRITICAL THINKING   (5)

c   i Name another stimulus that produces a growth response in plants. **(1)**

   ii Describe the ways roots and shoots respond to this stimulus **(2)**

(6)

   iii What is the benefit to the plant of these responses? **(2)**

Total 14 marks

SKILLS › ANALYSIS

The diagram below shows a flower.

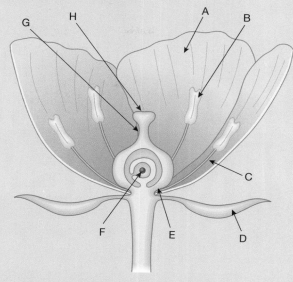

a Give the letter of the structure which:

   i produces pollen grains

   ii becomes the seed

   iii becomes the fruit wall.   **(3)**

b Write down two ways that you can see in the diagram that this flower is adapted for insect pollination.   **(2)**

c Give the letter of the structure which is:

   i the stigma

   ii the style

   iii the filament.   **(3)**

SKILLS › INTERPRETATION

d Explain the difference between:

   i pollination and fertilisation

   ii self-pollination and cross-pollination.   **(4)**

Total 12 marks

SKILLS › EXECUTIVE FUNCTION

6 When pollen grains land on the stigma of a plant they germinate (grow a pollen tube). Growth of the pollen tube is thought to be stimulated by sugars produced by the stigma.

If pollen grains are placed in certain solutions they will germinate. The drawing shows some grains seen through a microscope after 2 hours in a solution.

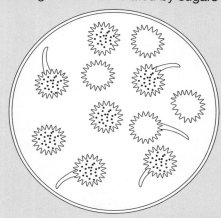

Design an investigation to find out if germination of pollen grains is stimulated by sucrose.

Your answer should include experimental details and be written in full sentences.

Total 6 marks

UNIT 4
ECOLOGY AND THE ENVIRONMENT

Ecology is the scientific study of the interactions between organisms and their surroundings. It analyses the interactions that living things have with each other and with their non-living, physical environment. In Chapter 14 we look at the components of ecosystems and these interactions. The activities of humans are having many serious effects on the world ecosystems, in particular as a consequence of agricultural practices and industrial pollution. Chapter 15 looks at these activities and considers how humans are damaging the planet.

14 ECOSYSTEMS

An ecosystem is a distinct, self-supporting system of organisms interacting with each other and with a physical environment. An ecosystem can be small, such as a pond, or large, such as a mangrove swamp or a large forest. This chapter looks at a variety of ecosystems and the interactions that happen within them.

LEARNING OBJECTIVES

- Understand the terms ecosystem, habitat, population and community

- Investigate the population size of an organism in two different areas using quadrats

BIOLOGY ONLY

- Understand the term biodiversity

- Investigate the distribution of organisms in their habitats and measure biodiversity usng quadrats

- Understand how biotic and abiotic factors affect the population size and distribution of organisms

- Understand the names given to different trophic levels, including producers, primary, secondary and tertiary consumers, and decomposers

- Understand the concepts of food chains, food webs, pyramids of number, pyramids of biomass and pyramids of energy transfer

- Understand the transfer of substances and energy along a food chain

- Understand why only about 10% of energy is transferred from one trophic level to the next

- Describe the stages in the carbon cycle, including respiration, photosynthesis, decomposition and combustion

BIOLOGY ONLY

- Describe the stages in the nitrogen cycle, including the roles of nitrogen-fixing bacteria, decomposers, nitrifying bacteria and denitrifying bacteria

▲ Figure 14.1 A pond is a small ecosystem.

▲ Figure 14.2 A mangrove swamp is a larger ecosystem.

THE COMPONENTS OF ECOSYSTEMS

Whatever their size, **ecosystems** usually have the same components:

- **producers** – plants which photosynthesise to produce food
- **consumers** – animals that eat plants or other animals
- **decomposers** – organisms that break down dead material and help to recycle nutrients
- the physical environment – all the non-biological components of the ecosystem; for example, the water and soil in a pond or the soil and air in a forest.

The living components of an ecosystem are called the *biotic* components. The non-living (physical) components are the *abiotic* components (compare these with **biotic** and **abiotic** *factors*, below).

An ecosystem contains a variety of habitats. A **habitat** is the place where an organism lives. For example, habitats in a pond ecosystem include the open water, the mud at the bottom of the pond, and the surface water.

All the organisms of a particular species found in an ecosystem at a certain time form the **population** of that species. All the immature frogs (tadpoles) swimming in a pond are a population of tadpoles; all the water lily plants growing in the pond make up a population of water lilies.

The populations of *all* species (animals, plants and other organisms) found in an ecosystem at a particular time form the **community**. Figure 14.3 illustrates the main components of a pond ecosystem.

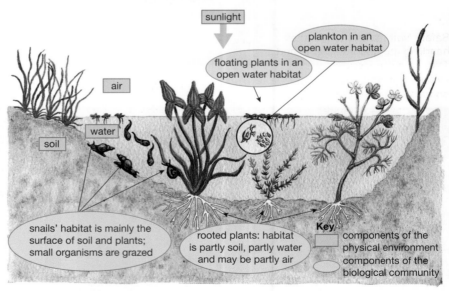

▲ Figure 14.3 A pond ecosystem

USING QUADRATS TO SAMPLE FROM A HABITAT

When an ecologist wants to know how many organisms there are in a particular habitat, it would not be possible for him to count them all. Instead, he is forced to count a smaller representative part of the population, called a *sample*. Sampling of plants, or animals that do not move much (such as snails), can be done using a sampling square called a **quadrat**. A quadrat is usually made from metal, wood or plastic. The size of quadrat you use depends on the size of the organisms being sampled. For example, to count plants growing on a school field, you could use a quadrat with sides 0.5 or 1 metre in length (Figure 14.4).

▲ Figure 14.4 A student sampling with a quadrat

It is important that sampling in an area is carried out *at random*, to avoid bias. For example, if you were sampling from a school field, but for convenience only placed your quadrats next to a path, this probably wouldn't give you a sample that was representative of the whole field! It would be a *biased* sample.

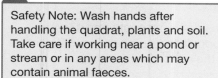

Safety Note: Wash hands after handling the quadrat, plants and soil. Take care if working near a pond or stream or in any areas which may contain animal faeces.

ACTIVITY 1

▼ PRACTICAL: USING QUADRATS TO COMPARE THE SIZE OF A PLANT POPULATION IN TWO AREAS OF A FIELD

One way that you can sample randomly is to place quadrats at coordinates on a numbered grid.

Imagine that there are two areas of a school field (A and B) that seem to contain different numbers of dandelion plants. Area A is more trampled than area B and looks like it contains fewer dandelions. This might lead you to propose the hypothesis: 'The dandelion population in area A is smaller than the dandelion population in area B'.

In area A, two 10-metre tape measures are arranged to form the sides of a square (Figure 14.5).

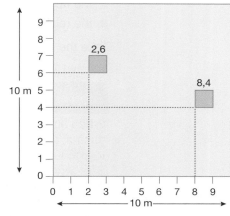

▲ Figure 14.5 A 10-m² grid with 1-m² quadrats positioned at coordinates 2,6 and 8,4.

A pair of random numbers is generated, using the random number function on a calculator. These numbers are used as coordinates to position the quadrat in the large square.

The numbers of dandelions in the quadrat are counted. The process is then repeated for nine more quadrats.

The tape measures are then moved to area B and the process repeated to sample from ten more quadrats in that part of the field.

The table below shows a set of results for a study like this.

Quadrat number	Number of dandelions in each quadrat in area A	Number of dandelions in each quadrat in area B
1	4	10
2	7	7
3	1	9
4	0	14
5	3	12
6	8	7
7	3	16
8	12	9
9	1	11
10	6	15

Calculate the mean number of dandelions per m² in each area. Do the results support the hypothesis that the population numbers are different? How could you improve the reliability of the results?

BIOLOGY ONLY

BIODIVERSITY

The amount of variation shown by species in an ecosystem is called the ecosystem's **biodiversity**. It is a combination of two measurements:

- the number of different species present (known as the *species richness*)
- the relative abundance of each species – their 'evenness' of numbers.

Take the example of the two 'communities' shown in Table 14.1

Table 14.1 Two 'communities' of organisms.

Species	Number of individuals of each species in community 1	Number of individuals of each species in community 2
A	10	1
B	10	1
C	10	1
D	10	1
E	10	46

Both communities contain the same number of species (5) and organisms (50) but community 2 is dominated by one species (E). Community 1 contains an even number of each species, so it has a higher biodiversity.

Some ecosystems such as tropical rainforests have a very high biodiversity (Figure 14.6). Other ecosystems are dominated by one species. This is shown by the pine forest plantations of northern Europe, which are dominated by one species of tree. The trees produce a very dense cover, or 'canopy'. Lack of light severely restricts the growth of other tree species and ground layer plants (Figure 14.7). The pine forest also provides a limited variety of habitats for animals.

▲ Figure 14.6 A tropical rainforest ecosystem contains thousands of different species of plants and animals and has a very high biodiversity.

▲ Figure 14.7 A pine forest has a low biodiversity

Biodiversity is generally a good thing for an ecosystem. Ecosystems with a high biodiversity are often more stable than ones with a low biodiversity. This is because an ecosystem that is dominated by one (or a few) species is more likely to be affected by any sort of ecological disaster. For example, if a new disease arose that wiped out the dominant tree species, this would have an impact on other species that relied on the tree for food and shelter. In a more diverse ecosystem other tree species might supply these resources.

ACTIVITY 2

▼ PRACTICAL: USING QUADRATS TO COMPARE THE BIODIVERSITY OF PLANTS IN TWO HABITATS

The sampling method used in Practical 17 can be modified to compare the biodiversity of plants in two habitats.

Imagine that there are two other areas of the field (C and D) that seem to contain different numbers of various plant species. A hypothesis about this observation might be: 'The diversity of plants in area C is higher than in area D'.

Ten 1 m² quadrat samples are taken in each area. The numbers of each plant species present in each quadrat are counted.

The table below shows a set of results. The student who carried out the investigation was only interested in comparing the broad-leaved plants in the two areas, and did not record the grasses present.

Plant species	Total from 10 quadrats in area C	Total from 10 quadrats in area D
dandelion	69	12
ribwort plantain	78	188
broad-leaved plantain	38	99
daisy	95	22
yarrow	44	12
creeping buttercup	30	5
white clover	49	7
common cat's ear	8	0
groundsel	5	0

Plot the results as two bar charts (total numbers of each species against species name). Use the same axis scales for each bar chart. Describe the biodiversity in the two areas. Do the results support the hypothesis that the two plant communities have a different biodiversity?

END OF BIOLOGY ONLY

INTERACTIONS IN ECOSYSTEMS

The organisms in an ecosystem are continually interacting with each other and with their physical environment. Interactions include the following.

- Feeding among the organisms – the plants, animals and decomposers are continually recycling the same nutrients through the ecosystem.
- Competition among the organisms – animals compete for food, shelter, mates, nesting sites; plants compete for carbon dioxide, mineral ions, light and water.
- Interactions between organisms and the environment – plants absorb mineral ions, carbon dioxide and water from the environment; plants also give off water vapour and oxygen into the environment; animals use materials from the environment to build shelters; the temperature of the environment can affect processes occurring in the organisms; processes occurring in organisms can affect the temperature of the environment (all organisms produce some heat).

REMINDER

Don't forget that plants take in carbon dioxide and give out oxygen only when there is sufficient light for photosynthesis to occur efficiently. When there is little light, plants take in oxygen and give out carbon dioxide. You should be able to explain the reasons for this (see Chapter 10).

BIOTIC AND ABIOTIC FACTORS

There are many factors that influence the numbers and distribution of organisms in an ecosystem. There are two types of factor – *biotic* and *abiotic*.

Biotic factors are biological. Many (but not all) involve feeding relationships. They include:

- availability of food and competition for food resources
- predation
- parasitism
- disease
- presence of pollinating insects
- availability of nest sites.

Abiotic factors are physical or chemical factors. They include:

- climate, such as light intensity, temperature and water availability
- hours of daylight
- soil conditions, such as clay content, nitrate level, particle size, water content and pH
- other factors specific to a particular habitat, such as salinity (salt content) in an estuary, flow rate in a river, or oxygen concentration in a lake
- pollution.

Clearly *which* factors affect population sizes and distribution of organisms will depend on the *type* of ecosystem. If you take the example of a river, some of the main abiotic factors could be:

- depth of water
- flow rate
- type of material at the bottom of the stream (stones, sand, mud etc.)
- concentration of minerals in the water
- pH
- oxygen concentration
- cloudiness of the water
- presence of any pollution.

The main biotic factors affecting animals in the river will be food supply, either from plants or other animals. But other factors are important too – large fish could not live in a shallow stream!

It is impossible to generalise about which factors are the most important. In a heavily polluted river all the organisms could be killed by the pollution, while in a clean river depth and flow rate might have a greater effect on the animals that could live there. The different factors may also affect one another. For example, a faster flow rate could mix the water with air, increasing the amount of dissolved oxygen.

The main factor affecting large ecosystems is climate, particularly temperature and rainfall. Climate is the reason why tropical rainforests are restricted to a strip near the equator of the Earth, while pine forests grow in the higher latitudes of the northern hemisphere.

KEY POINT

A **predator** is an animal that kills and eats another animal. A **parasite** is an organism (animal or plant) that lives in or on another organism (called its host) and gets its nutrition from the host. e.g. mosquitoes are parasites of humans (and other animals), and the human is the host of the mosquito.

KEY POINT

Large areas of the Earth dominated by a specific type of vegetation are called *biomes*. Temperate grassland and tropical rain forest are two examples of biomes. You could carry out some research to find out about these and other biomes.

FEEDING RELATIONSHIPS

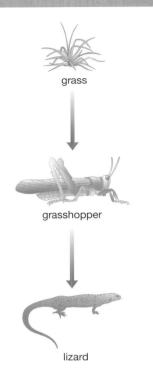

grass

grasshopper

lizard

▲ Figure 14.8 A simple food chain.

KEY POINT

Another way of describing the food chain is that it shows how energy is moved from one organism to another as a result of feeding. The arrows show the direction of energy flow.

The simplest way of showing feeding relationships within an ecosystem is a **food chain** (Figure 14.8).

In any food chain, the arrow (→) means 'is eaten by'. In the food chain illustrated, the grass is the **producer**. It is a plant so it can photosynthesise and produce food materials. The grasshopper is the **primary consumer**. It is an animal which eats the producer and is also a **herbivore**. The lizard is the **secondary consumer**. It eats the primary consumer and is also a **carnivore**. The different stages in a food chain (producer, primary consumer and secondary consumer) are called **trophic levels**.

Many food chains have more than three links in them. Here are two examples of longer food chains:

   filamentous algae → mayfly nymph → caddis fly larvae → salmon

In this freshwater food chain, the extra link in the chain makes the salmon a **tertiary consumer**.

   plankton → crustacean → fish → ringed seal → polar bear

In this marine (sea) food chain, the fifth link makes the polar bear a **quaternary consumer**. Because nothing eats the polar bear, it is also called the *top carnivore*.

Food chains are a convenient way of showing the feeding relationships between a few organisms in an ecosystem, but they oversimplify the situation. The marine food chain above implies that only crustaceans feed on plankton, which is not true. Some whales and other mammals also feed on plankton. For a fuller understanding, you need to consider how the different food chains in an ecosystem relate to each other. Figure 14.9 gives a clearer picture of the feeding relationships involved in a freshwater ecosystem in which salmon are the top carnivores. This is the **food web** of the salmon.

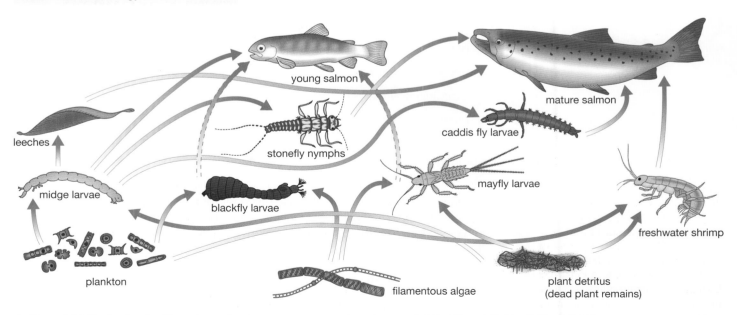

▲ Figure 14.9 The food web of the salmon. As you can see, young salmon have a slightly different diet to mature salmon.

This is still an oversimplification of the true situation, as some feeding relationships are still not shown. It does, however, give some indication of the interrelationships that exist between food chains in an ecosystem. With a little thought, you can predict how changes in the numbers of an organism in one food chain in the food web might affect those in another food chain. For example, if the leech population were to decline through disease, there could be several possible consequences:

■ the stonefly nymph population could increase as there would be more midge larvae to feed on

■ the stonefly nymph population could decrease as the mature salmon might eat more of them as there would be fewer leeches

■ the numbers could remain the same due to a combination of the above.

Although food webs give us more information than food chains, they don't give any information about how many, or what mass of organisms is involved. Neither do they show the role of the decomposers. To see this, we must look at other ways of presenting information about feeding relationships in an ecosystem.

ECOLOGICAL PYRAMIDS

Ecological pyramids are diagrams that represent the relative amounts of organisms at each trophic level in a food chain. There are two main types:

■ **pyramids of numbers**, which represent the numbers of organisms in each trophic level in a food chain, irrespective of their mass

■ **pyramids of biomass**, which show the total mass of the organisms in each trophic level, irrespective of their numbers.

Consider these two food chains:

a grass → grasshopper → frog → bird

b oak tree → aphid → ladybird → bird

Figures 14.8 and 14.9 show the pyramids of numbers and biomass for these two food chains.

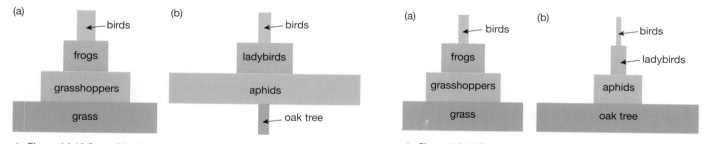

▲ Figure 14.10 Pyramids of numbers for two food chains   ▲ Figure 14.11 Pyramids of biomass for the two food chains

KEY POINT

Biomass is the total mass of organisms. If it refers to living organisms, this is called the *fresh biomass*. More commonly the *dry biomass* is used. This is the mass of plant or animal material after water has been removed, by drying in an oven. Dry biomass is a more reliable measure, since the water content of organisms (especially plants) varies with environmental conditions.

The two pyramids for the 'grass' food chain look the same – the numbers at each trophic level decrease. The *total* biomass also decreases along the food chain – the mass of *all* the grass plants in a large field would be more than that of *all* the grasshoppers which would be more than that of *all* the frogs, and so on.

The two pyramids for the 'oak tree' food chain look different because of the size of the oak trees. Each oak tree can support many thousands of aphids, so the numbers *increase* from first to second trophic levels. But each ladybird will need to eat many aphids and each bird will need to eat many ladybirds, so the numbers *decrease* at the third and fourth trophic levels. However, the total biomass *decreases* at each trophic level – the biomass of one oak tree is much greater than that of the thousands of aphids it supports. The total biomass of all these aphids is greater than that of the ladybirds, which is greater than that of the birds.

Suppose the birds in the second food chain are parasitised by nematode worms (small worms living in the bird's gut). The food chain now becomes:

oak tree → aphid → ladybird → bird → nematode worm

The pyramid of numbers now takes on a very strange appearance (Figure 14.12a) because of the large numbers of parasites on each bird. The pyramid of biomass, however, has a true pyramid shape because the total biomass (Figure 14.12b) of the nematode worms must be less than that of the birds they parasitise.

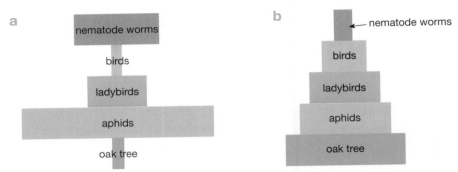

▲ Figure 14.12 (a) A pyramid of numbers and (b) a pyramid of biomass for the parasitised food chain.

WHY ARE DIAGRAMS OF FEEDING RELATIONSHIPS A PYRAMID SHAPE?

The explanation is relatively straightforward (Figure 14.11). When a rabbit eats grass, not all of the materials in the grass plant end up as rabbit! There are losses:

- some parts of the grass are not eaten (the roots for example)

- some parts are not digested and so are not absorbed – even though rabbits have a very efficient digestive system

- some of the materials absorbed form excretory products

- many of the materials are respired to release energy, with the loss of carbon dioxide and water.

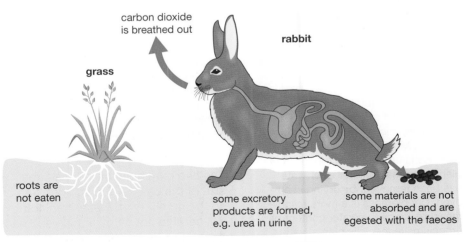

▲ Figure 14.13 Not all the grass eaten by a rabbit ends up as rabbit tissue.

In fact, only a small fraction of the materials in the grass ends up in new cells in the rabbit. Similar losses are repeated at each stage in the food chain, so smaller and smaller amounts of biomass are available for growth at successive trophic levels. The shape of pyramids of biomass reflects this.

Feeding is a way of transferring energy between organisms. Another way of modelling ecosystems looks at the energy flow between the various trophic levels.

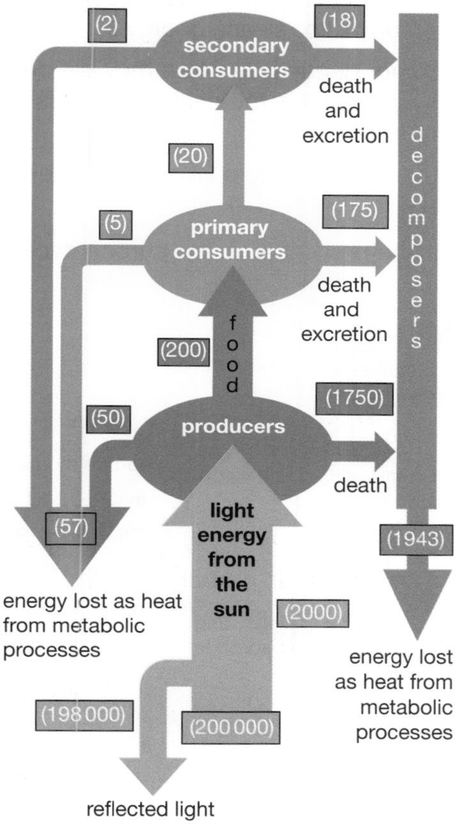

THE FLOW OF ENERGY THROUGH ECOSYSTEMS

This approach focuses less on individual organisms and food chains and rather more on energy transfer between trophic levels (producers, consumers and decomposers) in the whole ecosystem. There are a number of key ideas involved:

- Photosynthesis 'fixes' sunlight energy into chemicals such as glucose and starch.

- Respiration releases energy from organic compounds such as glucose.

- Almost all other biological processes (e.g. muscle contraction, growth, reproduction, excretion, active transport) use the energy released in respiration.

- If the energy released in respiration is used to produce new cells, then the energy remains 'fixed' in molecules in that organism. It can be passed on to the next trophic level through feeding.

- If the energy released in respiration is used for other processes then it will, once used, eventually escape as heat from the organism. Energy is therefore lost from food chains and webs at each trophic level.

Figure 14.14 is an energy flow diagram. It shows the main ways in which energy is transferred in an ecosystem. It also gives the amounts of energy transferred between the trophic levels of this particular (grassland) ecosystem.

As you can see, only about 10% of the energy entering a trophic level is passed on to the next trophic level. This explains why not many food chains have more than five trophic levels. Think of the food chain:

$$A \rightarrow B \rightarrow C \rightarrow D \rightarrow E$$

If we use the idea that only about 10% of the energy entering a trophic level is passed on to the next level, then, of the original 100% reaching A (a producer), 10% passes to B, 1% (10% of 10%) passes to C, 0.1% passes to D and only 0.001% passes to E. There just isn't enough energy left for another trophic level. In certain parts of the world, some marine food chains have six trophic levels because of the huge amount of light energy reaching the surface waters.

CYCLING NUTRIENTS THROUGH ECOSYSTEMS

The chemicals that make up our bodies have all been around before – probably many times! You may have in your body some carbon atoms that were part of the body of Mahatma Gandhi or were in carbon dioxide molecules breathed out by Winston Churchill. This constant recycling of substances is all part of the cycle of life, death and decay.

Microorganisms play a key role in recycling. They break down complex organic molecules in the bodies of dead animals and plants into simpler substances, which they release into the environment.

All figures given are kilojoules ($\times 10^5$) per m^2 per year.

▲ Figure 14.14 The main ways in which energy is transferred in an ecosystem. The amounts of energy transferred through 1 m^2 of a grassland ecosystem per year are shown in brackets.

THE CARBON CYCLE

Carbon is a component of all major biological molecules. Carbohydrates, lipids, proteins, DNA, vitamins and many other molecules all contain carbon. The following processes are important in cycling carbon through ecosystems.

■ Photosynthesis 'fixes' carbon atoms from carbon dioxide into organic compounds

■ Feeding and assimilation pass carbon atoms already in organic compounds along food chains

■ Respiration produces inorganic carbon dioxide from organic compounds (mainly carbohydrates) as they are broken down to release energy

■ Fossilisation – sometimes living things do not decay fully when they die due to the conditions in the soil (decay is prevented if it is too acidic) and fossil fuels (coal, oil, natural gas and peat) are formed

■ Combustion releases carbon dioxide into the atmosphere when fossil fuels are burned.

Figures 14.15 and 14.16 show the role of these processes in the carbon cycle in different ways.

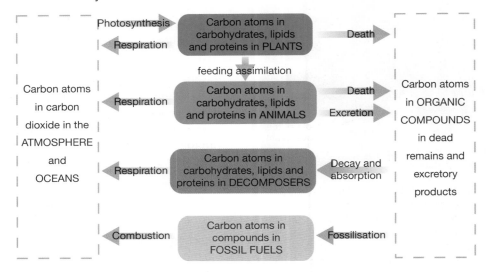

▲ Figure 14.15 The main stages in the carbon cycle

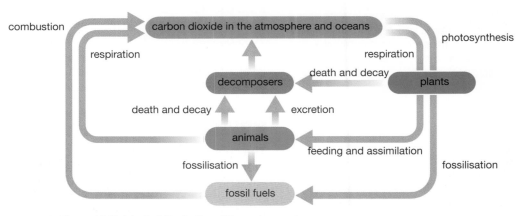

▲ Figure 14.16 A typical illustration of the carbon cycle

THE NITROGEN CYCLE

Some nitrogen-fixing bacteria are free-living in the soil. Others form associations with the roots of legumes (legumes are plants that produce seeds in a pod, like peas and beans). They form little bumps or 'root nodules' (Figure 14.17).

these nodules contain millions of nitrogen-fixing bacteria

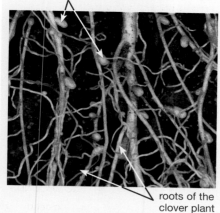

roots of the clover plant

▲ Figure 14.17 Root nodules on a clover plant.

Instead of entering the soil, the ammonia that the bacteria make by fixing nitrogen is passed to the plant which uses it to make amino acids. In return, the plant provides the bacteria with organic nutrients. This is an example of **mutualism**, where both organisms benefit from the relationship. This nitrogen fixation enriches the soil with nitrates when the plants die and are decomposed.

Nitrogen is an element that is present in many biological compounds, including proteins, amino acids, most vitamins, DNA and ATP. Like the carbon cycle, the nitrogen cycle involves feeding, assimilation, death and decay. Photosynthesis and respiration are not directly involved in the nitrogen cycle as these processes fix and release carbon, not nitrogen. The following processes are important in cycling nitrogen through ecosystems.

- Feeding and assimilation pass nitrogen atoms already in organic compounds along food chains.
- **Decomposition** by fungi and bacteria produces ammonia from the nitrogen in compounds like proteins, DNA and vitamins.
- The ammonia is oxidised first to nitrite and then to nitrate by **nitrifying bacteria**. This overall process is called nitrification.
- Plant roots can absorb the nitrates. They are combined with carbohydrates (from photosynthesis) to form amino acids and then proteins, as well as other nitrogen-containing compounds.

This represents the basic nitrogen cycle, but other bacteria carry out processes that affect the amount of nitrate in the soil that is available to plants, as follows:

- **Denitrifying bacteria** use nitrates as an energy source and convert them into nitrogen gas. Denitrification *reduces* the amount of nitrate in the soil.
- Free-living **nitrogen-fixing bacteria** in soil convert nitrogen gas into ammonia. This is used by the bacteria to make amino acids and proteins. When the bacteria die, their proteins decompose, releasing ammonia back to the soil.
- Nitrogen-fixing bacteria in **root nodules** (see margin box) also make ammonia but this is converted by the plant into amino acids and other organic nitrogen compounds. Death and decomposition of the plant returns the nitrogen to the soil as ammonia.

Figure 14.18 shows the role of these processes in the nitrogen cycle.

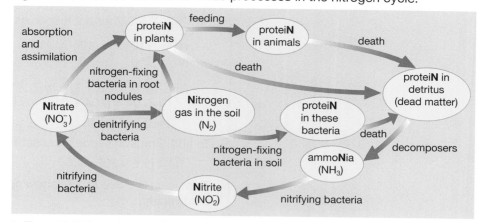

▲ Figure 14.18 The main stages in the nitrogen cycle

To remember if these organic compounds contains nitrogen, check to see if the letter **N** (the symbol for nitrogen) is present:

Protei**N**s, ami**N**o acids and **DNA** all contain nitrogen. Carbohydrates and fats don't – they have no **N**.

In addition to all the processes described in the biological part of the nitrogen cycle (Figure 14.18), there are other ways that nitrates are formed. Lightning converts nitrogen gas in the air into oxides of nitrogen. These dissolve in rainwater, enter the soil and are converted into nitrates by nitrifying bacteria. Humans also make nitrates industrially from nitrogen gas. These nitrates are mainly used as fertilisers (see Chapter 15.) because they increase the rate of growth of crops.

END OF BIOLOGY ONLY

LOOKING AHEAD – MEASURING BIODIVERSITY

Ecology is a very mathematical branch of biology. An example of this is that biodiversity can be measured by calculating a *diversity index.* One example is called Simpson's Index. One formula for Simpson's Index (*D*) is:

$$D = 1 - \sum \left(\frac{n}{N}\right)^2$$

In this formula:

N = the total number of individuals of all species

n = the total number of individuals of a particular species.

You calculate $\frac{n}{N}$ for each species, square each of these numbers, and then sum (Σ) all the squares. Subtracting this value from 1 gives you *D*. Values of *D* range from 0 to 1, where 0 represents a low diversity and 1 a high diversity.

Take the data you saw in Table 14.1:

Species	Number of individuals of each species in community 1	Number of individuals of each species in community 2
A	10	1
B	10	1
C	10	1
D	10	1
E	10	46

For community 1:

The total number of individuals of all species (*N*) = 50

For each species the total number of individuals (*n*) = 10.

So $\frac{n}{N} = \frac{10}{50} = 0.2$ for each species

$$\sum\left(\frac{n}{N}\right)^2 = \left(\frac{10}{50}\right)^2 + \left(\frac{10}{50}\right)^2 + \left(\frac{10}{50}\right)^2 + \left(\frac{10}{50}\right)^2 + \left(\frac{10}{50}\right)^2$$

$$= 0.2^2 + 0.2^2 + 0.2^2 + 0.2^2 + 0.2^2$$

$$= 0.04 + 0.04 + 0.04 + 0.04 + 0.04$$

$$= 0.20$$

To find Simpson's Index you subtract this value from 1:

$D = (1 - 0.20) = 0.80$

This value is close to 1.0, showing that community 1 has a high biodiversity.

Can you calculate Simpson's Index for community 2? If you do it carefully you should get the answer $D = 0.152$, showing that community 2 has a lower biodiversity.

EXTENSION WORK

If you are interested, you can do some background research about Simpson's Index. You will find that there are different formulas used for the Index.
It doesn't matter which one you use, as long as you are consistent.

CHAPTER QUESTIONS

More questions on ecosystems can be found at the end of Unit 4 on page 221.

SKILLS   CRITICAL THINKING

1 Which of the following terms is defined as 'all the organisms living in a particular place and their interactions with each other and with their environment'?

  A habitat   **B** population

  C community   **D** ecosystem

2 Which of the following is a biotic factor?

  A light intensity   **B**   food supply

  C pollution   **D**   temperature

3 The diagram below shows a food web containing four food chains.

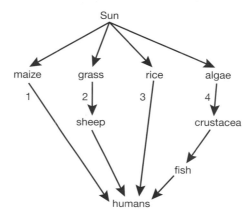

Which food chains are most efficient at using energy from the Sun?

  A 1 and 3   **B** 2 and 4   **C** 2 and 3   **D** 3 and 4

BIOLOGY ONLY

4 In the nitrogen cycle, which of the following conversions is carried out by nitrifying bacteria?

  A nitrogen gas to nitrates

  B nitrates to nitrogen gas

  C ammonia to nitrates

  D nitrates to ammonia

END OF BIOLOGY ONLY

5 a Explain what is meant by the terms habitat, community, environment and population.

  b What are the roles of plants, animals and decomposers in an ecosystem?

6 A marine food chain is shown below.

    plankton → small crustacean → krill → seal → killer whale

  a Which organism is **i** the producer, **ii** the secondary consumer?

  b What term best describes the killer whale?

  c Suggest why five trophic levels are possible in this case, when many food chains only have three or four.

7 Part *(a)* of the diagram shows a woodland food web. Part *(b)* shows a pyramid of numbers and a pyramid of biomass for a small part of this wood.

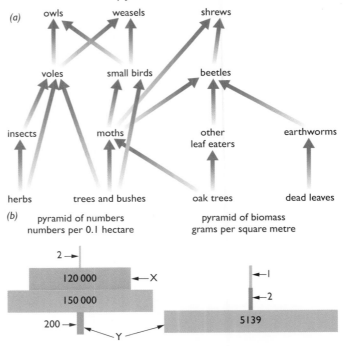

a Write out *two* food chains (from the food web) containing four organisms, both involving moths.

b Name *one* organism in the food web which is both a primary consumer and a secondary consumer.

c Suggest how a reduction in the dead leaves may lead to a reduction in the numbers of shrews.

d In part *(b)* of the diagram, explain why level Y is such a different width in the two pyramids.

BIOLOGY ONLY

8 The diagram shows part of the nitrogen cycle.

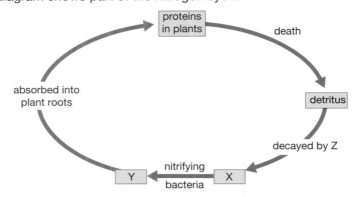

a What do X, Y and Z represent?

b Name the process by which plant roots absorb nitrates.

c What are nitrogen-fixing bacteria?

d Give *two* ways, not shown in the diagram, in which animals can return nitrogen to the soil.

9   In a year, 1 m² of grass produces 21 500 kJ of energy. The diagram below shows the fate of the energy transferred to a cow feeding on the grass.

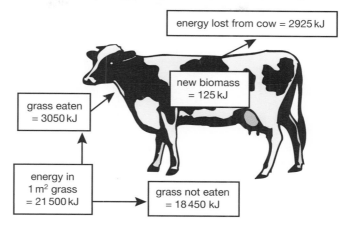

energy lost from cow = 2925 kJ

new biomass = 125 kJ

grass eaten = 3050 kJ

energy in 1 m² grass = 21 500 kJ

grass not eaten = 18 450 kJ

a   Calculate the energy efficiency of the cow from the following equation.

$$\text{Energy efficiency} = \frac{\text{energy that ends up as part of cow's biomas}}{\text{energy available}} \times 100$$

b   State two ways that energy is lost from the cow.

c   Suggest what may happen to the 18 450 kJ of energy in the grass that was not eaten by the cow.

10 Read the following description of the ecosystem of a mangrove swamp.

Pieces of dead leaves (detritus) from mangrove plants in the water are fed on by a range of crabs, shrimps and worms. These, in turn, are fed on by several species of fish, including young butterfly fish, angelfish, tarpon, snappers and barracuda. Mature snappers and tarpon are caught by fishermen as the fish move out from the swamps to the open seas.

a   Use the description to construct a food web of the mangrove swamp ecosystem.

b   Write out two food chains, each containing four organisms from this food web. Label each organism in each food chain as producer, primary consumer, secondary consumer or tertiary consumer.

c   Decomposers make carbon in the detritus available again to mangrove plants.

    i   In what form is this carbon made available to the mangrove plants?

    ii Explain how the decomposers make the carbon available.

END OF BIOLOGY ONLY

15 HUMAN INFLUENCES ON THE ENVIRONMENT

Humans have intelligence far beyond that of any other animal on Earth. This chapter looks at the ways in which we have used our intelligence to influence natural environments, and some of the problems we have caused.

the ozone layer is being destroyed in many places

the greenhouse effect means that heat is trapped in the Earth's atmosphere leading to global warming

smoke contains gases which contribute to acid rain and the greenhouse effect

acid rain forms when some of the products of burning fuels react with water in the atmosphere

deforestation increases the amount of carbon dioxide in the atmosphere and destabilises soils

exhaust fumes contain gases that contribute to acid rain

fertilisers in rivers cause algae to grow quickly which can lead to a lack of dissolved oxygen

▲ Figure 15.1 How the actions of humans influence the environment.

LEARNING OBJECTIVES

- Describe how glasshouses and polythene tunnels can be used to increase the yield of certain crops
- Understand the effects on crop yield of increased carbon dioxide and increased temperature in glasshouses
- Understand how the use of fertiliser can increase crop yield
- Understand the reasons for pest control and the advantages and disadvantages of using pesticides and biological control with crop plants

BIOLOGY ONLY
- Understand the methods used to farm large numbers of fish to provide a source of protein

- Understand that water vapour, carbon dioxide, nitrous oxide, methane and CFCs are greenhouse gases
- Understand how human activities contribute to greenhouse gases

- Understand how an increase in greenhouse gases results in an enhanced greenhouse effect and that this may lead to global warming and its consequences
- Understand the biological consequences of pollution of air by sulfur dioxide and carbon monoxide

BIOLOGY ONLY
- Understand the effects of deforestation, including leaching and soil erosion; and disturbance to evapotranspiration, the carbon cycle, and the balance of atmospheric gases

- Understand the biological consequences of pollution of water by sewage
- Understand the biological consequences of eutrophication caused by leached minerals from fertiliser

Since humans first appeared on Earth, our numbers have grown dramatically (Figure 15.2). The secret of our success has been our intelligence. Unlike other species, we have not adapted to one specific environment, we have changed many environments to suit our needs.

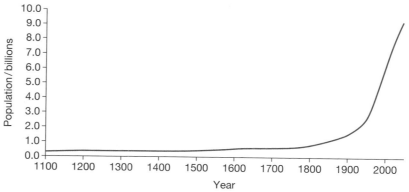

▲ Figure 15.2 The rate of increase of the human population has been particularly steep over the last 200 years. The curve has been extended to 2050, showing that the population is expected to reach 10 billion by that date..

As our numbers have grown, so has the sophistication of our technology. Early humans made tools from materials readily to hand. Today's technology involves much more complex processes. As a result, we produce ever-increasing amounts of materials that pollute our air, soil and waterways.

Early humans influenced their environment, but the enormous size of the population today and the extent of our industries mean that we affect the environment much more significantly. We make increasing demands on the environment for:

- food to sustain an ever-increasing population
- materials to build homes, schools and industries

- fuel to heat homes and power vehicles
- space in which to build homes, schools and factories, as well as for leisure facilities
- space in which to dump our waste materials.

MODERN AGRICULTURE – PRODUCING THE FOOD WE NEED

A modern farm is a sort of managed ecosystem. Many of the interactions are the same as in natural ecosystems. Crop plants depend on light and mineral ions from the soil as well as other factors in the environment. Stock animals (sheep, cattle, etc.) depend on crop plants for food (see Figure 15.3).

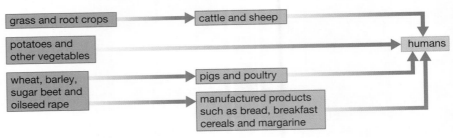

▲ Figure 15.3 A food web on a farm

Farmers must make a profit from their farms. To do this, they try to control the environment in such a way as to maximise the yield from crop plants and livestock.

IMPROVING YIELDS FROM CROP PLANTS

Table 15.1 summaries some factors that can be controlled by the farmer in order to maximise crop yield.

Table 15.1 Some ways the yield from crops can be improved.

Factor controlled	How it is controlled	Reason for controlling the factor
soil ions (e.g. nitrates)	adding fertilisers to the soil or growing in a hydroponic culture (Figure 15.4a)	extra mineral ions can be taken up and used to make proteins and other compounds for growth
soil structure	ploughing fields to break up compacted soil; adding manure to improve drainage and aeration of heavy, clay soils	good aeration and drainage allow better uptake of mineral ions (by active transport) and water
soil pH	adding lime (calcium salts) to acidic soils; few soils are too alkaline to need treatment	soil pH can affect crop growth as an unsuitable pH reduces uptake of mineral ions
carbon dioxide, light and heat	these cannot be controlled for field crops but in a glasshouse or polytunnel all can be altered to maximise yield of crops (Figure 15.4b); burning fuels produces heat and carbon dioxide	all may limit the rate of photosynthesis and the production of the organic substances needed for growth

DID YOU KNOW?

Soil pH can vary between about pH 3 to 8. Soil pH is tested using indicator kits or pH meters. Some soils are too acidic – they can be made more alkaline by adding lime.

a

▲ Figure 15.4 (a) A glasshouse maintains a favourable environment for plants. Crops grown by hydroponics in a glasshouse.

Glasshouses (otherwise known as 'greenhouses') and polytunnels can provide very controlled conditions for plants to grow (Figure 15.4). There are several reasons for this.

- The transparent walls of the glasshouse allow enough natural light for photosynthesis during the summer months, while additional lighting gives a 'longer day' during the winter.

▲ Figure 15.4 (b) Many crops are grown in large tunnels made of transparent polythene, called polytunnels.

■ The 'greenhouse effect' doesn't just happen to the Earth, but also in greenhouses! Short wavelength infrared radiation entering the glasshouse is absorbed and re-radiated as longer wavelength infrared radiation. This radiation cannot escape through the glass, so the glasshouse heats up (see the section on the Earth's greenhouse effect later in this chapter). The glasshouse also reduces convection currents that would cause cooling.

■ The glasshouse can be heated to raise the temperature if the outside temperature is too low.

■ If heaters use fossil fuels such as gas, this produces carbon dioxide and water vapour. The carbon dioxide is a raw material of photosynthesis. The water vapour maintains a moist atmosphere and reduces water loss by transpiration.

■ If the plants are grown in a hydroponic culture this provides *exactly* the right balance of mineral ions for the particular crop.

KEY POINT

By heating glasshouses to the optimum temperature for photosynthesis, a farmer can maximise his yield. Heating above this temperature is a waste of money as there is no further increase in yield.

CYCLING NUTRIENTS ON A FARM

In Chapter 14 you saw how the elements nitrogen and carbon are cycled in nature. On a farm, the situation is quite different, particularly with regard to the circulation of nitrogen.

Nitrates from the soil supply nitrogen that is needed to make proteins in plants. Some of these plants are crops that will be sold; others are used as food for the stock animals (fodder). When the crops are sold, the nitrogen in the proteins goes with them and is lost from the farm ecosystem. Similarly, when livestock is sold, the nitrogen in their proteins (gained from the fodder) goes with them and is lost from the farm ecosystem. To replace the lost nitrogen, a farmer usually adds some kind of fertiliser. The amount of fertiliser added must be carefully monitored to ensure the maximum growth and yield of the crop – using excess fertiliser wastes money.

Figure 15.5 summarises the circulation of nitrogen on a farm.

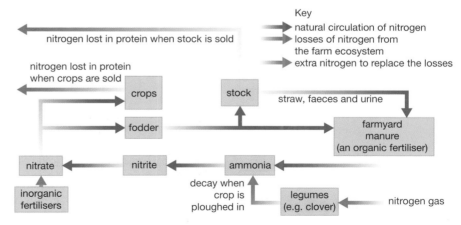

▲ Figure 15.5 The nitrogen cycle on a farm. The effects of denitrification and lightning (see Chapter 14) have been omitted.

FERTILISERS INCREASE CROP YIELD

There are two main types of fertilisers – *organic* and *inorganic*. Many organic fertilisers (such as farmyard manure) are made from the faeces of farm animals mixed with straw. Inorganic fertilisers are simply inorganic compounds such as potassium nitrate or ammonium nitrate, carefully formulated to provide a specific amount of nitrate (or some other ion) when applied according to the manufacturer's instructions.

Adding farmyard manure returns some nitrogen to the soil. But, as farmyard manure is made from livestock faeces and indigestible fodder, it can only replace a portion of the lost nitrogen. Most farmers apply inorganic fertilisers to replace the nitrates and other mineral ions lost. Whilst this can replace *all* the lost ions, it can also lead to pollution problems, which we will discuss later. Inorganic fertilisers do not improve soil structure in the way that organic fertilisers can, because they do not contain any decaying matter that is an essential part of the soil.

Another way to replace lost nitrates is to grow a legume crop (e.g. clover) in a field one year in four. Legumes have nitrogen-fixing bacteria in nodules on their roots (see Chapter 14). These bacteria convert nitrogen gas in the soil air to ammonium ions. Some of this is passed to the plants, which use it to make proteins. At the end of the season, the crop is ploughed back into the soil, and decomposers convert the nitrogen in the proteins to ammonia. This is then oxidised to nitrate by nitrifying bacteria and made available for next year's crop.

PEST CONTROL

Pests are organisms that reduce the yield of crop plants or stock animals. The 'yield' of a crop is the amount produced for sale. A pest can harm this in two ways:

- lowering the amount by reducing growth, e.g. by damaging leaves and reducing photosynthesis
- affecting the appearance or quality of a crop, making it unsuitable for sale (Figure 15.6).

> **DID YOU KNOW?**
> In Britain, about 30% of the potential maize crop is lost to weeds, insects and fungal diseases (Figure 15.6).

▲ Figure 15.6 Damage to a maize (corn) cob caused by the corn earworm caterpillar.

Any type of organism – plants, animals, bacteria, fungi or protoctists, as well as viruses – can be a pest. Pests can be controlled in a number of ways. Chemicals called **pesticides** can be used to kill them, or their numbers can be reduced by using **biological control** methods.

Pesticides are named according to the type of organism they kill:

- herbicides kill plant pests (they are weedkillers)
- insecticides kill insects
- fungicides kill fungi
- molluscicides kill snails and slugs.

> **DID YOU KNOW?**
> A weed is a plant that is growing where it is not wanted. Weeds can be controlled mechanically or chemically. Mechanical control involves physically removing the weeds. Chemical control uses herbicides.

A farmer uses pesticides to kill particular pests and improve the yield from the crops or livestock. Pests are only a problem when they are present in big enough numbers to cause economic damage – a few whiteflies in a tomato crop are not a problem; the real damage arises when there are millions of them. Whether or not a farmer uses pesticides is largely a decision based on cost. The increase in income due to higher yields must be set against the cost of the pesticides.

PROBLEMS WITH PESTICIDES

DID YOU KNOW?

Cultivating large areas of land with a single crop (a *monoculture*) encourages pests. Monocultures make harvesting the crop easier. But if a pest arrives, it can easily spread through the crop. During the winter the pest can lie dormant in the soil ready to attack next year's crop.

Crop rotation breaks the pest cycle. This is where a different crop is grown each year (Figure 15.7). When over-wintering pests emerge, their preferred crop is no longer there.

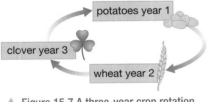

▲ Figure 15.7 A three-year crop rotation.

DID YOU KNOW?

DDT is a historical case, but there are problems with many modern pesticides. One recent example involves a class of insecticides called *neonicotinoids* (chemicals similar to nicotine). They have been linked to several serious ecological problems, including honey bee colony collapse disorder (CCD) and loss of birds due to a reduction in insect populations.

One problem with using pesticides is that a pest may develop *resistance* to the chemical. This happens through natural selection (see Chapter 19). It makes the existing pesticide useless, so that another must be found. Other problems are to do with the fact that pesticides can cause environmental damage. There are several reasons for this:

- they may be slow to decompose – they are *persistent* in the environment
- they build up in the tissues of organisms – *bioaccumulation*
- they build up and become more concentrated along food chains – *biomagnification*
- they kill other insects that are harmless, as well as helpful species, such as bees.

An ideal pesticide should:
- control the pest effectively
- be biodegradable, so that no toxic products are left in the soil or on crops
- be specific, so that only the pest is killed
- not accumulate in organisms
- be safe to transport, store and apply
- be easy to apply.

We can explain these problems by looking at one well-documented example – the insecticide DDT (dichlorodiphenyltrichloroethane).

DDT was invented in 1874, and first discovered to be an insecticide in 1939. In World War II it was used with great success to kill malaria carrying mosquitoes and the lice that carried typhus. Its use increased up until the 1960s, when we began to understand its harmful effects. It was banned from general use in the USA in 1972, and worldwide in 2004. Limited use of DDT is still allowed for control of insects that transmit disease, although this is still regarded as controversial.

DDT is a very effective insecticide, so why has it been banned?

- DDT is very persistent, remaining active in the environment for many years. If DDT is sprayed onto a field, around half will still be there ten years later. To make things worse, the missing half won't have degraded to harmless products – some will have broken down to form a similar compound called DDE, which is also a potent insecticide, and some will have spread to other habitats. DDT is carried all around the world by wind, and has been identified in polar ice caps and deserts, thousands of kilometres away from where it was applied.

- By the 1950s, many types of insect began to appear that were resistant to DDT. These insects had developed a genetic mutation that prevented them from being killed by the insecticide. While DDT continued to be used, the resistant insects had an advantage over the non-resistant ones. Their numbers increased with every generation since they were able to survive exposure to the pesticide. They reproduced, passing on their resistance genes to their offspring. This is an example of **natural selection** (see

Chapter 19). There are now hundreds of examples of pest species that have developed resistance to different insecticides.

■ DDT doesn't just kill pests. It will kill any type of insect, including harmless ones such as butterflies and useful species such as bees, as well as natural predators such as wasps, which might themselves kill the pest insects. Insecticides damage ecosystems by disrupting food chains.

■ DDT is very soluble in fats. When a herbivore feeds on plants that are contaminated with DDT, the insecticide is not broken down or excreted. Instead it becomes concentrated in the fatty tissues of the animal. This is called **bioaccumulation**. When a carnivore eats the herbivore this process is repeated, so that the insecticide builds up in concentration along the food chain. This is known as **biomagnification** (Figure 15.8). The levels at the top of the food chain may be toxic, leading to the death of the top carnivores in the chain. This can disrupt the food web of an ecosystem.

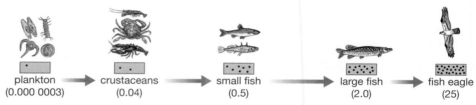

plankton (0.000 0003) → crustaceans (0.04) → small fish (0.5) → large fish (2.0) → fish eagle (25)

▲ Figure 15.8 Biomagnification of DDT in a food chain

BIOLOGICAL CONTROL

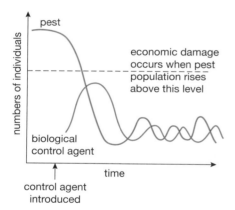

▲ Figure 15.9 Biological control.

Another option that may be used to control pests is biological control. Instead of using a toxic chemical, biological control uses another organism to reduce the numbers of a pest. We have already mentioned whiteflies as pests of tomatoes. One way of controlling them in large glasshouses is to introduce a parasite that will kill the whiteflies. A tiny wasp called *Encarsia* parasitises and kills their larvae, reducing the numbers of adult whitefly.

A problem with biological control is that it never fully gets rid of a pest. If the control organism killed *all* the pests, then it too would die out, as it would have no food supply. Biological control aims to reduce pest numbers to a level where they no longer cause significant economic damage (Figure 15.9).

There are several methods of biological control. Some examples include:

■ introducing a natural predator – ladybirds can be used to control the populations of aphids in orange groves

■ introducing a herbivore – a moth was introduced from South America to control the prickly pear cactus that was becoming a serious weed in grazing land in Australia

■ introducing a parasite – the wasp *Encarsia* is used to control whitefly populations in glasshouse tomato crops

■ introducing a pathogenic (disease-causing) microorganism – the myxomatosis virus was deliberately released in Australia to control the rabbit population

■ introducing sterile males – these mate with the females but no offspring are produced from these matings, so numbers fall

■ using pheromones – these are natural chemicals produced by insects to attract a mate. They are used to attract pests (either males or females) to traps. The pests are then destroyed, reducing the reproductive potential of the population. Male-attracting pheromones are used to control aphids (greenfly) in plum crops.

FISH FARMING

Fish is a good source of high quality protein. Over the last 60 years, the world's population has demanded an increasing amount of fish to eat. The tonnage of fish caught by ships has increased steadily, while the Earth's stock of many fish species has decreased dramatically.

Increasingly, fish farming (aquaculture) is meeting the need for fish as a food supply. Farming of fish and shellfish is the fastest-growing area of animal food production. In 1970 only 5% of the world's seafood was produced by aquaculture, in 2016 it was 50%. The most commonly farmed fish are various types of carp, catfish, tilapia, trout, salmon, cod, bream and sea bass, as well as various types of crustaceans, such as lobsters and prawns. They are not all used for human food: about one-quarter of farmed fish is used to make animal feed.

The fish are kept in densely stocked tanks or enclosures in rivers or lakes, or in sea cages (Figure 15.10). Fish farming has a number of advantages. The water quality can be monitored; for example the temperature, oxygen levels, water clarity and amount of chlorophyll in the water are measured. Large concentrations of chlorophyll give a warning of 'algal blooms' (see later in this chapter), which can be toxic to fish. Some conditions can be modified, e.g. air can be pumped into the enclosures to increase the amount oxygen dissolved in the water. The water is pumped through filtration units to remove the waste products of the fish.

The diet of the fish is also carefully controlled in both its quality and in the frequency of feeding. Enclosing the fish protects them against predators, and pesticides are used to kill parasites. Small fish may be eaten by larger members of their own species, so fish are regularly sorted by size and placed into different cages or tanks.

▲ Figure 15.10 Feeding the fish at a fish farm

Selective breeding programmes (see Chapter 20) can be used to improve the quality of the fish. For example, they are bred to produce faster growth and to be more 'placid' (less aggressive) than wild fish.

However, fish farming has been heavily criticised by environmentalists. In any intensive production system, the potential for the spread of disease is greater than normal because the animals are so close together. Antibiotics are often used to treat disease. This is a cause for concern because the antibiotics may not have degraded by the time the fish are eaten by humans, adding to the problem of antibiotic resistance in bacteria (see Chapter 19). Fish farms also cause a pollution problem, producing organic material from the animals' faeces and from food pellets, which can contaminate the waters outside the fish farm and cause eutrophication of the water (see page 217). The pesticides used to kill fish parasites may be highly toxic to other non-harmful species of invertebrates.

In fact, there is strong evidence that fish farming has a negative effect on 'wild' fish stocks. Carnivorous species like salmon and sea bass are fed with pellets made from other fish. They need to eat several kilograms of wild fish to produce 1 kilogram of farmed fish! The wild fish used for fishmeal are less marketable species, such as herring and sardines.

AIR POLLUTION

One definition of **pollution** is:

'Pollution is the contamination of the environment by harmful substances that are produced by the activities of humans'.

Human activities pollute the air with many gases. Some major examples that we will look at in this chapter are carbon dioxide, methane, carbon monoxide and sulfur dioxide.

CARBON DIOXIDE AND GLOBAL WARMING

KEY POINT

In any one year, there is a peak and a trough in the levels of carbon dioxide in the atmosphere

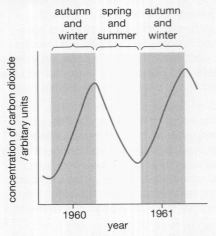

Figure 15.12 Seasonal fluctuations in carbon dioxide levels.

In the autumn and winter, trees lose their leaves. Without leaves they cannot photosynthesise and so do not absorb carbon dioxide. They still respire, which produces carbon dioxide, so in the winter months, they give out carbon dioxide and the level in the atmosphere rises. In the spring and summer, with new leaves and brighter sunlight, the trees photosynthesise faster than they respire. As a result, they absorb more carbon dioxide from the atmosphere than they produce, so the level decreases. However, because there are fewer trees overall, it doesn't quite return to the low level of the previous summer.

The levels of carbon dioxide have been rising for several hundred years. Over the last 100 years alone, the level of carbon dioxide in the atmosphere has increased by nearly 30%. This rise is mainly due to the increased burning of fossil fuels, such as coal, oil and natural gas, as well as petrol and diesel in vehicle engines. It has been made worse by cutting down large areas of tropical rainforest (see later in this chapter). These extensive forests have been called 'the lungs of the Earth' because they absorb such vast quantities of carbon dioxide and produce equally large amounts of oxygen. Extensive deforestation (see below) means that less carbon dioxide is being absorbed. Figure 15.11 shows changes in the level of carbon dioxide in the atmosphere (in parts per million) from 1960 to 2010.

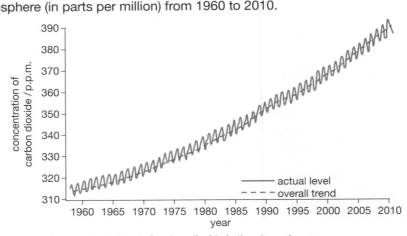

▲ Figure 15.11 Changes in the level of carbon dioxide in the atmosphere.

The increased levels of carbon dioxide and other gases contribute to global warming, or the enhanced greenhouse effect. It is important to understand that the 'normal' greenhouse effect occurs naturally – without it, more heat would be lost into space and the surface temperature of the Earth would be about 30 °C lower than it is today, and life as we know it would be impossible.

Carbon dioxide is one of the 'greenhouse gases' that are present in the Earth's upper atmosphere. Other greenhouse gases include water vapour (H_2O), methane (CH_4), nitrous oxide (N_2O) and chlorofluorocarbons (CFCs). Most greenhouse gases occur naturally, while some (like CFCs) are only produced by human activities.

The 'normal' greenhouse effect is shown in Figure 15.13..

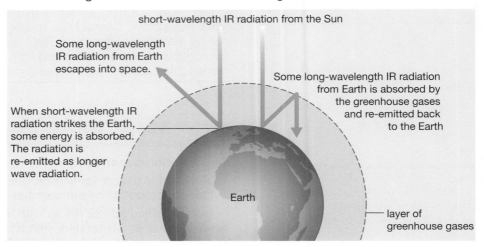

Figure 15.13 The Greenhouse effect

Short-wavelength infrared (IR) radiation from the Sun reaches the Earth. Some is absorbed by the Earth's surface and emitted again as longer-wavelength IR radiation. The greenhouse gases absorb and then re-emit some of this long-wavelength IR radiation, which would otherwise escape into space. This then heats up the surface of the Earth.

The problem is that human activities are polluting the atmosphere with *extra* greenhouse gases such as carbon dioxide. This is thought to be causing a rise in the Earth's surface temperature – the enhanced greenhouse effect, or global warming.

A rise in the Earth's temperature of only a few degrees would have many effects.

- Polar ice caps would melt and sea levels would rise.

- A change in the major ocean currents would result in warm water flowing into previously cooler areas.

- A change in global rainfall patterns could result. With a rise in temperature, there will be more evaporation from the surface of the sea, leading to more water vapour in the atmosphere and more rainfall in some areas. Other areas could experience a decrease in rainfall. Long-term climate change could occur.

- It could change the nature of many ecosystems. If species could not migrate quickly enough to a new, appropriate habitat, or adapt quickly enough to the changed conditions in their original habitat, they could become extinct.

- Changes in farming practices would be necessary as some pests became more abundant. Higher temperatures might allow some pests to complete their life cycles more quickly.

METHANE

Methane (CH_4) is an organic gas. It is produced when microorganisms ferment larger organic molecules to release energy. The most significant sources of methane are:

- decomposition of waste buried in the ground ('landfill sites'), by microorganisms

- fermentation by microorganisms in the rumen (stomach) of cattle and other ruminants

- fermentation by bacteria in rice fields.

Methane is a greenhouse gas, with effects similar to carbon dioxide. Although there is less methane in the atmosphere than carbon dioxide, each molecule has a much bigger greenhouse effect.

CARBON MONOXIDE

When substances containing carbon are burned in a limited supply of oxygen, carbon monoxide (CO) is formed. This happens when petrol and diesel are burned in vehicle engines. Exhaust gases contain significant amounts of carbon monoxide. It is a dangerous pollutant as it is colourless, odourless and tasteless and can be fatal. Haemoglobin binds more strongly with carbon monoxide than with oxygen. If a person inhales carbon monoxide for a period of time, more and more haemoglobin becomes bound to carbon monoxide and so cannot bind with oxygen. The person may lose consciousness and eventually die, as a result of a lack of oxygen reaching the cells, so that organs such as the heart and brain stop working.

SULFUR DIOXIDE

Sulfur dioxide (SO_2) is an important air pollutant. It is formed when fossil fuels are burned, and combines with water droplets in the air. It can be carried hundreds of miles in the atmosphere before falling as **acid rain**. Rain normally has a pH of about 5.5 – it is slightly acidic due to the carbon dioxide dissolved in it. Both sulfur dioxide and nitrogen oxides dissolve in rainwater to form a mixture of acids, including sulfuric acid and nitric acid. As a result, the rainwater is more acidic with a much lower pH than normal rain (Figure 15.14).

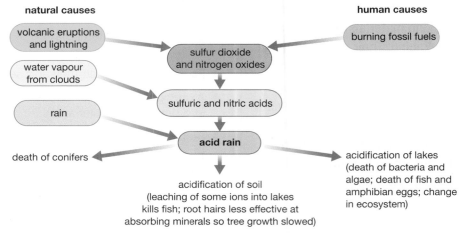

▲ Figure 15.14 The formation of acid rain and its effects on living organisms.

Lichens are small moss-like organisms. Some lichens are more tolerant of sulfur dioxide than others. In some countries, patterns of lichen growth can be used to monitor the level of pollution by sulfur dioxide. The different lichens are called *indicator species* as they 'indicate' different levels of sulfur dioxide pollution.

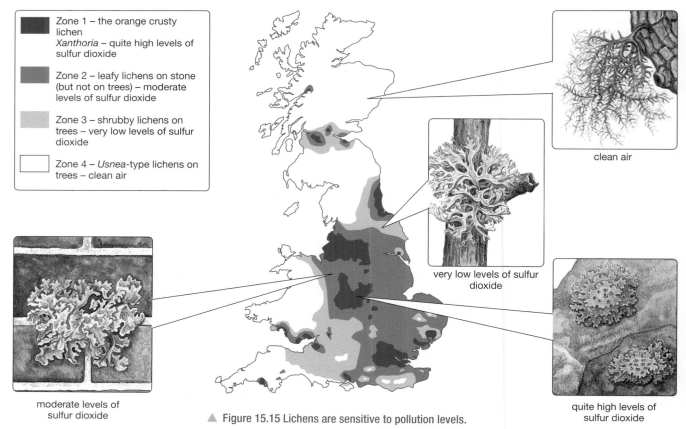

Zone 1 – the orange crusty lichen *Xanthoria* – quite high levels of sulfur dioxide

Zone 2 – leafy lichens on stone (but not on trees) – moderate levels of sulfur dioxide

Zone 3 – shrubby lichens on trees – very low levels of sulfur dioxide

Zone 4 – *Usnea*-type lichens on trees – clean air

clean air

very low levels of sulfur dioxide

moderate levels of sulfur dioxide

quite high levels of sulfur dioxide

▲ Figure 15.15 Lichens are sensitive to pollution levels.

The map in Figure 15.15 shows zones in Britain where different types of lichen are found.

BIOLOGY ONLY

DEFORESTATION

The last great natural forests of the world are the **tropical rainforests**. These form a belt around areas of the Earth near the equator, in South America, central Africa and Indonesia. They have a very high biodiversity. The rainforests are rapidly being destroyed by humans, in a process called deforestation. Deforestation is a consequence of the enormous growth of the human population. Every year, tens of thousands of hectares of rainforest are cut down to provide wood (timber) for building or other purposes, or to clear the land for farming (Figure 15.16). Much of the clearing is done by 'slash and burn' methods, where trees are cut down and burned, This adds to the carbon dioxide in the atmosphere and contributes to global warming. It also removes the trees, which would otherwise be absorbing carbon dioxide for photosynthesis. Deforestation is adding to global warming and climate change.

▲ Figure 15.16 Rainforest destroyed by the 'slash and burn' method.

▲ Figure 15.17 This river in Madagascar is full of silt from erosion caused by deforestation.

Apart from adding to global warming, there is a wide range of other problems caused by deforestation – some of these are listed below.

■ Destruction of habitats and reduced biodiversity. Rainforests are home to millions of species of plants, animals and other organisms. It has been estimated that 50–70% of species on Earth live in rainforests.

- Reduced soil quality. There are no trees and other plants to return minerals to the soil when they die, and no tree roots to hold the soil together. Crops planted in deforested areas rapidly use up minerals from the soil, and rain washes the minerals out (leaching).
- The soil is exposed due to lack of tree cover (canopy), and is blown or washed away (Figure 15.17). Soil may be washed into rivers, causing rising water levels and flooding of lowland areas.
- Deforestation may produce climate change. Trees are an important part of the Earth's water cycle, returning water vapour from the soil to the air by transpiration through their leaves. Cutting down the forests will upset the water cycle.
- In the past, rainforests have been a valuable source of many medicinal drugs, as well as species of plant that have been cultivated as crops. There are probably many undiscovered drugs and crop plants that will be lost with the deforestation.

You can see that there are many reasons why conservation of the remaining rainforests is urgently needed. Sometimes controlled replanting schemes (reforestation or re-afforestation) are carried out, allowing for *sustainable* timber production. 'Sustainable' production means replacing the trees that are removed and ensuring that there is no ecological damage to the environment.

Tackling the problem of clearance of trees for farming is a more complex issue. Large-scale cattle farming on deforested land is carried out mainly to supply meat for the burger industry, or palm oil for cosmetics, and you may feel these reasons are not ethical ones. The small-scale farming by farmers around the edges of rainforests is more understandable. These farmers are poor, and their livelihood and that of their family depends on this way of life. The only alternative would be to resettle the farmers and give them financial help to establish farms in other areas.

END OF BIOLOGY ONLY

WATER POLLUTION

Two major pollutants of freshwater are sewage and minerals from fertiliser.

POLLUTION OF WATER BY SEWAGE

Sewage is wet waste from houses, factories and farms. In developed countries where large-scale sewage treatment takes place, industrial and agricultural sewage is usually dealt with separately from household sewage. Household sewage consists of wastewater from kitchens and bathrooms and contains human urine and faeces, as well as dissolved organic and inorganic chemicals such as soaps and detergents. It is carried away in pipes called sewers, to be treated before it enters waterways such as rivers or the sea.

If sewage is discharged untreated into waterways, it produces two major problems:

- Aerobic bacteria in the water polluted by the sewage use up the dissolved oxygen in the water as they break down the organic materials. This reduction in the level of oxygen kills larger animals such as freshwater insects and fish.
- Untreated sewage contains pathogenic bacteria, which are a danger to human health (see Chapter 13).

Where untreated or 'raw' sewage enters a river, the level of oxygen in the water becomes very low as the aerobic bacteria and other microorganisms from the sewage decompose the organic matter. Only species that are adapted to

DID YOU KNOW?

Cutting down and burning rainforests is adding to carbon dioxide in the atmosphere. But does it affect oxygen? Books often talk about the rainforests as being the 'lungs of the world' because of the large amount of oxygen they produce. But scientists think that overall they don't have much effect on the world's oxygen levels. In a rainforest, the oxygen produced by the living plants is roughly balanced by the oxygen consumed by decomposers feeding on dead plant material.

live in low-oxygen conditions, such as anaerobic bacteria, can survive. As the water moves away from the outlet, it becomes oxygenated again as it mixes with clean water and absorbs oxygen from the air. The increase in dissolved oxygen levels allows more clean-water species to survive. Figure 15.18 shows the changes in oxygen content, numbers of clean-water animals, and numbers of polluted-water animals downstream from a sewage outlet.

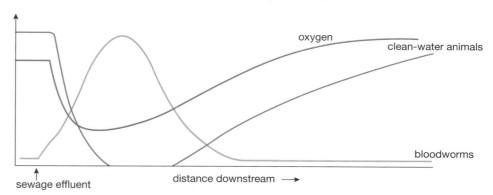

▲ Figure 15.18 Changes in oxygen levels and types of organism living downstream from a sewage outlet into a river.

The aim of sewage treatment is to remove solid and suspended organic matter and pathogenic microorganisms, so that cleaner waste can be discharged into waterways.

As with air pollution by sulfur dioxide, the level of pollution by organic material can be monitored by the presence or absence of indicator species. Figure 15.19 shows some of these.

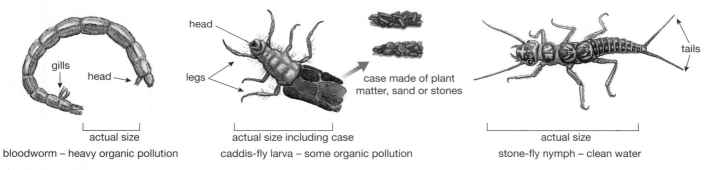

▲ Figure 15.19 Some freshwater animals will only live in very clean water, while others can survive in very polluted areas.

EUTROPHICATION

Eutrophication comes from a Greek word, meaning 'well-fed'. It refers to a situation where large amounts of nutrients enter a body of water, such as a river, lake or even the sea. The nutrients in question are inorganic mineral ions, usually nitrates or phosphates. Pollution by these minerals can have very harmful effects on an aquatic ecosystem.

There are two main sources of excess minerals:

■ from untreated or treated sewage

■ from artificial nitrate or phosphate fertilisers.

Eutrophication is often caused by the use of artificial fertiliser. Streams and rivers that run through agricultural land that have been treated with fertiliser can contain high concentrations of nitrate and phosphate. This is because nitrate is very soluble in water, and is easily washed out of the soil by rain, a process known as **leaching**. This is less of a problem with phosphate fertiliser, but phosphate is also washed into waterways by surface run-off of water.

The excess mineral ions stimulate the growth of all plants in the river or lake, but this is usually seen first as a rapid growth of algae, called an **algal bloom**. The algae can increase in numbers so rapidly that they form a thick scum on the surface of the water (Figure 15.20).

Figure 15.20 An algal bloom caused by fertiliser

REMINDER

The sequence of events following eutrophication is:

increase in mineral ions
↓
algal bloom
↓
death of algae
↓
decomposition by aerobic bacteria
↓
bacteria use up oxygen
↓
fish and other animals die

The algae soon start to die, and are decomposed by aerobic bacteria in the water. Because the bacteria respire aerobically, they use up the oxygen in the water. In addition, the algae block the light from reaching other rooted plants, further decreasing the oxygen produced by photosynthesis. The low oxygen levels can result in fish and other aerobic animals dying. In severe cases, the water becomes *anoxic* (containing very little oxygen) and smelly from gases like hydrogen sulfide and methane from the bacteria. By this stage only anaerobic bacteria can survive.

Rapid eutrophication is less likely when farmers use organic fertiliser (manure). The organic nitrogen-containing compounds in manure are less soluble and so are leached less quickly from the soil.

CHAPTER QUESTIONS

More questions on human influences on the environment can be found at the end of Unit 4 on page 221.

SKILLS ▷ CRITICAL THINKING

1   Which of the following is a desirable property for a pesticide?

    A It kills all insects

    B It does not show bioaccumulation

    C It is persistent in the environment

    D It is not biodegradable

SKILLS CRITICAL THINKING

2 The following gases are found in atmospheric air:

1. nitrogen

2. methane

3. carbon dioxide

4. water vapour

Which of these is/are *not* a greenhouse gas?

A 1 only   **B** 1 and 4

C 2 and 3   **D** 2 and 4

BIOLOGY ONLY

SKILLS CRITICAL THINKING

3 Which of the following is a consequence of deforestation?

A soil erosion

B increased water vapour in the air

C reduced CO_2 in the air

D increased biodiversity

END OF BIOLOGY ONLY

4 In a lake affected by eutrophication, which of the following causes the death of aquatic plants?

A Poisoning by excess minerals

B Lack of oxygen in the water

C Lack of light reaching the plants

D Lack of carbon dioxide in the water

SKILLS REASONING

5 Why are humans having much more of an impact on their environment now than they did 500 years ago?

SKILLS ANALYSIS

6 The graph shows the changing concentrations of carbon dioxide at Mauna Loa, Hawaii over a number of years.

a Describe the overall trend shown by the graph.

SKILLS REASONING

b Explain the trend described in a).

c In any one year, the level of atmospheric carbon dioxide shows a peak and a trough. Explain why.

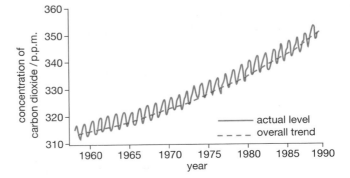

7 The diagram shows how the greenhouse effect is thought to operate.

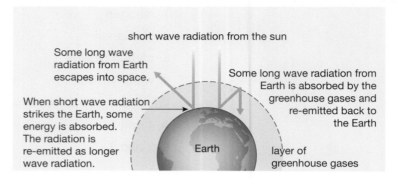

a Name two greenhouse gases.

b Explain one benefit to the Earth of the greenhouse effect.

c Suggest why global warming may lead to malaria becoming more common in Europe.

8 The diagram shows the profile of the ground on a farm either side of a pond.

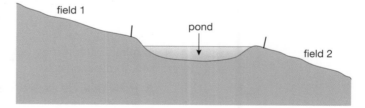

The farmer applied nitrate fertiliser to the two fields in alternate years. When he applied the fertiliser to Field 1, the pond often developed an algal bloom. This did not happen when fertiliser was applied to Field 2.

a Explain why an algal bloom developed when he applied the fertiliser to Field 1.

b Explain why no algal bloom developed when he applies the fertiliser to Field 2.

c Suggest why the algal bloom is greater in hot weather.

9 Some untreated sewage was accidentally discharged into a small river. A short time afterwards, a number of dead fish are seen downstream of the point of discharge. Explain, as fully as you can, how the discharge might have led to the death of the fish.

10 Some farmers use pesticides and fertilisers to improve crop yields. Those practising 'organic' farming techniques do not use any artificial products.

a Describe how the use of pesticides and fertilisers can improve crop yields.

b Explain how organic farmers can maintain fertile soil and keep their crops free of pests.

UNIT QUESTIONS

SKILLS ▶ INTERPRETATION

1 The diagram shows a simplified food web of a fish, the herring.

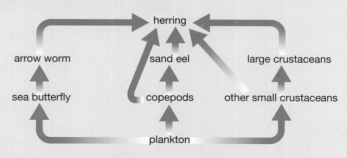

a   **i**   Write out a food chain from the above food web containing four organisms.   **(1)**

SKILLS ▶ ANALYSIS   4

     ii   From your food chain, name the primary consumer and secondary consumer.   **(2)**

     iii   Name one organism in the web that is both a secondary consumer and a tertiary consumer. Explain your answer.   **(2)**

b   The amount of energy in each trophic level has been provided for the following food chain. The units are kJ per m^2 per year

     plankton (8869) → copepod (892) → herring (91)

SKILLS ▶ INTERPRETATION   6

     i   Sketch a pyramid of energy for this food chain.   **(1)**

SKILLS ▶ PROBLEM SOLVING   7

     ii   Calculate the percentage of energy entering the plankton that passes to the copepod.   **(2)**

     iii   Calculate the percentage of energy entering the copepod that passes to the herring.   **(2)**

8

     iv   Calculate the amount of energy that enters the food chain per year if the plankton use 0.1% of the available energy.   **(2)**

SKILLS ▶ REASONING   6

     v   Explain two ways in which energy is lost in the transfer from the copepod to the herring.   **(2)**

Total 14 marks

SKILLS   INTERPRETATION

2   A farmer added nitrate fertiliser to his wheat crop to try to increase yield. The table below shows the wheat yield when he added different amounts of nitrate fertiliser.

Fertiliser added / kg per hectare	Wheat yield / tonnes per hectare
0	15
50	18
100	22
150	31
200	30
250	31

a   Use the information in the table to draw a line graph showing the effects of fertiliser on wheat yield.   **(4)**

SKILLS   PROBLEM SOLVING   **6**

b   What amount of fertiliser would you advise the farmer to use on his wheat crop? Explain your answer.   **(3)**

SKILLS   REASONING   **5**

c   Why is nitrate needed to help plants grow?   **(1)**

7

d   Excess nitrate can be washed into rivers. Explain the effects that this could have on the river ecosystem.   **(5)**

Total 13 marks

SKILLS   ANALYSIS, REASONING   **7**

3   Insecticides are used by farmers to control the populations of insect pests. New insecticides are continually being developed.

a   A new insecticide was trialled over three years to test its effectiveness in controlling an insect pest of potato plants. Three different concentrations of the insecticide were tested. Some results are shown in the table.

Concentration	Percentage of insect pest killed each year of insecticide		
	Year 1	Year 2	Year 3
1 (weakest)	95	72	18
2 (intermediate)	98	90	43
3 (strongest)	99	91	47

i   Describe, and suggest an explanation for, the change in the effectiveness of the insecticide over the three years.   **(3)**

ii   Which concentration would a farmer be most likely to choose to apply to potato crops? Explain your answer.   **(3)**

b The trials also showed that there was no significant bioaccumulation of the insecticide.

i What is bioaccumulation?   **(1)**

ii Give an example of bioaccumulation of an insecticide and describe its consequences.   **(2)**

iii Explain why it is particularly important that there is no bioaccumulation of this insecticide.   **(1)**

Total 10 marks

4 Carbon is passed through ecosystems by the actions of plants, animals and decomposers, in the carbon cycle. Humans influence the carbon cycle more than other animals.

a Explain the importance of plants in the carbon cycle.   **(2)**

b Describe two human activities that have significant effects on the world carbon cycle.   **(2)**

c The graph shows the activity of decomposers acting on the bodies of dead animals under different conditions.

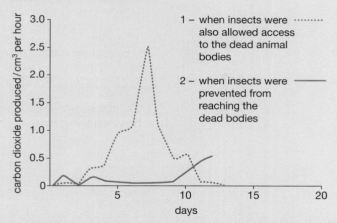

1 – when insects were also allowed access to the dead animal bodies

2 – when insects were prevented from reaching the dead bodies

i Explain why carbon dioxide production was used as a measure of the activity of the decomposers.   **(2)**

ii Describe and explain the changes in decomposer activity when insects were also allowed access to the dead bodies (curve 1)   **(3)**

iii Describe two differences between curves (1) and (2). Suggest an explanation for the differences you describe.   **(4)**

Total 13 marks

5 In natural ecosystems, there is competition between members of the same species as well as between different species.

a Explain how competition between members of the same species helps to control population growth.   **(3)**

b Crop plants must often compete with weeds for resources. Farmers often control weeds by spraying herbicides (weedkillers).

i Name two factors that the crop plants and weeds may compete for and explain the importance of each.   **(4)**

ii Farmers usually prefer to spray herbicides on weeds early in the growing season. Suggest why.   **(2)**

c Two species of the flour beetle, *Tribolium*, compete with each other for flour. Both are parasitised by a protozoan. The graphs show the changes in numbers of the two species over 900 days when the parasite is absent and when it is present.

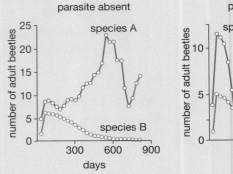
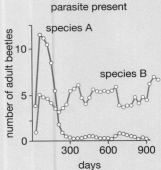

i Which species is the most successful when the parasite is absent? Justify your answer. (2)

ii What is the effect of the parasite on the relative success of the two beetles? Suggest an explanation for your answer. (4)

Total 15 marks

6 The table gives information about the pollutants produced in extracting aluminium from its ore (bauxite) and in recycling aluminium.

Pollutants	Amount / g per kg aluminium produced	
air	extraction from bauxite	recycling aluminium
sulfur dioxide	88 600	886
nitrogen oxides	139 000	6760
carbon monoxide	34 600	2440
Water		
dissolved solids	18 600	575
suspended solids	1600	175

a Calculate the percentage reduction in sulfur dioxide pollution by recycling aluminium. (2)

b Explain how extraction of aluminium from bauxite may contribute to the acidification of water hundreds of miles from the factory where the aluminium is extracted. (3)

c Suggest two reasons why there may be little plant life in water near an extraction. (4)

Total 9 marks

SKILLS ▸ INTERPRETATION   7

7 Rust fungi infect the leaves of many crop plants.

a Explain two ways in which an infection by rust fungi can reduce
   crop yield.   (2)

b Infections of rust fungi can be controlled by spraying the crops with a
   fungicide. In an investigation into the effectiveness of this method of control,
   a field of wheat was divided into two plots. As soon as the first signs of
   infection appeared, one plot was sprayed with fungicide. The other plot was
   left unsprayed.

   The relative amount of infection in the two plots was monitored over 50
   days. The table shows the results of the investigation.

Day	Relative amount of infection / arbitrary units	
	Unsprayed (control)	Sprayed plants
0	0	3
10	17	8
20	32	10
30	48	15
40	62	17
50	65	18

i   Plot a line graph of these data.   (5)

SKILLS ▸ ANALYSIS   6

ii From your graph, find the relative infection of sprayed and unsprayed
    plants on day 24.   (2)

SKILLS ▸ EXECUTIVE FUNCTION

iii Explain why it is necessary to have a control group.   (2)

SKILLS ▸ ANALYSIS

iv The investigation was repeated every year for five years. Each year
    seeds were kept for planting the following year. On day 30 in Year 5,
    the relative amounts of infection were 46 (unsprayed) and 38 (sprayed).
    Compare these results with Year 1 and suggest an explanation for the
    difference.   (3)

Total 14 marks

UNIT 5 VARIATION AND SELECTION

The genetic code is contained within a chemical in the nucleus of the cell, called DNA. DNA is organised into units called genes. The discovery of the structure of DNA in 1953 was the beginning of a new discipline called molecular biology. Molecular biology looks at how the genetic code in the DNA affects the function of the cell. Molecular biology is closely linked with genetics, which concerns the visible effects of the genes on living organisms . This Unit starts with a look at how DNA forms the genetic code, and how it is passed from cell to cell when a cell divides. Chapter 18 deals with genetics – how the DNA shows itself in organisms. Chapter 19 describes the theory that underpins all of biology – Darwin's theory of evolution by natural selection. The last chapter in this Unit describes the principles of selective breeding.

16 CHROMOSOMES, GENES AND DNA

This chapter looks at the structure and organisation of genetic material, namely chromosomes, genes and DNA.

LEARNING OBJECTIVES

■ Understand that the nucleus of a cell contains chromosomes on which genes are located

■ Understand that a gene is a section of a molecule of DNA that codes for a specific protein

BIOLOGY ONLY

■ Describe a DNA molecule as two strands coiled to form a double helix, the strands being linked by a series of paired bases: adenine (A) with thymine (T), and cytosine (C) with guanine (G)

■ Understand that an RNA molecule is single stranded and contains uracil (U) instead of thymine (T)

■ Describe the stages of protein synthesis (transcription and translation), including the role of mRNA, ribosomes, tRNA, codons and anticodons

■ Understand that mutation is a rare, random change in genetic material that can be inherited

BIOLOGY ONLY

■ Understand how a change in the DNA can affect the phenotype by altering the sequence of amino acids in a protein

■ Understand how most genetic mutations have no effect on the phenotype, some have a small effect and rarely do they have a significant effect

■ Understand that the incidence of mutations can be increased by exposure to ionising radiation (for example, gamma rays, x-rays and ultraviolet rays) and some chemical mutagens (for example, chemicals in tobacco)

■ Know that in human cells the diploid number of chromosomes is 46 and the haploid number is 23

■ Understand that the genome is the entire DNA of an organism

■ Understand how genes exist in alternative forms called alleles, which give rise to different inherited characteristics.

DID YOU KNOW?

DNA is short for deoxyribonucleic acid. It gets the 'deoxyribo' part of its name from the sugar in the DNA molecule – deoxyribose, a sugar containing five carbon atoms.

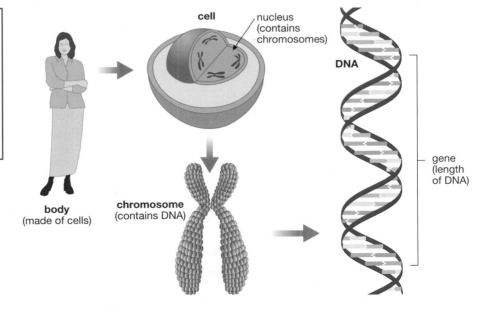

cell

nucleus (contains chromosomes)

DNA

gene (length of DNA)

body (made of cells)

chromosome (contains DNA)

▲ Figure 16.1 Our genetic make-up

The chemical that is the basis of inheritance in nearly all organisms is **DNA**. DNA is usually found in the nucleus of a cell, in the **chromosomes** (Figure 16.1). A small section of DNA that determines a particular feature is called a **gene**. Genes determine features by instructing cells to produce particular proteins which then lead to the development of the feature. So a gene can also be described as a section of DNA that codes for a particular protein.

DNA can replicate (make an exact copy of) itself. When a cell divides by mitosis (see Chapter 17), each new cell receives exactly the same type and amount of DNA. The cells formed are *genetically identical*.

BIOLOGY ONLY

THE STRUCTURE OF DNA

WHO DISCOVERED IT?

James Watson and Francis Crick, working at Cambridge University, discovered the structure of the DNA molecule in 1953 (Figure 16.2(a)). Both were awarded the Nobel prize in 1962 for their achievement. However, the story of the first discovery of the structure of DNA goes back much further than this. Watson and Crick were only able to propose the structure of DNA because of the work of others – Rosalind Franklin (Figure 16.2(b)) had been researching the structure of a number of substances using a technique called X-ray diffraction.

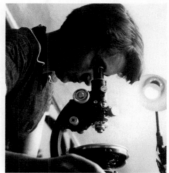

▲ Figure 16.2 (a) Watson and Crick with their double helix model

▲ Figure 16.2 (b) Rosalind Franklin (1920–1958)

Watson and Crick were able to use her results, together with other material, to propose the now-familiar double helix structure for DNA. Rosalind Franklin died of cancer in 1958 and so was unable to share in the award of the Nobel Prize.

A molecule of DNA is made from two strands of molecules called *nucleotides* (Figure 16.3). Each nucleotide contains a sugar molecule (deoxyribose), a phosphate group and a nitrogen-containing group called a **base.** There are four bases, adenine (A), thymine (T), cytosine (C) and guanine (G) (Figure 16.3).

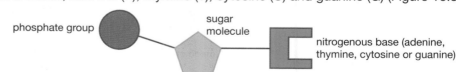

▲ Figure 16.3 The structure of a single nucleotide.

Notice that, in the two strands (see Figure 16.4), nucleotides with adenine are always opposite nucleotides with thymine, and cytosine is always opposite guanine. Adenine and thymine are *complementary* bases, as are cytosine and guanine. Complementary bases always link or 'bind' with each other and never with any other base. This is known as the *base-pairing rule*.

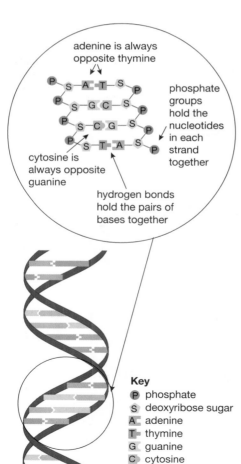

adenine is always opposite thymine

phosphate groups hold the nucleotides in each strand together

cytosine is always opposite guanine

hydrogen bonds hold the pairs of bases together

Key
- **P** phosphate
- **S** deoxyribose sugar
- **A** adenine
- **T** thymine
- **G** guanine
- **C** cytosine

▲ Figure 16.4 Part of a molecule of DNA.

DNA is the only chemical that can replicate itself exactly. Because of this, it is able to pass genetic information from one generation to the next as a 'genetic code'.

DNA REPLICATION

When a cell is about to divide (see Mitosis, Chapter 17) it must first make an exact copy of each DNA molecule in the nucleus. This process is called **replication**. As a result, each cell formed receives exactly the same amount and type of DNA. Figure 16.5 summarises this process.

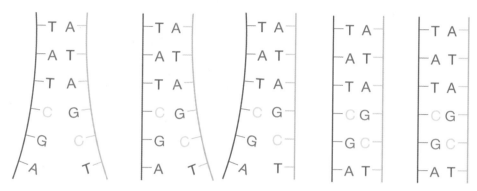

1 The polynucleotide strands of DNA separate.

2 Each strand acts as a template for the formation of a new strand of DNA.

3 DNA polymerase assembles nucleotides into two new strands according to the base-pairing rule.

4 Two identical DNA molecules are formed – each contains a strand from the parent DNA and a new complementary strand.

▲ Figure 16.5 How DNA replicates itself.

DID YOU KNOW?

A 'template' is a pattern that can be used to make something. For example a dress template is a paper pattern for cutting out the material of a dress.

THE GENETIC CODE

Only one of the strands of a DNA molecule actually codes for the manufacture of proteins in a cell. This strand is called the **template strand.** The other strand is called the non-template strand.

Many of the proteins manufactured are enzymes, which go on to control processes within the cell. Some are structural proteins, e.g. keratin in the skin or myosin in muscles. Other proteins have particular functions, such as haemoglobin and some hormones.

Proteins are made of chains of amino acids. A sequence of *three* bases in the template strand of DNA codes for *one* amino acid. For example, the base sequence TGT codes for the amino acid cysteine. Because three bases are needed to code for one amino acid, the DNA code is a *triplet code*. The sequence of triplets that codes for *all* the amino acids in a protein is a gene (Figure 16.6).

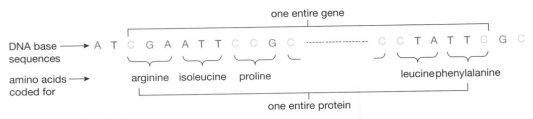

▲ Figure 16.6 The triplet code.

The triplets of bases that code for individual amino acids are the same in all organisms. The base sequence TGT codes for the amino acid cysteine in humans, bacteria, bananas, monkfish, or in any other organism you can think of – the DNA code is a *universal* code.

THE STAGES OF PROTEIN SYNTHESIS

DID YOU KNOW?

Ribose and deoxyribose are very similar in structure. Ribose contains an extra oxygen atom. Similarly, the bases uracil and thymine are very similar in structure.

DID YOU KNOW?

There is a third type of RNA called ribosomal RNA (rRNA). Ribosomes are made of RNA and protein.

DNA stays in the nucleus, but protein synthesis takes place in the cytoplasm. This means that for proteins to be made, the genetic code must be copied, and then transferred out of the nucleus to the cytoplasm. This is carried out by a different kind of nucleic acid called **ribonucleic acid (RNA)**.

There are three main differences between DNA and RNA:

■ DNA is a double helix, RNA is a single strand

■ DNA contains the sugar deoxyribose, RNA contains ribose

■ RNA contains the base uracil (U) instead of thymine (T).

Two types of RNA take part in protein synthesis:

■ **Messenger RNA (mRNA)**, which forms a copy of the DNA code

■ **Transfer RNA (tRNA)**, which carries amino acids to the **ribosomes** to make the protein.

Protein synthesis takes place in two stages, called **transcription** and **translation**.

TRANSCRIPTION

Transcription happens in the nucleus. In a chromosome, part of the DNA double helix unwinds and 'unzips', so that the two strands separate, exposing the bases along the template strand (Figure 16.7).

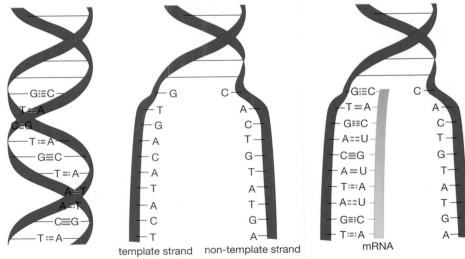

template strand   non-template strand   mRNA

▲ Figure 16.7 (a) The DNA double helix, showing some base pairs, (b) the two strands of the DNA have separated and unwound, (c) a short length of mRNA has formed.

An RNA nucleotide consists of a ribose group, a phosphate and one of four bases (C, G, A or U).

The template strand of the DNA forms a framework upon which a molecule of mRNA is formed. The building blocks of the mRNA are RNA nucleotides. They line up alongside the template strand according to the complementary base-pairing rules (Table 16.1).

Table 16.1 Base-pairing rules in transcription

Base on DNA	Base on mRNA
G	C
C	G
T	A
A	U

One at a time, the RNA nucleotides link up to form an mRNA molecule. They form bonds between their ribose and phosphate groups, joining together to make the sugar–phosphate backbone of the molecule. When a section of DNA corresponding to a protein has been transcribed, the mRNA molecule leaves the DNA and passes out of the nucleus to the cytoplasm. It leaves through pores (holes) in the nuclear membrane. The DNA helix then 'zips up' again. Because of complementary base pairing the triplet code of the DNA is converted into a triplet code in the mRNA.

TRANSLATION

Converting the code in the mRNA into a protein is called translation. It takes place at the ribosomes. By this stage the code consists of sets of three bases in the mRNA (e.g. AUG, CCG, ACA). These triplets of bases are called **codons.** Each codon codes for a particular amino acid, e.g. CCU codes for the amino acid proline, and AUG codes for methionine.

The mRNA molecule attaches to a ribosome. Now the tRNA molecules begin their part in the process. Each has an **anticodon** of three bases at one end of the molecule, which is complementary to a particular codon on the mRNA. At the other end of the tRNA molecule is a site where a specific amino acid can attach (Figure 16.8). In other words there is a particular tRNA molecule for each type of amino acid. The tRNA molecule carries its amino acid to the ribosome, where its specific anticodon pairs up with the three bases of the corresponding mRNA codon.

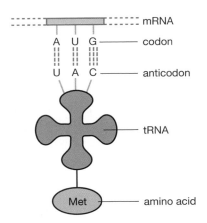

▲ Figure 16.8 A tRNA molecule with the anticodon UAC, carrying the amino acid methionine.

This interaction between mRNA and tRNA is the basis of translation. The process is shown in Figure 16.9.

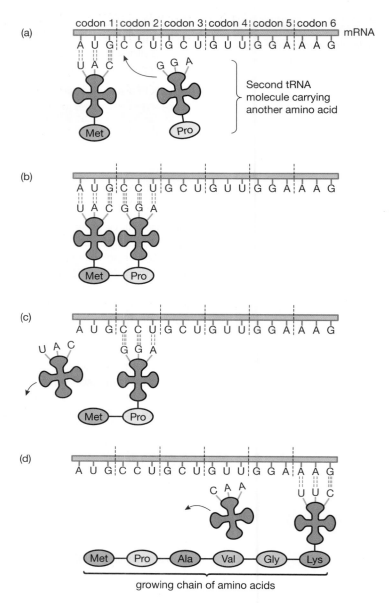

▲ Figure 16.9 Interaction between mRNA and tRNA is the basis of translation

It happens as follows:

- The first tRNA to bind at the mRNA does so at the 'start codon', which always has the base sequence AUG. This codes for the amino acid methionine.

- Another tRNA brings along a second amino acid. The anticodon of the second tRNA binds to the next codon on the mRNA.

- A bond forms between the methionine and the second amino acid.

- The first tRNA molecule is released and goes off to collect another amino acid.

- More tRNA molecules arrive at the mRNA and add their amino acids to the growing chain, forming a protein.

- At the end of the chain a 'stop codon' tells the translation machinery that the protein is complete, and it is released.

REMINDER

Summary of protein synthesis:

The order of bases in the template strand of the DNA forms the genetic code. The code is converted into the sequence of bases in the mRNA. In the cytoplasm the sequence of mRNA bases is used to determine the position of amino acids in a protein.

There are 20 different amino acids, so there need to be at least 20 different codons (and 20 different anticodons). In fact there are more than this, because some amino acids use more than one triplet code. For example the mRNA codons GGU, GGC, GGA and GGG all code for the amino acid glycine.

GENE MUTATIONS – WHEN DNA MAKES MISTAKES

KEY POINT

Double Award students only need to know that a mutation is a rare, random change in the genetic material that can be inherited. Students studying Biology will need to understand this section.

(a) ATT TCC GTT ATC

duplication here

ATT TTC CGT TAT C

extra T becomes first base of next triplet

(b) ATT TCC GTT ATC

deletion here

ATT CCG TTA TC

replaced by first base of next triplet

(c) ATT TCC GTT ATC

original base

ATG TCC GTT ATC

substituted base

(d) ATT TCC GTT ATC

inversion here

ATT CCT GTT ATC

▲ Figure 16.10 Gene mutations (a) duplication, (b) deletion, (c) substitution, (d) inversion.

A **mutation** is a change in the DNA of a cell. It can happen in individual genes or in whole chromosomes. Sometimes, when DNA is replicating, mistakes are made and the wrong nucleotide is used. The result is a gene mutation and it can alter the sequence of the bases in a gene. In turn, this can lead to the gene coding for the wrong amino acid and therefore, the wrong protein. There are several ways in which gene mutations can occur (Figure 16.10).

In *duplication*, Figure 16.10 (a), the nucleotide is inserted twice instead of once. Notice that the entire base sequence is altered – each triplet after the point where the mutation occurs is changed. The whole gene is different and will now code for an entirely different protein.

In *deletion*, Figure 16.10 (b), a nucleotide is missed out. Again, the entire base sequence is altered. Each triplet after the mutation is changed and the whole gene is different. Again, it will code for an entirely different protein.

In *substitution*, Figure 16.10 (c), a different nucleotide is used. The triplet of bases in which the mutation occurs is changed and it *may* code for a different amino acid. If it does, the structure of the protein molecule will be different. This may be enough to produce a significant alteration in the *functioning* of a protein or a total lack of function. However, the new triplet may not code for a different amino acid as most amino acids have more than one code. (In this case, the protein will have its normal structure and function.)

In *inversions*, Figure 16.10 (d), the sequence of the bases in a triplet is reversed. The effects are similar to substitution. Only one triplet is affected and this may or may not result in a different amino acid and altered protein structure.

Mutations that occur in body cells, such as those in the heart, intestines or skin, will only affect that particular cell. If they are very harmful, the cell will die and the mutation will be lost. If they do not affect the functioning of the cell in a major way, the cell may not die. If the cell then divides, a group of cells containing the mutant gene is formed. When the organism dies, however, the mutation is lost with it; it is not passed to the offspring. Only mutations in the gametes or in the cells that divide to form gametes can be passed on to the next generation. This is how genetic diseases begin.

Sometimes a gene mutation can be an advantage to an individual. For example, as a result of random mutations, insects can become resistant to insecticides (see Chapter 15). Resistant insects obviously have an advantage over non-resistant types when an insecticide is being used. They will survive the insecticide treatment and reproduce. Their offspring will be resistant and so the proportion of resistant individuals will increase generation after generation. This is an example of **natural selection.** Bacteria can become resistant to antibiotics in a similar way (see Chapter 19).

Gene mutations are random events that occur in all organisms. The rate at which they occur can be increased by a number of agents called **mutagens**. Mutagens include:

■ ionising radiation (such as ultraviolet light, X-rays and gamma rays)

■ chemicals including mustard gas and nitrous oxide, and many of the chemicals in cigarette smoke and the tar from cigarettes.

END OF BIOLOGY ONLY

THE STRUCTURE OF CHROMOSOMES

Each chromosome contains one double-stranded DNA molecule. The DNA is folded and coiled so that it can be packed into a small space. The DNA is coiled around proteins called **histones** (Figure 16.11).

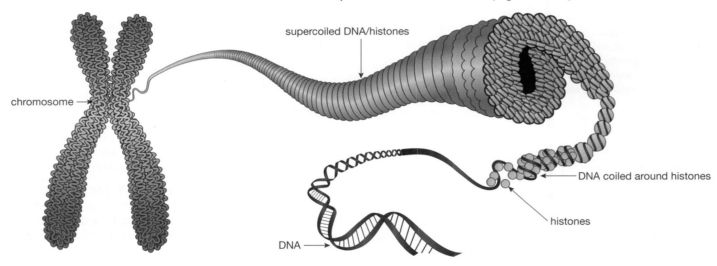

supercoiled DNA/histones

chromosome

DNA coiled around histones

histones

DNA

▲ Figure 16.11 The structure of a chromosome

Because a chromosome contains a particular DNA molecule, it will also contain the genes that make up that DNA molecule. Another chromosome will contain a different DNA molecule, and so will contain different genes.

HOW MANY CHROMOSOMES?

Nearly all human cells contain 46 chromosomes. The photographs in Figure 16.12 show the 46 chromosomes from the body cells of a human male.

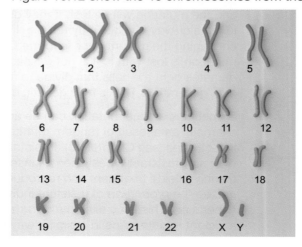

◀ Figure 16.12 A man's chromosomes. One of each of the 22 homologous pairs are shown, along with the X and Y sex chromosomes. A woman's chromosomes are the same, except that she has two X chromosomes. A picture of all the chromosomes in a cell is called a karyotype.

REMINDER

Red blood cells have no nucleus and therefore no chromosomes. (The lack of a nucleus means there is more room for carrying oxygen.)

KEY POINT

The X and the Y chromosomes are the **sex chromosomes**. They determine whether a person is male or female (see Chapter 18).

The chromosomes are not arranged like this in the cell. The original photograph has been cut up and chromosomes of the same size and shape 'paired up'. The cell from the male has 22 pairs of chromosomes and two that do not form a pair – the X and Y chromosomes. A body cell from a female has 23 matching pairs including a pair of X chromosomes.

Pairs of matching chromosomes are called **homologous pairs**. They carry genes for the same features, and these genes are arranged at the same positions and sequence along the chromosome (Figure 16.13). Cells with chromosomes in pairs like this are **diploid** cells.

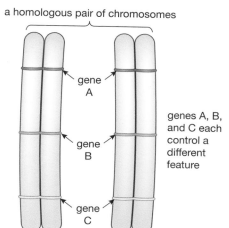

a homologous pair of chromosomes

gene A

gene B

gene C

genes A, B, and C each control a different feature

▲ Figure 16.13 Both chromosomes in a homologous pair have the same sequence of genes.

Not all human cells have 46 chromosomes. Red blood cells have no nucleus and so have none. Sex cells have only 23 – just half the number of other cells. They are formed by a cell division called **meiosis** (see Chapter 17). Each cell formed has one chromosome from each homologous pair, and one of the sex chromosomes. Cells with only half the normal diploid number of chromosomes, and therefore only half the DNA content of other cells, are **haploid** cells.

When two gametes fuse in **fertilisation**, the two nuclei join to form a single diploid cell (a **zygote**). This cell has, once again, all its chromosomes in homologous pairs and two copies of every gene. It has the normal DNA content.

HOW MANY GENES?

The entire DNA of an organism (the amount present in a diploid cell) is known as its **genome**. The human genome is made up of about 3.2 billion base pairs. One of the surprise discoveries of modern molecular biology is that only a small fraction of the genome consists of protein-coding genes. For example, the human genome contains about 20 000-25 000 genes coding for proteins, which is only about 1.5% of the total DNA. The rest have other functions, or functions yet to be discovered! (See the 'Looking ahead' feature at the end of this chapter.)

GENES AND ALLELES

Genes are sections of DNA that control the production of proteins in a cell. Each protein contributes towards a particular body feature. Sometimes the feature is visible, such as eye colour or skin pigmentation. Sometimes the feature is not visible, such as the type of haemoglobin in red blood cells or the type of blood group antigen on the red blood cells.

Some genes have more than one form. For example, the genes controlling several facial features have alternative forms, which result in alternative forms of the feature (Figure 16.14).

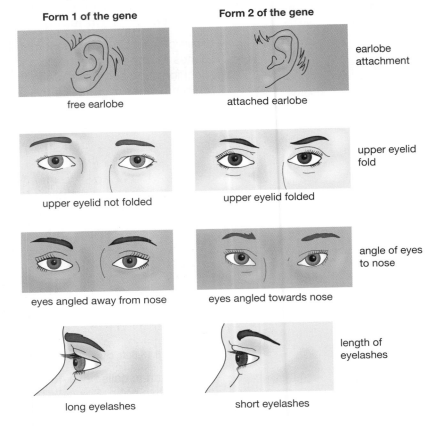

Figure 16.14 The alternate forms of four facial features

The gene for earlobe attachment has the forms 'attached earlobe' and 'free earlobe'. These different forms of the gene are called **alleles**. Homologous chromosomes carry genes for the same features in the same sequence, but the alleles of the genes may not be the same (Figure 16.15). The DNA in the two chromosomes is not quite identical.

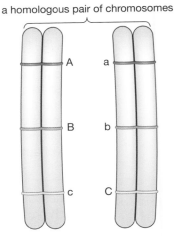

Figure 16.15 A and a, B and b, and C and c are different alleles of the same gene. They control the same feature but code for different expressions of that feature.

Each cell with two copies of a chromosome also has two copies of the genes on those chromosomes. Suppose that, for the gene controlling earlobe attachment, a person has one allele for attached earlobes and one for free earlobes. What happens? Is one ear free and the other attached? Are they both partly attached? Neither. In this case, both earlobes are free. The 'free' allele is **dominant.** This means that it will show its effect, whether or not the allele for 'attached' is present. The allele for 'attached' is called **recessive**. The recessive allele will only show up in the appearance of the person if there is no dominant allele present. You will find out more about how genes work in Chapter 18.

LOOKING AHEAD – REGULATING GENES

You have seen how 'normal' gene coding for proteins (known as structural genes) make up only 1.5% of the genome. Some of the rest of the genome is DNA that regulates the action of the structural genes, switching them on and off. One way this happens is in regions of the DNA called operons.

An operon is a group of structural genes headed by a non-coding length of DNA called an operator, along with another sequence of DNA called a promoter. The promoter starts transcription by binding to an enzyme called RNA polymerase. Close to the promoter is a regulatory gene, which codes for a protein called a repressor. The repressor can bind with the operator, preventing the promoter from binding with RNA polymerase, and stopping transcription (Figure 16.16).

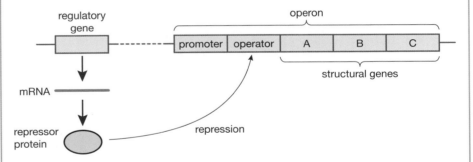

▲ Figure 16.16 An operon is a group of structural genes linked to an operator and a promoter. It is under the control of a regulatory gene.

The structural genes are in groups because they are related – e.g. they code for different enzymes in a metabolic pathway. Operons were first discovered in bacteria. At first we thought they only existed in prokaryotes, but molecular biologists have now found them in eukaryotic cells too.

CHAPTER QUESTIONS

More questions on DNA can be found at the end of Unit 5 on page 277.

BIOLOGY ONLY

SKILLS CRITICAL THINKING

1 Which of the following are components of DNA?

A deoxyribose, uracil and phosphate

B ribose, adenine and guanine

C deoxyribose, phosphate and adenine

D ribose, thymine and cytosine

2 Which of the following is the function of transfer RNA?

A transporting amino acids

B coding for the order of amino acids

C transcription of the DNA

D translation of the DNA

3 The base sequence for the same length of DNA before and after a gene mutation was as follows:

Before mutation: ATT TCC GTT ATC CGG

After mutation: ATT CCG TTA TCC GGA

Which type of mutation took place?

A duplication   B deletion

C substitution   D inversion

END OF BIOLOGY ONLY

4 How many chromosomes are there in the body cells of a man?

A 23 pairs + XX   B 23 pairs + XY

C 22 pairs + XX   D 22 pairs + XY

BIOLOGY ONLY

SKILLS INTERPRETATION

5 The diagram represents part of a molecule of DNA.

a Name the parts labelled 1, 2, 3, 4 and 5.

SKILLS CRITICAL THINKING

b What parts did James Watson, Frances Crick and Rosalind Franklin play in discovering the structure of DNA?

SKILLS INTERPRETATION

c Use the diagram to explain the base-pairing rule.

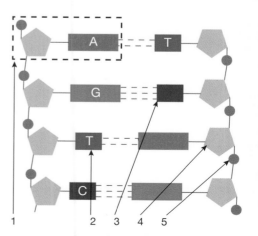

END OF BIOLOGY ONLY

6 a What is:

   i   a gene

   ii an allele?

b Describe the structure of a chromosome.

c How are the chromosomes in a woman's skin cells:

   i   similar to

   ii different from those in a man's skin cells?

BIOLOGY ONLY

7 DNA is the only molecule capable of replicating itself. Sometimes mutations occur during replication.

a Draw a flow diagram to describe the process of DNA replication.

b Explain how a single gene mutation can lead to the formation of a protein in which:

   i   many of the amino acids are different from those coded for by the non-mutated gene

   ii only one amino acid is different from those coded for by the non-mutated gene.

8 Below is a base sequence from part of the template strand of a DNA molecule.

TAC CTC GGT CAT CCC

a How many amino acids are coded for by this base sequence?

b The sequence of the coding strand was transcribed to form mRNA. Write down the base sequence of this mRNA.

c Write down the corresponding base sequence of the *non-template* strand of the DNA.

d Copy and complete this description of the next stage in protein synthesis:

The mRNA base sequence is converted into the amino acid sequence of a protein during a process called _____ .
The mRNA sequence consists of a triplet code. Each triplet of bases is called a _____ . Reading of the mRNA base sequence begins at a _____ and ends at a _____ . Molecules of tRNA carrying an amino acid bind to the mRNA at an organelle called the _____ .

END OF BIOLOGY ONLY

17 CELL DIVISION

Growth and reproduction are two characteristics of living things. Both involve cell division, which is the subject of this chapter.

LEARNING OBJECTIVES

- Understand how division of a diploid cell by mitosis produces two cells that contain identical sets of chromosomes

- Understand that mitosis occurs during growth, repair, cloning and asexual reproduction

- Understand how division of a cell by meiosis produces four cells, each with half the number of chromosomes, and that this results in the formation of genetically different haploid gametes

- Understand how random fertilisation produces genetic variation of offspring

- Understand that variation within a species can be genetic, environmental or a combination of both.

In most parts of the body, cells need to divide so that organisms can grow and replace worn out or damaged cells. The cells that are produced in this type of cell division should be exactly the same as the cells they are replacing. This is the most common form of cell division.

Only in the sex organs is cell division different. Here, some cells divide to produce gametes (sex cells), which contain only half the original number of chromosomes. This is so that when male and female gametes fuse together (fertilisation) the resulting cell (called a **zygote**) will contain the full set of chromosomes and can then divide and grow into a new individual.

Human body cells are **diploid** – they have 46 chromosomes in 23 homologous pairs. The gametes, with 23 chromosomes (one copy of each homologous chromosome), are **haploid** cells.

There are two kind of cell division: **mitosis** and **meiosis**. When cells divide by mitosis, two cells are formed. These have the same number and type of chromosomes as the original cell. Mitosis forms all the cells in our bodies except the gametes.

When cells divide by meiosis, four cells are formed. These have only half the number of chromosomes of the original cell. Meiosis forms gametes

KEY POINT

Meiosis is sometimes called a *reduction division*. This is because it produces cells with only half the number of chromosomes of the original cell.

MITOSIS

When a 'parent' cell divides it produces 'daughter' cells. Mitosis produces two daughter cells that are genetically identical to the parent cell – both daughter cells have the same number and type of chromosomes as the parent cell.

To achieve this, the dividing cell must do two things.

- It must copy each chromosome before it divides. This involves the DNA replicating and more proteins being added to the structure. Each daughter cell will then be able to receive a copy of each chromosome (and each molecule of DNA) when the cell divides.

■ It must divide in such a way that each daughter cell receives one copy of every chromosome. If it does not do this, both daughter cells will not contain all the genes.

(a) prophase

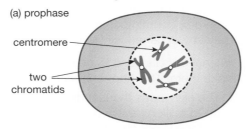

Before mitosis the DNA replicates and the chromosomes form two exact copies called chromatids. During the first stage of mitosis (prophase) the chromatids become visible, joined at a centromere. The nuclear membrane breaks down.

(b) metaphase

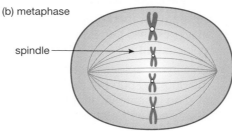

During metaphase a structure called the spindle forms. The chromosomes line up at the 'equator' of the spindle, attached to it by their centromeres.

(c) anaphase

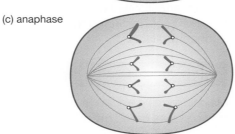

During anaphase, the spindle fibres shorten and pull the chromatids to the opposite ends ('poles') of the cell. The chromatids separate to become the chromosomes of the two daughter cells.

(c) telophase

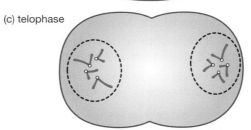

In the last stage (telophase) two new nuclei form at the poles of the cell. The cytoplasm starts to divide to produce two daughter cells. Both daughter cells have a copy of each chromosome from the parent cell.

▲ Figure 17.1 The stages of mitosis. For simplicity the cell shown contains two homologous pairs of chromosomes (one long pair, one short). (You do not need to remember the names of the stages.)

A number of distinct stages occur when a cell divides by mitosis. These are shown in Figure 17.1. Figure 17.2 is a photograph of some cells from the root tip of an onion. Cells in this region of the root divide by mitosis to allow growth of the root.

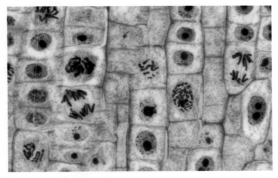

▲ Figure 17.2 Cells in the root tip of an onion dividing by mitosis. Can you identify any of the stages shown in Figure 17.1?

Each daughter cell formed by mitosis receives a copy of every chromosome, and therefore every gene, from the parent cell. Each daughter cell is genetically identical to the others. All the cells in our body (except the gametes) are formed by mitosis from the zygote (single cell formed at fertilisation). They all, therefore, contain copies of all the chromosomes and genes of that zygote. They are all genetically identical.

Whenever cells need to be replaced in our bodies, cells divide by mitosis to make them. This happens more frequently in some regions than in others.

- The skin loses thousands of cells every time we touch something. This adds up to millions every day that need replacing. A layer of cells beneath the surface is constantly dividing to produce replacements.

- Cells are scraped off the lining of the gut as food passes along. Again, a layer of cells beneath the gut lining is constantly dividing to produce replacement cells.

- Cells in our spleen destroy worn out red blood cells at the rate of 100 000 000 000 per day! These are replaced by cells in the bone marrow dividing by mitosis. In addition, the bone marrow forms all our new white blood cells and platelets.

- Cancer cells also divide by mitosis. The cells formed are exact copies of the parent cell, including the mutation in the genes that makes the cells divide uncontrollably.

MEIOSIS

Meiosis forms gametes. It is a more complex process than mitosis and takes place in two stages called meiosis I and meiosis II, resulting in four haploid cells. Each daughter cell is genetically different from the other three and from the parent cell.

During meiosis the parent cell must do two things:

- It must copy each chromosome so that there is enough genetic material to be shared between the four daughter cells

- It must divide twice, in such a way that each daughter cell receives just one chromosome from each homologous pair.

These processes are summarised in Figure 17.4. Figure 17.3 shows cells in the anther of a flower dividing by meiosis.

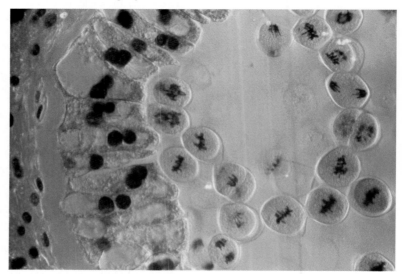

▲ Figure 17.3 Photomicrograph of an anther showing cells dividing by meiosis.

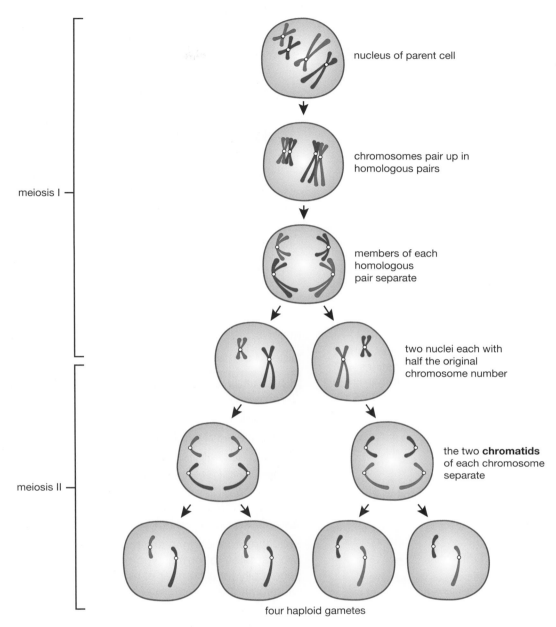

nucleus of parent cell

chromosomes pair up in homologous pairs

members of each homologous pair separate

two nuclei each with half the original chromosome number

the two **chromatids** of each chromosome separate

four haploid gametes

meiosis I

meiosis II

▲ Figure 17.4: The stages of meiosis. For simplicity the parent cell contains only two homologous pairs of chromosomes (one long pair, one short). To help you to see what happens, one member of each pair is coloured red and one blue. The cell membrane is shown, but the nuclear membrane has been omitted. A spindle forms during each division, but these have also been omitted for clarity.

There are two main events during meiosis:

- during the first division, one chromosome from each homologous pair goes into each daughter cell
- during the second division, the chromosome separates into two parts. One part goes into each daughter cell.

The gametes formed by meiosis don't all have the same combinations of alleles – there is *genetic variation* in the cells. During the two cell divisions of meiosis, the chromosomes of each homologous pair are shared between the two daughter cells independently of each of the other homologous pairs. This allows for much possible genetic variation in the daughter cells (Figure 17.5).

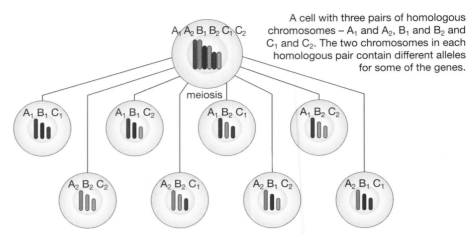

A cell with three pairs of homologous chromosomes – A_1 and A_2, B_1 and B_2 and C_1 and C_2. The two chromosomes in each homologous pair contain different alleles for some of the genes.

As a result of the two divisions of meiosis, each sex cell formed contains one chromosome from each homologous pair. This gives eight combinations. As A_1 and A_2 contain different alleles (as do B_1 and B_2, and C_1 and C_2) the eight possible sex cells will be genetically different.

▲ Figure 17.5 How meiosis produces variation

Table 17.1 summarises the similarities and differences between mitosis and meiosis.

Table 17.1 Comparison of meiosis and mitosis.

Feature of the process	Mitosis	Meiosis
Chromosomes are copied before division begins	Yes	Yes
Number of cell divisions	One	Two
Number of daughter cells produced	Two	Four
Daughter cells are haploid or diploid	Diploid	Haploid
Genetic variation in the daughter cells	No	Yes

SEXUAL REPRODUCTION AND VARIATION

Sexual reproduction in any multicellular organism involves the fusion of two gametes to form a zygote. The offspring from sexual reproduction vary genetically for a number of reasons. One reason is because of the huge variation in the gametes. Another reason is because of the random way in which fertilisation takes place. In humans, any one of the billions of sperm formed by a male during his life could, potentially, fertilise any one of the thousands of ova formed by a female.

This variation applies to both male and female gametes. So, just using our 'low' estimate of about 8.5 million different types of human gametes means that there can be 8.5 million different types of sperm and 8.5 million different types of ova. When fertilisation takes place, any sperm could fertilise any ovum. The number of possible combinations of chromosomes (and genes) in the zygote is 8.5 million × 8.5 million = 7.2 x 10^{13}, or 72 trillion! And remember, this is using our 'low' number!

This means that every individual is likely to be genetically unique. The only exceptions are identical twins. Identical twins are formed from the *same* zygote – they are sometimes called *monozygotic* twins. When the zygote divides by mitosis, the two *genetically identical* cells formed do not 'stay together'. Instead, they separate and each cell behaves as though it were an individual zygote, dividing and developing into an embryo (Figure 17.6). Because they have developed from genetically identical cells (and, originally, from the same

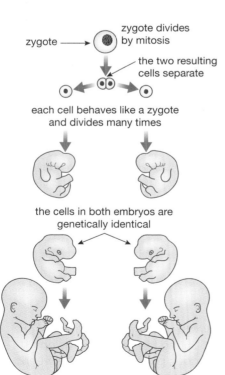

▲ Figure 17.6 How identical twins are formed

zygote), the embryos (and, later, the children and the adults they become) will be genetically identical.

Non-identical twins develop from different zygotes and so are not genetically identical.

Seeds are made by sexual reproduction in plants. Each seed contains an embryo, which results from a pollen grain nucleus fusing with an egg cell nucleus. Embryos from the same plant will vary genetically because they are formed by different pollen grains fertilising different egg cells and so contain different combinations of genes.

ASEXUAL REPRODUCTION AND CLONING

KEY POINT

Cloning is a process that produces a group of genetically identical offspring (a **clone**) from part of the parent organism. Gametes are not involved.

Plant breeders have known for a long time that sexual reproduction produces variation. They realised that if a plant had some desirable feature, the best way to get more of that plant was not to collect and plant its seeds, but to **clone** it in some way. Modern plant-breeding techniques allow the production of many thousands of identical plants from just a few cells of the original (see Chapter 20).

When organisms reproduce asexually, there is no fusion of gametes. A part of the organism grows and somehow breaks away from the parent organism. The cells it contains were formed by mitosis, so contain exactly the same genes as the parent. Asexual reproduction produces offspring that are genetically identical to the parent, and genetically identical to each other.

Asexual reproduction is common in plants (see Chapter 13). For example, flower bulbs grow and divide asexually each season to produce more bulbs. Asexual reproduction also occurs in some animals (see Chapter 9).

GENES AND ENVIRONMENT BOTH PRODUCE VARIATION

DID YOU KNOW?

Short pea plants are called 'dwarf' varieties. The genetics controlling the height of pea plants is described in Chapter 18.

There are two varieties of pea plants that are either tall or short. This difference in height is due to the genes they inherit. There are no 'intermediate height' pea plants. However, all the tall pea plants are not *exactly* the same height and neither are all the short pea plants *exactly* the same height. Figure 17.7 illustrates the variation in height of pea plants.

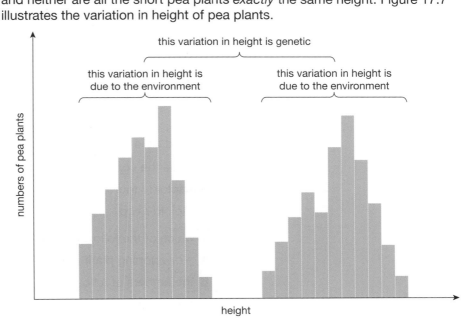

▲ Figure 17.7 Bar chart showing variation in height of pea plants.

Several environmental factors can influence the height of the plants.

■ They may not all receive the same amount of light and so some will not photosynthesise as well as others.

■ They may not all receive the same amount of water and mineral ions from the soil – this could affect the manufacture of a range of substances in the plant.

■ They may not all receive the same amount of carbon dioxide. Again, some plants will not photosynthesise as well as others.

Similar principles apply in humans. Identical twins have the same genes, and often grow up to look very alike (although not quite identical). Also, they often develop similar talents. However, identical twins never look *exactly* the same. This is especially true if, for some reason, they grow up apart. The different environments affect their physical, social and intellectual development in different ways.

CHAPTER QUESTIONS

SKILLS   CRITICAL THINKING

More questions on cell division can be found at the end of Unit 5 on page 277.

1 A species of mammal has 32 chromosomes in its muscle cells. Which row in the table below shows the number of chromosomes in the mammal's skin cells and sperm cells?

	Skin cells	Sperm cells
A	32	32
B	16	16
C	16	32
D	32	16

2 Consider the following statements about reproduction in plants:

1. Large numbers of offspring are quickly produced

2. There is little genetic variation in the offspring.

3. A mechanism such as wind or insects is not needed for pollination.

Which of the above are advantages of *asexual* reproduction in plants?

A 1 and 2   C 2 and 3

B 1 and 3   D 1, 2 and 3

3 Mitosis results in two _____ cells, while meiosis results in _____ haploid cells

Choose the correct pair of answers to fill the gaps in the sentence:

A haploid / four   C diploid / four

B diploid / two   D haploid / two

4 In which of the following does meiosis occur?

A a developing plant embryo

B the anthers of a flower

C the skin of a mammal.

D the tip of a shoot of a plant

5 Cells can divide by mitosis or by meiosis.

   a Give one similarity and two differences between the two processes.

   b Do cancer cells divide by mitosis or meiosis? Explain your answer.

   c Why is meiosis sometimes called a reduction division?

6 Daffodils reproduce sexually by forming seeds and asexually by forming bulbs. Explain why:

   a the bulbs formed from a single daffodil plant produce plants very similar to each other and to the parent plant

   b the seeds formed by a single daffodil plant produce plants that vary considerably.

7 The diagram shows two cuttings. They were both taken from the same clover plant and planted in identical soil. Both were left for several days to become established in their new pots. Some nitrogen-fixing bacteria were then added to the pot labelled 'inoculated'. The other pot was left untreated and labelled 'not inoculated'. The diagram shows the plants three weeks after the treatment to the 'inoculated' pot.

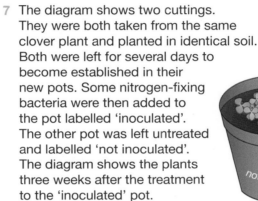

   a What is the name given to the part of the experiment represented by the untreated pot of seeds?

   b Why were cuttings from the same plant used rather than seeds from the same plant?

   c What does this experiment suggest about the influence of genes and the environment on variation in the height of clover plants?

8 Some cells divide by mitosis, others divide by meiosis. For each of the following examples, say whether mitosis or meiosis is involved. In each case, give a reason for your answers.

   a Cells in the testes dividing to form sperm.

   b Cells in the lining of the small intestine dividing to replace cells that have been lost.

   c Cells in the bone marrow dividing to form white blood cells.

   d Cells in an anther of a flower dividing to form pollen grains.

   e A zygote dividing to form an embryo.

9 Variation in organisms can be caused by the environment as well as by the genes they inherit. For each of the following examples, state whether the variation described is likely to be genetic, environmental or both. In each case, give a reason for your answers.

   a Humans have brown, blue or green eyes.

   b Half the human population is male, half is female.

   c Cuttings of hydrangea plants grown in soils with different pH values develop flowers with slightly different colours.

SKILLS CRITICAL THINKING, REASONING ⬤5

d Some pea plants are tall; others are dwarf. However, the tall plants are not exactly the same height and neither are all the dwarf plants the same height.

e People in some families are more at risk of heart disease than people in other families. However, not every member of the 'high risk' families have a heart attack and some members of the 'low risk' families do.

SKILLS ANALYSIS, INTERPRETATION ⬤9

10 In an investigation into mitosis, the distance between a chromosome and the pole (end) of a cell was measured.

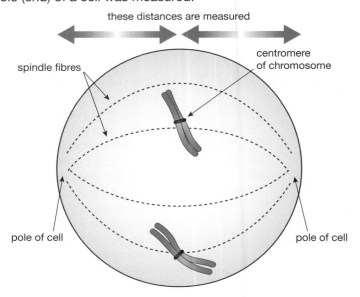

The graph shows how these distances changed during mitosis.

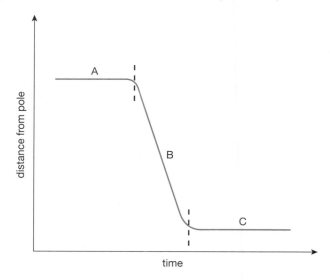

a Describe two events that occur during stage A.

b Explain what is happening during stage B.

c Describe two events that occur during stage C.

18 GENES AND INHERITANCE

How and why do we inherit features from our parents? This chapter answers these questions by looking at the work of Gregor Mendel and how he has helped us to understand the mysteries of inheritance.

LEARNING OBJECTIVES

- Understand that genes exist in alternative forms called alleles which give rise to differences in inherited characteristics

- Understand the meaning of the terms dominant, recessive, homozygous, heterozygous, phenotype and genotype

- Describe patterns of monohybrid inheritance using genetic diagrams

- Predict probabilities of outcomes from monohybrid crosses

- Understand how to interpret family pedigrees

BIOLOGY ONLY

- Understand the meaning of the term codominance

- Understand how the sex of a person is controlled by one pair of chromosomes, XX in a female and XY in a male

- Describe the determination of the sex of offspring at fertilisation, using a genetic diagram

- Understand that most phenotypic features are the result of polygenic inheritance rather than single genes

The groundbreaking research that uncovered the rules of how genes are inherited was carried out by Gregor Mendel and published in 1865. The rules of inheritance are now known as 'Mendelian genetics' in his honour.

GREGOR MENDEL

Gregor Mendel was a monk who lived in a monastery in Brno in what is now the Czech Republic (Figure 18.1). He became interested in the science of heredity, and carried out hundreds of breeding experiments using pea plants. From his research Mendel was able to explain the laws governing inheritance.

Mendel found that for every feature or 'character' he investigated:

- a 'heritable unit' (what we now call a **gene**) is passed from one generation to the next

- the heritable unit (gene) can have alternative forms (we now call these different forms **alleles**)

- each individual must have two alternative forms (alleles) per feature

- the gametes only have one of the alternative forms (allele) per feature

- one allele can be dominant over the other.

Mendel used these ideas to predict outcomes of cross-breeding or 'crosses' between plants, which he tested in his breeding experiments. He published his results and ideas in 1865 but few people took any notice, and his work went unrecognised for many years. It wasn't until 1900 that other biologists working on inheritance rediscovered Mendel's work and realised its importance. In 1903, the connection between the behaviour of genes in Mendelian genetics and the behaviour of chromosomes in meiosis was noticed and the science of genetics was established.

Mendel noticed that many of the features of pea plants had two alternative forms. For example, plants were either tall or very short (called a 'dwarf' variety); they either had purple or white flowers; they produced yellow seeds or green seeds. There were no intermediate forms, no pale purple flowers or green/yellow seeds or intermediate height plants. Figure 18.2 shows some of the contrasting features of pea plants that Mendel used in his breeding experiments.

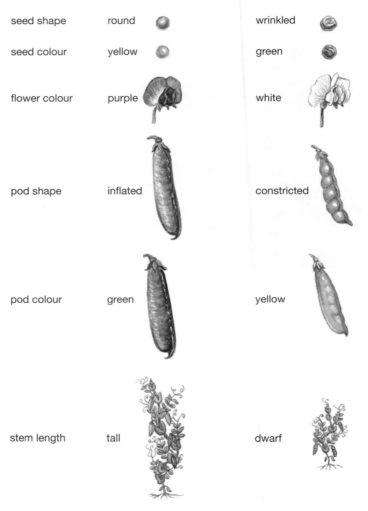

▲ Figure 18.2 Some features of pea plants used by Mendel in his breeding experiments.

Mendel decided to investigate, systematically, the results of cross breeding plants that had contrasting features. These were the 'parent plants', referred to as 'P' in genetic diagrams. He transferred pollen from one experimental plant to another. He also made sure that the plants could not be self-fertilised.

He collected all the seeds formed, grew them and noted the features that each plant developed. These plants were the first generation of offspring, called the F_1 **generation**. He did not cross-pollinate these plants, but allowed them to self-fertilise. Again, he collected the seeds, grew them and noted the features that each plant developed. These plants formed the second generation of offspring or F_2 **generation**. When Mendel used pure-breeding tall and pure-breeding dwarf plants as his parents, he obtained the results shown in Figure 18.3.

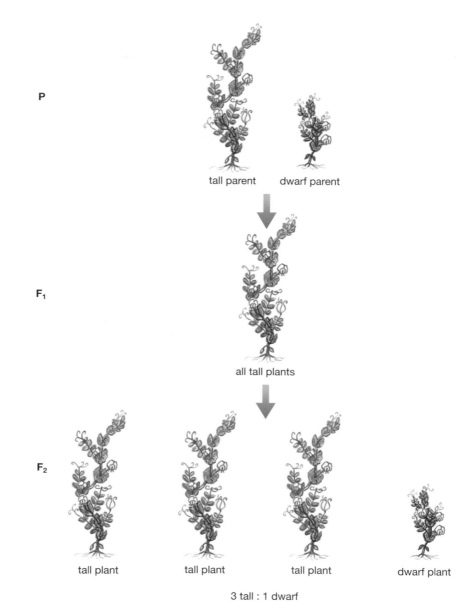

P

tall parent   dwarf parent

F₁

all tall plants

F₂

tall plant   tall plant   tall plant   dwarf plant

3 tall : 1 dwarf

▲ Figure 18.3 A summary of Mendel's results from breeding tall pea plants with dwarf pea plants.

Mendel obtained similar results when he carried out breeding experiments using plants with other pairs of contrasting characters (Figure 18.4). He noticed two things in particular.

■ All the plants of the F₁ generation were of one type. This type was not a blend of the two parental features, but one or the other. For example, when tall and dwarf parents were crossed, all the F₁ plants were tall.

■ There was always a 3:1 ratio of types in the F₂ generation. Three-quarters of the plants in the F₂ generation were of the type that appeared in the F₁ generation. One-quarter showed the other parental feature. For example, when tall F₁ plants were crossed, three-quarters of the F₂ plants were tall and one-quarter were dwarf.

Mendel was able to use his findings to work out how features were inherited, despite having no knowledge of chromosomes, genes or meiosis. Nowadays we can use our understanding of these ideas to explain Mendel's results.

- Each feature is controlled by a gene, which is found on a chromosome.

- There are two copies of each chromosome and each gene in all body cells, except the gametes.

- The gametes have only one copy of each chromosome and each gene (i.e. one allele).

- There are two alleles of each gene.

- One allele is **dominant** over the other allele, which is **recessive**.

- When two different alleles (one dominant and one recessive) are in the same cell, only the dominant allele is expressed (shown in the appearance of the organism).

- An individual can have two dominant alleles, two recessive alleles or a dominant allele and a recessive allele in each cell.

We can use the cross between tall and dwarf pea plants as an example (Figure 18.4). In pea plants, there are tall and dwarf alleles of the gene for height. We will use the symbol T for the tall allele and t for the dwarf allele. The term **genotype** describes the alleles each cell has for a certain feature (e.g. TT). The **phenotype** is the feature that results from the genotype (e.g. a tall plant).

KEY POINT

Normally, we use the first letter of the dominant feature to represent the gene, with a capital letter indicating the dominant allele and a lower case letter the recessive allele. Tall is dominant to dwarf in pea plants, so we use T for the allele for tall and t for dwarf.

KEY POINT

It is the accepted practice in genetics diagrams to show the gene present in a gamete as a letter in a circle.

phenotype of parents	tall	dwarf	Both parents are pure breeding. The tall parent has two alleles for tallness in each cell. The dwarf parent has two alleles for dwarfness in each cell. Because each has two copies of just one allele, we say that they are **homozygous** for the height gene.
genotype of parents	TT	tt	

gametes (sex cells)   (T)   (t)   The sex cells are formed by meiosis. As a result, they only have one allele each.

genotype of F_1 Tt   The F_1 plants have one tall allele and one dwarf allele. We say that they are **heterozygous** for the height gene.

phenotype of F_1 all tall   The plants are tall because the tall allele is dominant.

The F_1 plants are allowed to self-fertilise.

gametes from the F_1 plants   male gametes female gametes   The sex cells are formed by meiosis and so only have one allele. Because the F_1 plants are heterozygous, half of the gametes carry the T allele and half carry the t allele.

(T) or (t)   (T) or (t)

genotypes of F_2

	female gametes	
	(T)	(t)
(T)	TT	Tt
(t)	Tt	tt

male gametes

The diagram opposite is called a **Punnett square**. It allows you to work out the results from a genetic cross. Write the genotypes of one set of sex cells across the top of the square and those of the other sex cells down the side. Then combine the alleles in the two sets of gametes; the squares represent the possible fertilisations.

1 TT : 2 Tt : 1 tt   You can now work out the ratio of the different genotypes.

phenotypes of F_2 3 tall : 1 dwarf

▲ Figure 18.4 Results of crosses using true-breeding tall and dwarf pea plants.

It is important to remember that in genetic crosses, ratios such as 3:1 are *predicted* ratios. In breeding experiments the *actual* numbers of offspring are unlikely to *exactly* fit a 3:1 ratio.

Imagine you flip a coin 20 times. The most likely outcome is that you will get 10 heads and 10 tails. However, you wouldn't be surprised to get, by chance, 11 heads and 9 tails, or 8 heads and 12 tails. The same principle applies to the outcome of a breeding experiment.

For example, one of Mendel's experiments produced 787 tall plants and 277 dwarf plants. This is a ratio of 2.84:1, not quite the expected 3:1. The reason for this is that there are a number of factors that affect survival of the plants – some pollen may not fertilise some ova, some seedlings may die before they mature, and so on. These are unpredictable or 'chance' events. The numbers that Mendel found were statistically close enough to the expected 3:1 ratio, and he found the same thing when he repeated his experiments with other characteristics.

WORKING OUT GENOTYPES – THE TEST CROSS

You cannot tell just by looking at a tall pea plant whether it is **homozygous** (TT) or **heterozygous** (Tt). Both these genotypes would appear equally tall because the tall allele is dominant.

It would help if you knew the genotypes of its parents. You could then write out a genetic cross and perhaps work out the genotype of your tall plant. If you don't know the genotypes of the parents, the only way you can find out is by carrying out a breeding experiment called a **test cross**.

In a test cross, the factor under investigation is the unknown genotype of an organism showing the dominant phenotype. A tall pea plant could have the genotype TT or Tt. You must control every other possible variable *including the genotype of the plant you breed it with*. The only genotype you can be *certain* of is the genotype of plants showing the *recessive* phenotype (in this case dwarf plants). They *must* have the genotype tt.

In this example, you must breed the 'unknown' tall pea plant (TT or Tt) with a dwarf pea plant (tt). You can write out a genetic cross for both possibilities (TT ´ tt and Tt ´ tt) and *predict* the outcome for each (Figure 18.5). You can then compare the result of the breeding experiment with the predicted outcome, to see which result matches the prediction most closely.

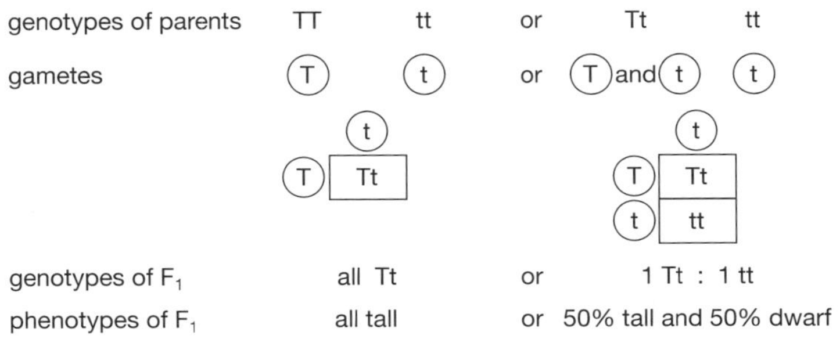

genotypes of parents	TT	tt	or	Tt	tt

gametes   (T)   (t)   or   (T) and (t)   (t)

	t
T	Tt

	t
T	Tt
t	tt

genotypes of F₁   all Tt   or   1 Tt : 1 tt

phenotypes of F₁   all tall   or   50% tall and 50% dwarf

▲ Figure 18.5 A test cross

From our crosses we would expect:

- *all* the offspring to be tall if the tall parent was homozygous (TT)

- *half* the offspring to be tall and *half* to be dwarf if the tall parent was heterozygous (Tt).

WAYS OF PRESENTING GENETIC INFORMATION

Writing out a genetic cross is a useful way of showing how genes are passed through one or two generations, starting from the parents. To show a family history of a genetic condition requires more than this. We can use a diagram called a **pedigree**.

Polydactyly is an inherited condition in which a person develops an extra digit (finger or toe) on the hands and feet. It is determined by a dominant allele. The recessive allele causes the normal number of digits to develop.

If we use the symbol D for the polydactyly allele and d for the normal-number allele, the possible genotypes and phenotypes are:

- DD – person has polydactyly (has two dominant polydactyly alleles)

- Dd – person has polydactyly (has a dominant polydactyly allele and a recessive normal allele)

- dd – person has the normal number of digits (has two recessive, normal-number alleles).

We don't use P and p to represent the alleles as you would expect, because the letters P and p look very similar and could easily be confused. The pedigree for polydactyly is shown in Figure 18.6.

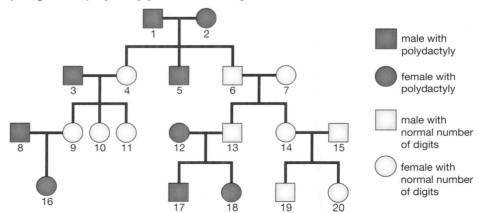

▲ Figure 18.6 A pedigree showing the inheritance of polydactyly in a family.

We can extract a lot of information from a pedigree. In this case:

- there are four generations shown (individuals are arranged in four horizontal lines)

- individuals 4, 5 and 6 are children of individuals 1 and 2 (a family line connects each one directly to 1 and 2)

- individual 4 is the first-born child of 1 and 2 (the first-born child is shown to the left, then second born to the right of this, then the third born and so on)

- individuals 3 and 7 are not children of 1 and 2 (no family line connects them directly to 1 and 2)

- individuals 3 and 4 are father and mother of the same children – as are 1 and 2, 6 and 7, 8 and 9, 12 and 13, 14 and 15 (a horizontal line joins them).

It is usually possible to work out which allele is dominant from a pedigree. You look for a situation where two parents show the same feature and at least one child shows the contrasting feature. In Figure 18.6, individuals 1 and 2 both have polydactyly, but children 4 and 6 do not. There is only one way to explain this:

■ the normal alleles in 4 and 6 can only have come from their parents (1 and 2), so 1 and 2 must both carry normal alleles

■ 1 and 2 show polydactyly, so they must have polydactyly alleles as well

■ if they have both polydactyly alleles *and* normal alleles but show polydactyly, the polydactyly allele must be the dominant allele.

Now that we know which allele is dominant, we can work out most of the genotypes in the pedigree. All the people with the normal number of digits *must* have the genotype dd (if they had even one D allele, they would show polydactyly). All the people with polydactyly must have at least one polydactyly allele (they must be either DD or Dd).

From here, we can begin to work out the genotypes of the people with polydactyly. To do this we need to remember that people with the normal number of digits must inherit one 'normal-number' allele from each parent, and also that people with the normal number of digits will pass on one 'normal-number' allele to each of their children.

From this we can say that any person with polydactyly who has children with the normal number of digits must be heterozygous (the child must have inherited one of their two 'normal-number' alleles from this parent), and also that any person with polydactyly who has one parent with the normal number of digits must also be heterozygous (the 'normal-number' parent can only have passed on a 'normal-number' allele). Individuals 1, 2, 3, 16, 17 and 18 fall into one or other of these categories and must be heterozygous.

We can now add this genetic information to the pedigree. This is shown in Figure 18.7.

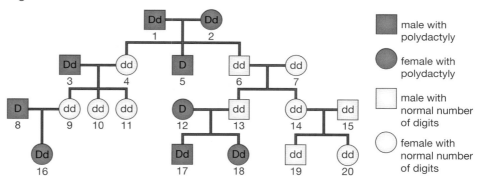

▲ Figure 18.7 A pedigree showing the inheritance of polydactyly in a family, with details of genotypes added.

We are still uncertain about individuals 5, 8 and 12. They could be homozygous or heterozygous. For example, individuals 1 and 2 are both heterozygous. Figure 18.8 shows the possible outcomes from a genetic cross between them. Individual 5 could be any of the outcomes indicated by the shading. It is impossible to distinguish between DD and Dd.

genotypes of parents	Dd	Dd
gametes	D and d	D and d

female gametes

genotypes of children

	D	d
male gametes D	DD	Dd
d	Dd	dd

▲ Figure 18.8 Possible outcomes from a genetic cross between two parents, both heterozygous for polydactyly.

▲ Figure 18.9 Flower colours in snapdragons are caused by a gene showing codominance.

CODOMINANCE

So far, all the examples of genetic crosses that we have seen involve *complete* dominance, where one dominant allele completely hides the effect of a second, or recessive allele. However, there are many genes with alleles that *both* contribute to the phenotype. If two alleles are expressed in the same phenotype, they are called **codominant**. For example, snapdragon plants have red, white or pink flowers (Figure 18.9).

If a plant with red flowers is crossed with one that has white flowers, all the plants resulting from the cross will have pink flowers. The appearance of a third phenotype shows that there is codominance. We can represent the alleles for flower colour with symbols:

R = allele for red flower

W = allele for white flower.

Figure 18.10 shows the cross between the parent plants. Note that the alleles for red and white flowers are given *different* letters, since one is not dominant over the other.

genotypes of parent plants	RR	WW
gametes	all Ⓡ	all Ⓦ

genotypes of offspring
all RW

	Ⓡ	Ⓡ
Ⓦ	RW	RW
Ⓦ	RW	RW

▲ Figure 18.10 Crossing red-flowered snapdragons with white-flowered plants produces a third phenotype, pink.

When pink-flowered plants are crossed together, all three phenotypes reappear, in the ratio 1 red : 2 pink : 1 white (Figure 18.11).

genotypes of parent plants	RW	RW
gametes	Ⓡ all Ⓦ	Ⓡ all Ⓦ

genotypes of offspring
1RR : 2RW : 1WW

	Ⓡ	Ⓦ
Ⓡ	RR	RW
Ⓦ	RW	WW

▲ Figure 18.11 Crossing pink-flowered snapdragons

In fact, *most* genes do not show complete dominance. Genes can show a range of dominance, from complete dominance as in tall and dwarf pea plants through to equal dominance as in the snapdragon flowers, where the new phenotype is halfway between the other two.

SEX DETERMINATION

Our sex – whether we are male or female – is not under the control of a single gene. It is determined by the X and Y chromosomes – the **sex chromosomes**. As well as the 44 non-sex chromosomes, there are two X chromosomes in all cells of females (except the egg cells) and one X and one Y chromosome in all cells of males (except the sperm). Our sex is effectively determined by the presence or absence of the Y chromosome. The full chromosome complement of male and female is shown in Figure 16.12 on page 234.

The inheritance of sex follows the pattern shown in Figure 18.12. In any one family, however, this ratio may well not be met. Predicted genetic ratios are usually only met when large numbers are involved. The overall ratio of male and female births in all countries is 1 : 1.

phenotypes of parents	male	female
genotypes of parents	XY	XX
gametes	X and Y	X

female gametes

		X
male gametes	X	XX
	Y	XY

ratio of genotypes   50% XX : 50% XY

ratio of phenotypes   50% female : 50% male

▲ Figure 18.12 Determination of sex in humans

POLYGENIC INHERITANCE

All of the genetic crosses that you have seen in this chapter have been examples of inheritance involving single genes. The reason for this is that it is easier to draw genetics diagrams and explain what is happening if we start by considering alleles of a single gene. However, many characteristics are controlled by two or more genes working together. This is called **polygenic inheritance**.

A good example is human skin colour. Darker skins contain greater amounts of a black pigment called melanin. This is controlled by several genes, which act together to determine the amount of melanin in the skin. Each gene has alleles that promote melanin production and alleles which do not. This produces a wide range of phenotypes (Figure 18.13).

▲ Figure 18.13 Skin colour depends on the amount of melanin in the skin. It is a result of polygenic inheritance.

Other human characteristics determined by several genes (**polygenes**) are human height and body mass (weight).

CHAPTER QUESTIONS

More questions on chromosomes, genes and inheritance can be found at the end of Unit 5 on page 277.

SKILLS CRITICAL THINKING

1   Which of the following is true of dominant alleles?

   A they are only expressed if present as a pair

   B they determine the most favourable of a pair of alternative features

   C they are inherited in preference to recessive alleles

   D a dominant allele is expressed if present with a recessive allele

SKILLS PROBLEM SOLVING

2   In pea plants, the allele for purple petals is dominant to the allele for white petals. A plant heterozygous for petal colour was crossed with a plant with white petals. What would be the ratio of genotypes in the offspring?

   A 1:1   **B** 2:1

   C 1:0   **D** 3:1

3   The allele for yellow coat colour in mice (Y) is dominant to the allele for non-yellow coat colour (y).

   Mice with the genotype yy have non-yellow coats.

   Mice with the genotype Yy have yellow coats.

   Mice with the genotype YY die as embryos.

   Two heterozygous mice were crossed. What is the probability that a surviving mouse in the F1 generation will be yellow?

   A 0.00   **B** 0.25

   C 0.50   **D** 0.67

BIOLOGY ONLY

4   Alleles B and b are codominant. Two heterozygous individuals were crossed. What would be the expected ratio of phenotypes in the F_1 generation?

   A 1:1   **B** 3:1

   C 1:2:1   **D** 1:1:1:1

END OF BIOLOGY ONLY

5   Predict the *ratios* of offspring from the following crosses between tall/dwarf pea plants.

   a   TT × TT

   b   TT × Tt

   c   TT × tt

   d   Tt × Tt

   e   Tt × tt

   f   tt × tt.

6 In cattle, a pair of alleles controls coat colour. The allele for black coat colour is dominant over the allele for red coat colour. The genetic diagram represents a cross between a pure-breeding black bull and a pure-breeding red cow. B = dominant allele for black coat colour; b = recessive allele for red coat colour.

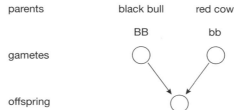

a i What term describes the genotypes of the pure-breeding parents?

ii Explain the terms dominant and recessive.

b i What are the genotypes of the gametes of each parent?

ii What is the genotype of the offspring?

c Cows with the same genotype as the offspring were bred with bulls with the same genotype.

i What genetic term describes this genotype?

ii Draw a genetic diagram to work out the ratios of:

the genotypes of the offspring

the phenotypes of the offspring.

7 In nasturtiums, a single pair of alleles controls flower colour.

The allele for red flower colour is dominant over the allele for yellow flower colour. The diagram represents the results of a cross between a pure-breeding red-flowered nasturtium and a pure-breeding yellow-flowered nasturtium. R = dominant allele for red flower colour; r = recessive allele for yellow flower colour.

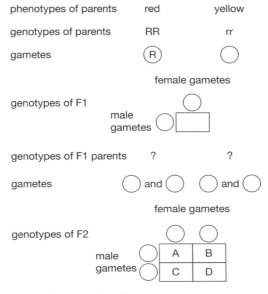

a Copy and complete the genetic diagram.

b What are the colours of the flowers of A, B, C and D?

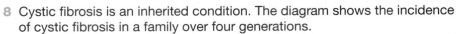

SKILLS ANALYSIS

8 Cystic fibrosis is an inherited condition. The diagram shows the incidence of cystic fibrosis in a family over four generations.

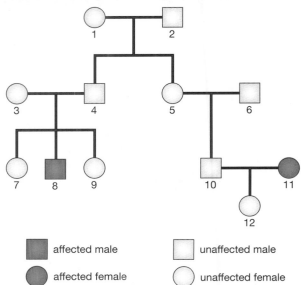

a What evidence in the pedigree suggests that cystic fibrosis is determined by a recessive allele?

b What are the genotypes of individuals 3, 4 and 11? Explain your answers.

c Draw genetic diagrams to work out the probability that the next child born to individuals 10 and 11 will

SKILLS INTERPRETATION

   i   be male,

   ii suffer from cystic fibrosis.

SKILLS REASONING

9 In guinea pigs, the allele for short hair is dominant to that for long hair.

a Two short-haired guinea pigs were bred and their offspring included some long-haired guinea pigs. Explain these results.

b How could you find out if a short-haired guinea pig was homozygous or heterozygous for hair length?

BIOLOGY ONLY

SKILLS CRITICAL THINKING

10 When two different alleles of a gene are expressed in the same phenotype, they are called codominant. Coat colour in shorthorn cattle is controlled by a codominant gene. 'Red' cattle crossed with 'white' cattle produce offspring which all have a pale brown coat, called roan.

a Explain the terms gene, allele and phenotype.

b Draw genetic diagrams to show the possible genotypes of offspring resulting from a cross between:

SKILLS INTERPRETATION

   i   a red bull and a white cow

   ii   a red bull and a roan cow

   iii a roan bull and a roan cow.

SKILLS PROBLEM SOLVING

c For each of the crosses in (b) state the ratio of the phenotypes you would expect from the cross.

END OF BIOLOGY ONLY

19 NATURAL SELECTION AND EVOLUTION

Over millions of years, life on this planet has evolved from its simple beginnings into the vast range of organisms present today. This has happened by a process called natural selection.

LEARNING OBJECTIVES

- Explain Darwin's theory of evolution by natural selection
- Understand how resistance to antibiotics can increase in bacterial populations.

▲ Figure 19.1 Charles Darwin (1809–1882).

The meaning of '**evolution**' is that species of animals and plants are not fixed in their form, but change over time. It is not a new idea. For thousands of years philosophers have discussed this theory. By the beginning of the nineteenth century there was overwhelming evidence for evolution, and many scientists had accepted that it had taken place. What was missing was an understanding of the *mechanism* by which evolution could have occurred.

The person who proposed the mechanism for evolution that is widely accepted today was the English biologist Charles Darwin (Figure 19.1). He called the mechanism **natural selection**.

THE WORK OF CHARLES DARWIN

Charles Darwin was the son of a country doctor. He did not do particularly well at school or university and was unable to decide on a profession. His father is supposed to have said: 'you're good for nothing but shooting guns and rat-catching … you'll be a disgrace to yourself and all of your family'. He was wrong – Darwin went on to become one of the most famous scientists of all time!

At the age of 22, Charles Darwin became the unpaid biologist aboard the survey ship HMS *Beagle*, which left England for a five-year voyage in 1831 (Figure 19.2).

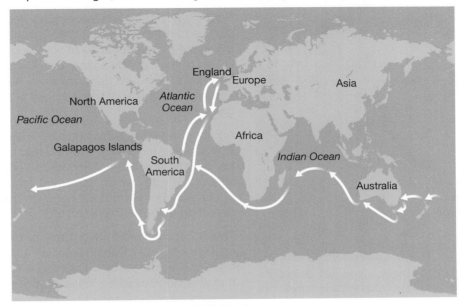

▲ Figure 19.2 The five-year journey of HMS *Beagle*

DID YOU KNOW?

A fossil is the remains of an animal or plant that lived thousands or millions of years ago, preserved in sedimentary rocks. Fossils are formed when minerals replace the materials in bone and tissue, creating a replica of the original organism in the rock.

KEY POINT

The phrase 'survival of the fittest' does not mean *physical* fitness, but *biological* fitness. This depends on how well adapted an organism is to its environment so that it is successful in *reproducing*. A good way of putting it is; 'survival of the individuals that will leave most offspring in later generations'.

▲ Figure 19.3 Darwin's ideas were unpopular and many newspapers of the time made fun of them.

During the voyage, Darwin collected hundreds of specimens and made many observations about the variety of organisms and the ways in which they were adapted to their environments. He gained much information, in particular, from the variety of life forms in South America and the Galapagos Islands. Darwin was influenced by the work of Charles Lyell who was, at the time, laying the foundations of modern geology. Lyell was using the evidence of rock layers to suggest that the surface of the Earth was constantly changing. The layers of sediments in rocks represented different time periods. Darwin noticed that the fossils found in successive layers of rocks often changed slightly through the layers. He suggested that life forms were continually changing – evolving.

On his return to England, Darwin began to evaluate his data and wrote several essays, introducing the ideas of natural selection. He arrived at his theory of natural selection from observations made during his voyage on HMS *Beagle* and from deductions made from those observations. Darwin's observations were that:

■ organisms generally produce more offspring than are needed to replace them – a single female salmon can release 5 million eggs per year; a giant puffball fungus produces 40 million spores

■ despite this over-reproduction, stable, established populations of organisms generally remain the same size – the seas are not overflowing with salmon, and we are not surrounded by lots of giant puffball fungi!

■ members of the same species are not identical – they show variation.

He made two important deductions from these observations.

■ From the first two observations he deduced that there is a 'struggle for existence'. Many offspring are produced, yet the population stays the same size. There must be competition for resources and many individuals must die.

■ From the third observation he deduced that, if some offspring survive whilst others die, those organisms best suited to their environment would survive to reproduce. Those less suited will die. This gave rise to the phrase 'survival of the fittest'.

Notice a key phrase in the second deduction – the best-suited organisms survive *to reproduce*. This means that those characteristics that give the organism a better chance of surviving will be passed on to the next generation. Fewer of the individuals that are less suited to the environment survive to reproduce. The next generation will have more of the type that is better adapted and fewer of the less well adapted type. This will be repeated in each generation.

Another naturalist, Alfred Russell Wallace, had also studied life forms in South America and Indonesia and had reached the same conclusions as Darwin. Darwin and Wallace published a scientific paper on natural selection jointly, although it was Darwin who went on to develop the ideas further. In 1859, he published his now famous book *On the Origin of Species by Means of Natural Selection* (usually shortened to *The Origin of Species*).

This book changed forever the way in which biologists think about how species originate. Darwin went on to suggest that humans could have evolved from ape-like ancestors. For this he was ridiculed, largely by people who had misunderstood his ideas (Figure 19.3). He also carried out considerable research in other areas of biology, such as plant tropisms (see Chapter 12).

When Darwin proposed his theory of natural selection, he did not know about genes and how they control characteristics. Gregor Mendel had yet to publish his work on inheritance, and as you have seen, the significance of Mendel's work was not recognised until 1903.

The theory of natural selection proposes that some factor in the environment 'selects' which forms of a species will survive to reproduce. Forms that are not well adapted will not survive.

The following is a summary of how we think natural selection works:

1. there is variation within the species
2. changing conditions in the environment (called a selection pressure) favours one particular form of the species (which has a selective advantage)
3. the frequency of the favoured form increases (it is selected for) under these conditions (survival of the fittest)
4. the frequency of the less well adapted form decreases under these conditions (it is selected against).

As you have seen, many gene mutations are harmful, and cells that carry them will not usually survive. Some mutations are 'neutral' and if they arise in the gametes, may be passed on without affecting the survival of the offspring. However, a few mutations can actually be beneficial to an organism. Beneficial mutations are the 'raw material' that are ultimately the source of new inherited variation.

SOME EXAMPLES OF HOW NATURAL SELECTION MIGHT HAVE WORKED

THE HOVERFLY

(a)

(b)

▲ Figure 19.4 Two insects showing 'warning colouration'. (a) A wasp, which has a sting. (b) A harmless hoverfly.

Figure 19.4 shows two species of insect: a wasp and a hoverfly. Wasps can defend themselves against predators using a sting. They also have a body with yellow and black stripes. This is called a 'warning colouration'. Predators such as birds soon learn that these colours mean that wasps have the sting, and they avoid attacking them.

Hoverflies do not have a sting. However, they have an appearance that is very like a wasp, with similar yellow and black stripes – they are 'mimics' of wasps. Predators treat hoverflies as if they do have a sting.

Clearly, mimicking a wasp is an advantage to the hoverfly. How could they have evolved this appearance? We can explain how it could have happened by natural selection.

The selection pressure was predation by birds and other animals. Among the ancestors of present-day hoverflies there would have been variation in colours. As a result of mutations, some hoverflies gained genes that produced stripes on their bodies. These insects were less likely to be eaten by predators than hoverflies without the stripes – they had a *selective advantage*.

Since the hoverflies with stripes were more likely to survive being eaten, they were more likely to reproduce, and would pass on the genes for stripes to their offspring. This process continued over many generations. Gradually more mutations and selection for 'better' stripes took place, until the hoverflies evolved the excellent warning colouration that they have today.

Note that perfect stripes didn't have to appear straight away. Even a slight stripy appearance could give a small selective advantage over hoverflies without stripes. This would be enough to result in an increase in stripy hoverflies in the next generation.

▲ Figure 19.5 A polar bear hunting on the Arctic sea ice.

The polar bear lives in the Arctic, inhabiting landmasses and sea ice covering the waters within the Arctic Circle (Figure 19.5). It is a large predatory carnivore, mainly hunting seals. One way the bear hunts is to wait near holes in the ice where seals come up to breathe. It also silently approaches and attacks seals that are resting on the ice.

Polar bears have many **adaptations** that suit them to their habitat. These include:

- a thick layer of white fur, which reduces heat loss and acts as camouflage in the snow

- wide, large paws. These help with walking in the snow, and are used for swimming

- strong, muscular legs – a bear can swim continuously in the cold Arctic waters for days

- nostrils that close when the bear is swimming under water

- a large body mass. Polar bears are the largest bears on Earth. An adult male averages 350 to 550 kilograms, and the record is over 1000 kilograms. This large size results in the animal having a small surface area to volume ratio, which reduces heat loss

- a 10 centimetre thick layer of insulating fat under the skin

- a well developed sense of smell – used to detect the bear's prey

- bumps on the pads of the paws to provide grip on the ice

- short, powerful claws, which also provide grip, and are needed for holding the heavy prey.

The polar bear is thought to have evolved from a smaller species, the brown bear, about 150,000 years ago. How did it evolve its adaptations for life in the Arctic? Let's consider just one of the adaptations, the thick white fur.

There are two main selection pressures in favour of thick white fur. The first is the need for insulation to reduce heat loss. The polar bear often has to survive temperatures of –30°C, and temperatures in the Arctic can fall as low as –70°C. The second is camouflage; white fur camouflages the animal against the snow so that it can approach its prey unseen and then attack it.

Among the brown bears that were the ancestors of the polar bear there would have been variations in fur length and colour. When some of these bears came to live in colder, more northerly habitats, those individuals with longer and paler fur would have had a selective advantage over others with shorter, darker fur. Any gene mutations that produced long, pale fur increased this advantage. Bears with these genes were less likely to die from the cold, or from lack of food. As a result, well-adapted bears were more likely to reproduce and pass on their genes. Over many thousands of years more mutations and selection for long, white fur produced the adaptation we see in the polar bear today. The same process of natural selection is thought to have happened to bring about the other adaptations shown by the polar bear.

DID YOU KNOW?

Polar bear fur appears to be white, but in fact the individual hairs are actually transparent. The white colour results from light being refracted through the clear hair strands.

CAN WE OBSERVE NATURAL SELECTION IN ACTION?

Most animals and plants reproduce slowly, so it takes a long time for natural selection to have an observable effect. To observe natural selection happening we can study organisms that reproduce quickly, such as bacteria or insects.

ANTIBIOTIC RESISTANCE IN BACTERIA

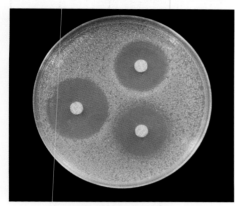

▲ Figure 19.6 This photo shows a colony of bacteria growing on a petri dish of nutrient agar. The circular discs contain different antibiotics. The discs have clear areas around them, where the bacteria have been killed by the antibiotic diffusing out from the discs.

Antibiotics are chemicals that kill or reduce the growth of microorganisms (Figure 19.6). They are used in medicine mainly to treat bacterial infections, although a few antibiotics are effective against fungal pathogens. Antibiotics do not work on viruses, so they are no use in treating any disease caused by a virus.

Natural antibiotics are produced by bacteria and fungi. They give a microorganism an advantage over other microorganisms when competing for nutrients and other resources, since the antibiotic kills the competing organisms.

Alexander Fleming discovered the first antibiotic in 1929. It is made by the mould *Penicillium*, and is called **penicillin**. Penicillin kills bacteria, and was first used to treat bacterial infections in the 1940s. Since then other natural antibiotics have been discovered, and many more have been chemically synthesised in laboratories. The use of antibiotics has increased dramatically, particularly over the last 20 years. We now almost expect to be given an antibiotic for even the most minor of ailments. This can be dangerous, as it leads to the development of bacterial resistance to antibiotics, so that the antibiotics are no longer effective in preventing bacterial infection.

> **DID YOU KNOW?**
> A particularly worrying example of a resistant bacterium is MRSA. MRSA stands for methicillin-resistant *Staphylococcus aureus*. It is sometimes called a 'super bug' because it is resistant to many antibiotics including methicillin (a type of penicillin). It is a particular problem in hospitals where it is responsible for many difficult-to-treat infections.

Resistance starts when a random mutation gives a bacterium resistance to a particular antibiotic. In a situation where the antibiotic is widely used, the new resistant bacterium has an advantage over non-resistant bacteria of the same type. The resistant strain of bacterium will survive and multiply in greater numbers than the non-resistant type. Bacteria reproduce very quickly – the generation time of a bacterium (the time it takes to divide into two daughter cells) can be as short as 20 minutes. This means that there could be 72 generations in a single day, producing a population of millions of resistant bacteria.

Resistant bacteria will not be killed by the antibiotic, meaning the antibiotic is no longer effective in controlling the disease.

Bacterial resistance to antibiotics was first noticed in hospitals in the 1950s, and has grown to be a major problem today. The resistant bacteria have a selective advantage over non-resistant bacteria – they are 'fitter'. In effect, the bacteria have evolved as a result of natural selection.

Doctors are now more reluctant to prescribe antibiotics. They know that by using them less, the bacteria with resistance have less of an advantage and will not become as widespread.

PESTICIDE RESISTANCE IN INSECTS

Just as pathogenic bacteria can become resistant to antibiotics, insect pests can develop resistance to insecticides. The powerful insecticide DDT was first used in the 1940s (see Chapter 15). By the 1950s many species of insect (e.g. mosquitoes) appeared to be resistant to DDT. The resistant insects had developed a gene mutation that stopped them being killed by the insecticide.

> **HINT**
> Some people talk about bacteria becoming *immune* to antibiotics. This is a misunderstanding. Immunity happens in people – we become immune to microorganisms that infect us, as a result of the immune response.
>
> Bacteria become *resistant* to antibiotics.

While DDT continued to be used, the resistant insects had a selective advantage over the non-resistant ones. They survived to breed, so that with each generation the numbers of resistant insects in the population increased.

The same thing has happened with modern insecticides. There are now hundreds of examples of insect pests that have developed resistance to different insecticides.

CHAPTER QUESTIONS

More questions on natural selection and evolution can be found at the end of Unit 5 on page 277.

SKILLS CRITICAL THINKING

1 Which of the following statements is/are correct?

    1. Antibiotics are made by bacteria

    2. Antibiotics kill bacteria

    3. Antibiotics do not work on viruses

A 1 and 2

B 2 only

C 2 and 3

D 1, 2 and 3

2 Which of the following best describes the meaning of biological 'fitness'

A a measure of an organism's ability to survive in different habitats

B a measure of the reproductive success of an organism

C a measure of the relative health of an organism

D a measure of the strength of an organism

3 What is the source of genetic variation?

A mutations

B mitosis

C selection pressures

D changes in the environment

4 Why is natural selection easy to observe in bacteria?

A they are very small

B they can be killed by antibiotics

C they are composed of simple cells

D they reproduce very quickly

5 a What does the term 'survival of the fittest' mean?

   b Which two biologists arrived at the same idea concerning the 'survival of the fittest' at the same time?

6 Antibiotics are chemicals that are used to kill pathogens.

   a What is a pathogen?

   b Name two types of organism that make natural antibiotics.

SKILLS REASONING

   c Some antibiotics are no longer effective in killing pathogens. Use your understanding of natural selection to explain why.

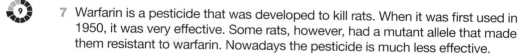

SKILLS REASONING

7 Warfarin is a pesticide that was developed to kill rats. When it was first used in 1950, it was very effective. Some rats, however, had a mutant allele that made them resistant to warfarin. Nowadays the pesticide is much less effective.

a Use the ideas of natural selection to explain why warfarin is much less effective than it used to be.

b Suggest what might happen to the number of rats carrying the allele for warfarin resistance, if warfarin were no longer used. Explain your answer.

SKILLS ANALYSIS

8 In the Galapagos Islands, Charles Darwin identified a number of species of birds, now known as Darwin's finches. He found evidence to suggest that they had all evolved from one ancestral type, which had colonised the islands from South America. The main differences between the finches were their beaks. The diagram shows some of the beak types and that of the likely ancestral finch.

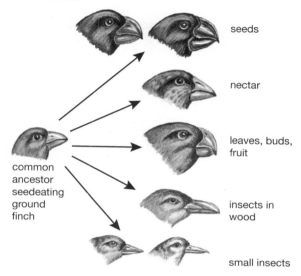

a Explain how the seed-eating finches are adapted to their environment.

b Explain how the finches that eat insects and live in woodland are adapted to their environment.

SKILLS REASONING

c Use the information in the diagram to help you explain how the common ancestor could have evolved into the different type of finches.

SKILLS CRITICAL THINKING

9 Read the description below and answer the questions that follow.

Natural selection happens when a selection pressure favours individuals with particular characteristics, so that they have a selective advantage.

Some plants growing in areas contaminated by waste from mines have developed a tolerance to toxic metals such as lead and copper. They are able to grow on polluted soil, while non-tolerant plants are killed by the metals in the soil.

a How did the new tolerant varieties of plants arise?

SKILLS REASONING

b With reference to this example, explain the terms

i selection pressure

ii selective advantage

iii natural selection

SKILLS REASONING

c When metal-tolerant plants are grown on uncontaminated soil, they are out-competed by non-tolerant plants. Suggest a reason for this.

20 SELECTIVE BREEDING

Ever since humans became farmers they have been selectively breeding animals and plants. This chapter looks at traditional methods of selective breeding and modern developments involving cloned organisms.

LEARNING OBJECTIVES

- Understand how selective breeding can be used to produce plants with desired characteristics

- Understand how selective breeding can be used to produce animals with desirable characteristics

BIOLOGY ONLY

- Describe the process of micropropagation (tissue culture) in which explants are grown *in vitro*

BIOLOGY ONLY

- Understand how micropropagation can be used to produce commercial quantities of genetically identical plants with desirable characteristics

- Describe the stages in the production of cloned mammals involving the introduction of a diploid nucleus from a mature cell into an enucleated egg cell, illustrated by Dolly the sheep

- Understand how cloned transgenic animals can be used to produce human proteins

About 12 000 years ago, the human way of life changed significantly. Humans began to grow plants and keep animals for milk and meat. They became farmers rather than hunters. This change first took place in the Middle East. Similar changes took place a little later in the Americas (where potatoes and maize were being grown) and in the Far East (where rice was first cultivated).

In the Middle East, humans first grew the cereal plants wheat and barley, and kept sheep and goats. Later, their livestock included cattle and other animals. Cultivating crops and keeping stock animals made it possible for permanent settlements to appear – life in villages began. Because of the more certain food supply, there was spare time, for the first time ever, for some people to do things other than hunt for food.

Ever since the cultivation of the first wheat and barley and the domestication of the first stock animals, humans have tried to obtain bigger yields from them. They cross-bred different maize plants (and barley plants) to obtain strains that produced more grain. They bred sheep and goats to give more milk and meat – selective breeding had begun. Today, animals and plants are bred for more than just food. For example, animals are used to produce a range of medicines and for research into the action of drugs.

Selective breeding is best described as the breeding of only those individuals with desirable features. It is sometimes called 'artificial selection', as human choice, rather than environmental factors, is providing the selection pressure (compare this with natural selection, described in Chapter 19).

The methods used today for selective breeding are very different from those used only 50 years ago. Modern gene technology makes it possible to create a new strain of plant within weeks, rather than years.

TRADITIONAL SELECTIVE BREEDING

PLANTS

Traditionally, farmers have bred crop plants of all kinds to obtain increased yields. Probably the earliest example of selective breeding was the cross-breeding of strains of wild wheat. The aim was to produce wheat with a much increased yield of grain and with shorter, stronger stems (Figure 20.1). This wheat was used to make bread.

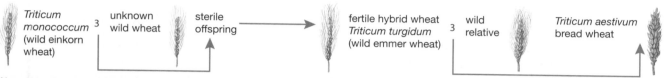

Triticum monococcum (wild einkorn wheat) 3 unknown wild wheat → sterile offspring → fertile hybrid wheat *Triticum turgidum* (wild emmer wheat) 3 wild relative → *Triticum aestivum* bread wheat

1 About 11 000 years ago, two strains of wild wheat were cultivated by farmers. Initially, all attempts at cross-breeding to produce wheats with a better yield gave only sterile offspring.

2 About 8000 years ago, a fertile hybrid wheat appeared from these two wild wheats. This was called emmer wheat and had a much higher yield than either of the original wheats.

3 The emmer wheat was cross-bred with another wild wheat to produce wheat very similar to the wheats used today to make bread. This new wheat had an even bigger yield and was much easier to 'process' to make flour.

▲ Figure 20.1 Modern wheat is the result of selective breeding by early farmers.

> **DID YOU KNOW?**
>
> The production of modern bread wheats by selective breeding is probably one of the earliest examples of producing genetically modified food. Each original wild wheat species had 14 chromosomes per cell. The wild emmer hybrid had 28 chromosomes per cell. Modern bread wheat has 42 chromosomes per cell. Selective breeding has modified the genetic make-up of wheat.

Other plants have been selectively bred for certain characteristics. *Brassica* is a genus of cabbage-like plants. One species of wild brassica (*Brassica olera*) was selectively bred to give several strains, each with specific features (see Figure 20.2). Some of the strains had large leaves, others had large flower heads, and others produced large buds.

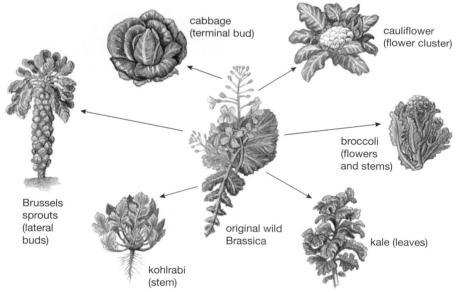

cabbage (terminal bud)

cauliflower (flower cluster)

broccoli (flowers and stems)

Brussels sprouts (lateral buds)

original wild Brassica

kale (leaves)

kohlrabi (stem)

▲ Figure 20.2 Selectively breeding the original wild brassica plants to enhance certain features has produced several familiar vegetables.

Selective breeding has produced many familiar vegetables. Besides the ones produced from *Brassica*, selective breeding of wild *Solanum* plants has produced the many strains of potatoes that are eaten today. Carrots and parsnips are also the result of selective breeding programmes. Crop plants are bred to produce strains that:

- give higher yields
- are resistant to certain diseases (the diseases would reduce the yields)
- are resistant to certain insect pest damage (the damage would reduce the yield)
- are hardier (so that they survive in harsher climates or are productive for longer periods of the year)
- have a better balance of nutrients in the crop (for example, plants that contain more of the types of amino acids needed by humans).

Figure 20.3 shows a field of potato plants. Some have been bred to be resistant to insect pests, while others were not selectively bred in this way.

DID YOU KNOW?

Plant breeders do not just breed plants for food. Nearly all garden flowers are the result of selective breeding. Breeders have selected flowers to have a particular size, shape, colour and fragrance. Roses and orchids are among the most selectively bred of our garden plants.

▲ Figure 20.3 Selective breeding can reduce damage by pests. The plants on the right have been bred to be resistant to a fungal pest. Plants on the left are not resistant to the pest.

ANIMALS

Farmers have bred stock animals for similar reasons to the breeding of crops. They have selected for animals that:

- produce more meat, milk or eggs
- produce more fur or better quality fur
- produce more offspring
- show increased resistance to diseases and parasites.

Again, like crop breeding, breeding animals for increased productivity has been practised for thousands of years. A stone tablet found in Iran appears to record the results of breeding domesticated donkeys. It was dated at over 5000 years old.

For many thousands of years, the only way to improve livestock was to mate a male and a female with the features that were desired in the offspring. In cattle, milk yield is an important factor and so high-yielding cows would be bred with bulls from other high-yielding cows.

Since about 1950, the technique of **artificial insemination** (**AI**) has become widely available. Bulls with many desirable features are kept and semen is obtained from them. The semen is diluted, frozen and stored. Farmers can buy quantities of this semen to inseminate their cows. the semen is transferred into the cow's uterus using a syringe. AI makes it possible for the semen from one prize bull to be used to fertilise many thousands of cows.

Modern sheep are domesticated wild sheep, and cows have been derived from wild aurochs. Just think of all the varieties of dogs that now exist. All these have been derived from one ancestral type. This original 'dog' was a domesticated wolf (Figure 20.4). In domesticating the wolf, humans gained an animal that was capable of herding stock animals. The sheepdog has all the same instincts as the wolf except the instinct to kill. This has been selectively 'bred out'.

▲ Figure 20.4 The many different breeds of dog all originate from a common ancestor – the wolf.

BIOLOGY ONLY

MODERN SELECTIVE BREEDING

CLONING PLANTS

The term cloning describes any procedure that produces genetically identical offspring. Taking cuttings of plants and growing them is a traditional cloning technique (Figure 20.5).

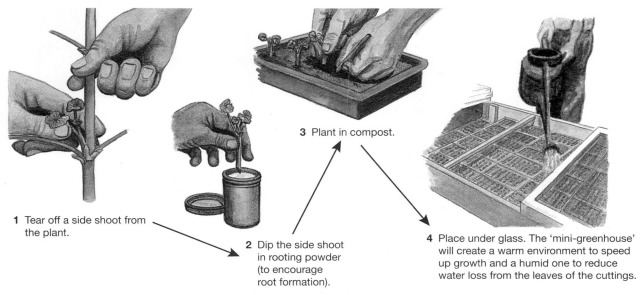

1 Tear off a side shoot from the plant.

2 Dip the side shoot in rooting powder (to encourage root formation).

3 Plant in compost.

4 Place under glass. The 'mini-greenhouse' will create a warm environment to speed up growth and a humid one to reduce water loss from the leaves of the cuttings.

▲ Figure 20.5 Taking stem cuttings

All the cuttings contain identical genes as they are all parts of the same parent plant. As they grow, they form new cells by mitosis, copying the genes in the existing cells exactly. The cuttings develop into a group of genetically identical plants – a **clone**. Any differences will be due to the environment. Many garden flowers have traditionally been propagated this way.

Some modern cloning techniques are essentially the same as taking cuttings – removing pieces of a plant and growing them into new individuals. The technology, however, is much more sophisticated. By using the technique of **micropropagation**, thousands of plants can quickly be produced from one original (Table 20.1).

Table 20.1 The main stages in micropropagation

Stages	Illustrations
The tips of the stems and side shoots are removed from the plant to be cloned. These parts are called **explants**. The explants are trimmed to a size of about 0.5–1 mm, and surface-sterilised to kill any microorganisms. They are then placed in a sterile agar medium that contains nutrients and plant hormones to encourage growth (Figure 20.6). More explants can be taken from the new shoots that form on the original ones. This can be repeated until there are enough to supply the demand.	 ▲ Figure 20.6 Explants growing in a culture medium.
The explants with shoots are transferred to another culture medium containing a different balance of plant hormones to induce root formation (Figure 20.7).	 ▲ Figure 20.7 Explants forming roots.
When the explants have grown roots, they are transferred to greenhouses and transplanted into compost (Figure 20.8). They are then gradually acclimatised to normal growing conditions. The atmosphere in the greenhouse is kept very moist to reduce water loss from the young plants. Because of the amount of water vapour in the air, they are often called 'fogging greenhouses'.	 ▲ Figure 20.8 Young plants being grown in compost in a greenhouse.

DID YOU KNOW?

Many strains of bananas are infertile. They are now commonly reproduced by micropropagation. Other plants produced this way include lilies, orchids and agave plants.

There are many advantages to propagating plants in this way:

■ large numbers of genetically identical plants can be produced rapidly

■ species that are difficult to grow from seed or from cuttings can be propagated by this method

■ plants can be produced at any time of the year

■ large numbers of plants can be stored easily (many can be kept in cold storage at the early stages of production and then developed as required)

■ genetic modifications can be introduced into thousands of plants quickly, after modifying only a few plants.

CLONING ANIMALS

We have been able to clone plants by taking cuttings for thousands of years. It is now possible to make genetically identical copies of animals. The first, and best-known, example of this is the famous cloned sheep, Dolly.

Dolly was cloned from a body cell of an adult sheep (Figure 20.9). Scientists first took an ovum (egg cell) from a donor sheep and removed its nucleus, producing an *enucleated* cell. They then took cells from the mammary (milk-producing) gland of a second 'parent' sheep (Dolly's genetic mother) and cultured them in a special solution that kept them alive but stopped them growing. Next they placed one of the mammary gland cells next to the enucleated cell and fused the two cells together using an electric current. The nucleus from the mammary gland cell was now inside the enucleated cell.

They monitored the resulting cell for several days to watch its development. When it started to divide they transferred the embryo into the uterus of another sheep, a 'surrogate mother', to complete its development. After 148 days, Dolly was born. Dolly's cells contained exactly the same genetic information as those of the body cell from the 'parent' sheep.

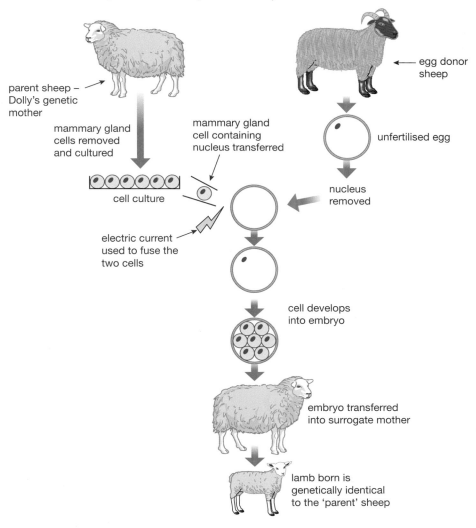

▲ Figure 20.9 How 'Dolly' was produced.

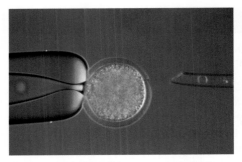

▲ Figure 20.10 Inserting a mammary gland cell into an egg cell that has had its nucleus removed.

Dolly was only produced after many unsuccessful attempts. Since then, the procedure has been repeated using more sheep and other animals. Some of the animals produced are born deformed. Some do not survive to birth. Biologists believe that these problems occur because the genes that are transferred to the egg are 'old genes'. These genes came from an animal that had already lived for several years and from cells specialised to do things other than produce sex cells. It will take much more research to make the technique reliable.

> **DID YOU KNOW?**
>
> In 2016 (20 years after the birth of Dolly) the University of Nottingham in the UK announced that they had cloned several more sheep, including four called Debbie, Denise, Dianna and Daisy. These four animals were derived from the same cell line that produced Dolly, so were exact genetic copies of her. At the time of publication of the research, the sheep were nine years old and had developed no age-related health problems.

USING CLONED ANIMALS TO MAKE PROTEINS

Cloning animals has special value if the animal produces an important product. Sheep have been genetically modified to produce several human proteins (see Chapter 22). One of these is a protein called alpha-1-antitrypsin, which is used to treat conditions such as emphysema and cystic fibrosis. The genetically modified sheep secrete the protein in their milk. Cloning sheep like these would allow production of much more of this valuable protein. Polly, the first cloned, genetically modified sheep, was born a year after Dolly.

> **DID YOU KNOW?**
>
> Animals that have had genes transferred from other species are called *transgenic* animals.

END OF BIOLOGY ONLY

CHAPTER QUESTIONS

More questions on selective breeding can be found at the end of Unit 5 on page 277.

SKILLS ▸ CRITICAL THINKING

1 Which of the following is another term for selective breeding?

   A cloning

   B natural selection

   C artificial insemination

   D artificial selection

SKILLS CRITICAL THINKING

2 Which of the following best describes a clone?

A a transgenic organism

B an offspring where the genetic material in every cell is identical to that of both parents

C an offspring where the genetic material in every cell is identical to that of one parent

D a type of sheep

3 Which of the following best describes artificial insemination?

A transplanting an embryo in the uterus

B taking sperm and placing it directly in the uterus

C selectively breeding healthy animals

D fertilisation of an egg in a test tube

4 Which of the following is the best description of how the first cloned mammal (Dolly the sheep) was produced?

A A mammary gland cell was fused with an enucleated egg, grown into an embryo and allowed to develop in a surrogate mother.

B A surrogate mother had an egg removed, which was fused with a mammary gland cell and allowed to develop into an embryo.

C An egg was placed in an enucleated mammary gland cell, grown into an embryo and allowed to develop in a surrogate mother.

D An egg was placed in the mammary gland of a donor sheep and allowed to develop into an embryo.

5 a How is selective breeding similar to natural selection?

b How is selective breeding different from natural selection?

SKILLS REASONING

6 Selective breeding of crop plants often aims to increase the yield of the crop.

a Describe, and explain the reasons for, three other aims of selective breeding programmes in crop plants.

b Describe two advantages of micropropagation over the more traditional technique of taking cuttings.

c Explain why plants produced by micropropagation will be genetically identical to each other and to the parent plant.

7 The diagram shows some of the features of a cow that might be used as a basis for a breeding programme.

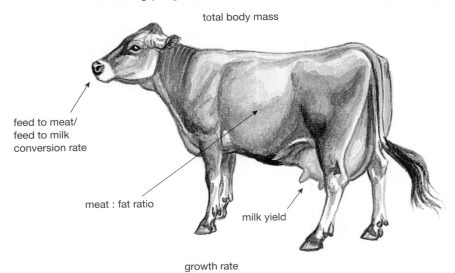

total body mass

feed to meat/ feed to milk conversion rate

meat : fat ratio

milk yield

growth rate

a Which features would you consider important in a breeding programme for dairy cattle?

b Assume that you had all the techniques of modern selective breeding available to you. Describe how you would set about producing a herd of high-yielding beef cattle.

8 The diagram shows the results of a breeding programme to improve the yield of maize (sweetcorn).

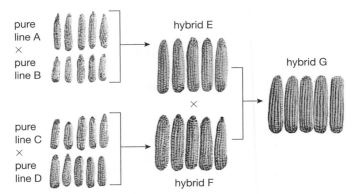

pure line A × pure line B

hybrid E

pure line C × pure line D

hybrid F

hybrid G

a Describe the breeding procedure used to produce hybrid G.

b Describe three differences between the corn cobs of hybrid G and those of C.

c How could you show that the differences between hybrid G and hybrid C are genetic?

9 Carry out some research and write an essay about the benefits and concerns of selective breeding of animals. You should write about one side of A4.

UNIT QUESTIONS

SKILLS ▶ REASONING

1 For natural selection to operate, some factor has to exert a 'selection pressure'. In each of the following situations, identify both the selection pressure and the likely result of this selection pressure.

a Near old copper mines, the soil becomes polluted with copper ions that are toxic to most plants. **(2)**

b In the Serengeti of Africa, wildebeest are hunted by lions. **(2)**

c A farmer uses a pesticide to try to eliminate pests of a potato crop. **(2)**

(Total 6 marks)

BIOLOGY ONLY

SKILLS ▶ CRITICAL THINKING

2 Micropropagation produces thousands of genetically identical plants. Small 'explants' from the parent plant are grown in culture media.

a Outline the main stages in micropropagation. **(4)**

SKILLS ▶ REASONING

b Explain why the plants formed by micropropagation are genetically identical. **(2)**

c In some cases, the explants used contain only a few cells, without roots or shoots. Plant hormones are added to the culture media to encourage root and shoot formation. Two of these hormones are called kinetin and auxin. The diagram shows the effects of using different concentrations of the two hormones on root and shoot growth of the explants.

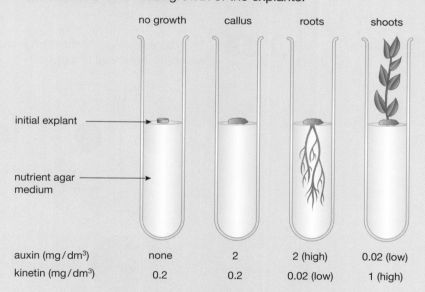

	no growth	callus	roots	shoots
auxin (mg/dm³)	none	2	2 (high)	0.02 (low)
kinetin (mg/dm³)	0.2	0.2	0.02 (low)	1 (high)

 i What is the effect of adding kinetin or auxin without any other hormone? **(2)**

ii Describe how you would treat these explants to produce first shoots and then roots. **(3)**

SKILLS ▶ CRITICAL THINKING

d Explain one advantage and one disadvantage of micropropagation. **(2)**

(Total 13 marks)

END OF BIOLOGY ONLY

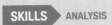 ANALYSIS

3 PTC (phenylthiocarbamide) is a chemical that to some people has a very bitter taste, while other people cannot taste it at all. The diagram shows the inheritance of PTC tasting in a family.

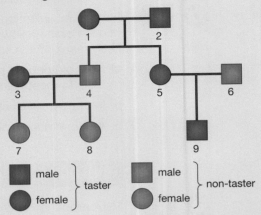

SKILLS REASONING

SKILLS ANALYSIS

SKILLS INTRPRETATION

a What evidence in the diagram suggests that the allele for PTC tasting is dominant? **(2)**

b Using T to represent the tasting allele and t to represent the non-tasting allele, give the genotypes of individuals 3 and 7. Explain how you arrived at your answers. **(4)**

c Why can we not be sure of the genotype of individual 5? **(2)**

d If individuals 3 and 4 had another child, what is the chance that the child would be able to taste PTC? Construct a genetic diagram to show how you arrived at your answer. **(4)**

(Total 12 marks)

SKILLS ANALYSIS

4 The diagrams A to F show an animal cell during cell division. The diploid number of this cell is eight.

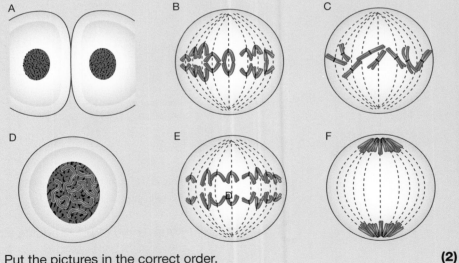

a Put the pictures in the correct order. **(2)**

b Is the cell going through mitosis or meiosis? Explain your answer. **(2)**

 CRITIACL THINKING

c What is the diploid number of a human cell? **(1)**

d Describe two differences between mitosis and meiosis. **(2)**

(Total 7 marks)

BIOLOGY ONLY

SKILLS CRITICAL THINKING **8**

5 The following flowchart shows how Dolly the sheep was cloned.

> **1** Cell taken from the mammary gland of sheep A and grown in culture in the laboratory for six days.
>
> **2** Unfertilised egg taken from sheep B. Nucleus of the egg removed.
>
> **3** Cell from sheep A fused with the empty egg by an electrical spark.
>
> **4** Embryo from stage 3 transferred to the uterus of sheep C, which acts as a surrogate mother.
>
> **5** Surrogate mother gives birth to Dolly.

a Where did scientists get the DNA to put into the unfertilised egg from sheep B? **(1)**

6

b How does the nucleus removed from an egg differ from the nucleus of an embryo? **(1)**

SKILLS ANALYSIS **8**

c Dolly is genetically identical to another sheep in the diagram. Which one? **(1)**

SKILLS CRITICAL THINKING **8**

d Give two ways in which this method is different from the normal method of reproduction in sheep. **(2)**

e Suggest two advantages of producing animal clones. **(2)**

(Total 7 marks)

7

6 Copy and complete the following passage about genes:

A gene is a section of a molecule known as _____ .
The molecule is found within the _____ of a cell,
within thread-like structures called _____ .
The strands of the molecule form a double helix joined by paired
bases. The base adenine is always paired with its complementary base
_____ , and the base cytosine is paired with
_____ . During the process of transcription,
the order of bases in one strand of the molecule is used to form
_____, which carries the code for making proteins
out to the cytoplasm.

(Total 6 marks)

SKILLS PROBLEM SOLVING **9**

7 In a section of double-stranded DNA there are 100 bases, of which 30 are cytosine (C).

Calculate the total number of each of the following in this section of DNA:

a complementary base pairs

b guanine (G) bases

c thymine (T) bases

d adenine (A) bases

e deoxyribose sugar groups

(Total 5 marks)

END OF BIOLOGY ONLY

UNIT 6
MICROORGANISMS AND GENETIC MODIFICATION

You have seen how some microorganisms are essential decomposers in ecosystems and recycle nutrients, while others are pathogens, causing diseases of animals and plants. There are a few species of microorganisms that are grown by humans in order to make useful products. In this unit we look at this last group, and consider traditional and modern biotechnology. In addition, microorganisms are sometimes genetically engineered to produce new products. This is described in the final chapter, along with some examples of genetically modified plants and animals.

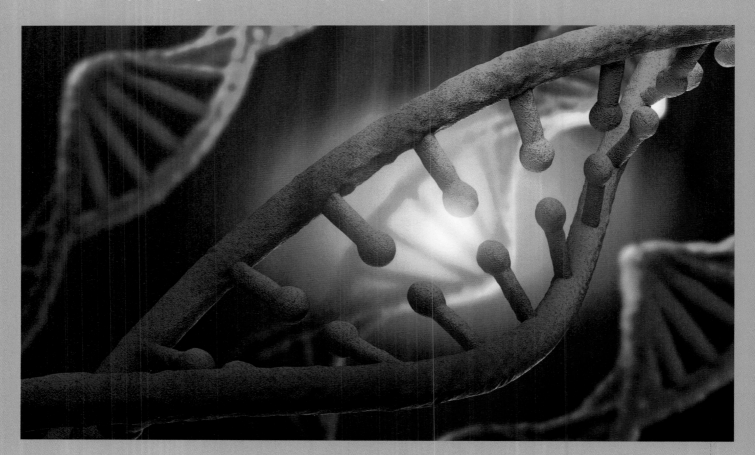

21 USING MICROORGANISMS

In this chapter we look at microorganisms that are grown in order to make products that are of use to humans.

LEARNING OBJECTIVES

- Understand the role of yeast in the production of food, including bread

- Investigate the rate of anaerobic respiration by yeast in different conditions

- Understand the role of bacteria (*Lactobacillus*) in the production of yoghurt

- Understand the use of an industrial fermenter and explain the need to provide suitable conditions for the growth of microorganisms in the fermenter, including aseptic precautions, nutrients, optimum temperature and pH, oxygenation, and agitation.

WHAT ARE MICROORGANISMS?

Microorganisms are living things that you can only see with the help of a microscope. The 'bodies' of most microorganisms are made of a single cell, although sometimes millions of cells are gathered together to form a colony. The colony of cells may then be visible to the human eye.

Microorganisms have critical roles to play in recycling the waste products of organisms, as well as recycling the organisms themselves when they die. Many types of microorganisms are studied because they cause disease in animals and plants. On the other hand, humans have made use of the great reproductive capacity of microorganisms to make useful products, such as food, drink and medicines.

There are several groups that we call microorganisms, including protoctists, bacteria, viruses and some fungi (see Chapter 2).

Figure 21.1 shows a few examples of the many types of microorganisms.

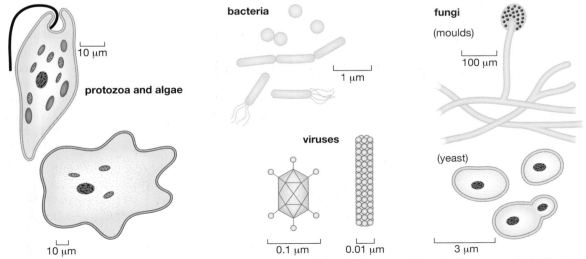

▲ Figure 21.1 Some examples of microorganisms. They are not drawn to the same scale. Notice the range of size, as shown by the scale bar alongside each organism. One micrometre (1 µm) is a millionth of a metre, or a thousandth of a millimetre.

FERMENTATION AND BIOTECHNOLOGY

Fermentation and **fermenters** are terms that you will encounter if you are involved in the growth of microorganisms. What do these words mean? Many microorganisms respire anaerobically (see Chapter 1), and originally, fermentation meant any anaerobic respiration process involving microorganisms, such as the fermentation of sugars by yeast. Nowadays, the word is used more generally to define other metabolic processes carried out by microorganisms, many of which are aerobic. Fermentation is normally used to make a useful product.

This overlaps with the definition of **biotechnology**, which means using any organisms (but mainly microbes) to make products that are useful to humans. Although the word itself is relatively new, humans have been using biotechnology processes for thousands of years without knowing it. Since ancient times, fermentation by yeast has been used to produce bread. Yoghurt is made by the action of bacteria on milk, and other bacteria and moulds are used in cheese manufacture.

Our ancestors used biotechnology to make products like these, but they did not understand *how* they were made and had no idea of the existence of microorganisms. Nowadays we understand what is happening when fermentation takes place, and can use biotechnology to produce not just foods but also a huge range of products, from medicines like penicillin to chemicals such as enzymes and fuels.

Modern biotechnology also allows us to alter the genes of microorganisms so that they code for new products. This is called **genetic engineering**. It is a topic that you will read about in Chapter 22.

TRADITIONAL BIOTECHNOLOGY – MAKING FOOD AND DRINKS

When yeast cells are deprived of oxygen, they respire anaerobically, breaking sugar down into ethanol and carbon dioxide:

glucose → ethanol + carbon dioxide

This process has been used for thousands of years to make bread, using a species of yeast called *Saccharomyces cerevisiae*.

MAKING DRINKS

Wine is made by using yeast to ferment sugars in grape juice. Commercial wine production takes place in large containers called vats which prevent air reaching the wine and ensure conditions remain anaerobic. Homemade wine is produced in small-scale fermenters fitted with an 'airlock', which allows carbon dioxide to escape but prevents the entry of oxygen (Figure 21.2). The alcohol increases in concentration until it kills the yeast cells, at which point fermentation stops.

Beer is made from barley. Barley contains starch rather than sugars so the starch needs to be broken down first. This happens by allowing the barley seeds to germinate. When they start to germinate they produce the enzyme amylase, which breaks down starch into the sugar maltose. Amylase breaks down starch into the sugar maltose. Later, the maltose from the seeds is fermented by yeast in a large open vat (Figure 21.3).

▲ Figure 21.2 A small fermenter for producing homemade wine. The U-shaped tube at the top is a water-filled airlock that prevents oxygen entering the fermenter but allows carbon dioxide to escape.

▲ Figure 21.3 A froth forming on the surface of the beer as yeast ferments the sugar to alcohol and carbon dioxide. The carbon dioxide in the froth prevents oxygen entering the mixture from the air – keeping conditions anaerobic.

MAKING BREAD

Yeast is also used to make bread. Wheat flour and water are mixed together and yeast added, forming the bread dough. Enzymes from the original cereal grains break down starch to sugars, which are respired by the yeast. Extra sugar may be added at this stage. In bread-making, the yeast begins by respiring aerobically, producing water and carbon dioxide. The carbon dioxide makes the dough rise. When the air runs out, conditions become anaerobic, so the yeast begins to respire anaerobically making ethanol and more carbon dioxide.

Later, when the dough is baked in the oven, the gas bubbles expand. This gives the bread a light, cellular texture (Figure 21.4). Baking also kills the yeast cells and evaporates any ethanol from the fermentation.

a   **b**

▲ Figure 21.4 (a) The 'holes' in this bread were produced by bubbles of carbon dioxide released from the respiration of the yeast. (b) Bread that is made without yeast is called unleavened bread. What is the difference in texture and appearance between leavened and unleavened bread?

ACTIVITY 1

▼ PRACTICAL: INVESTIGATING THE RATE OF ANAEROBIC RESPIRATION IN YEAST

! Safety Note: Wear eye protection and avoid skin contact with the indicator.

Some simple apparatus and materials can be used to investigate the rate of anaerobic respiration in yeast.

A small amount of water is gently boiled in a boiling tube to remove any air that is dissolved in the water. The water is allowed to cool, and a small amount of sugar (glucose or sucrose) is dissolved in the water. Finally, a little yeast is added and the mixture is stirred.

The apparatus is set up as shown in Figure 21.5.

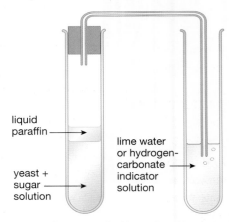

liquid paraffin

yeast + sugar solution

lime water or hydrogen-carbonate indicator solution

▲ Figure 21.5 Apparatus to test for carbon dioxide produced by anaerobic respiration in yeast.

A thin layer of liquid paraffin is added to the surface of the mixture, using a pipette. The boiled water ensures that there is no oxygen in the mixture, and the layer of paraffin stops any oxygen diffusing in from the air. A control apparatus is set up. This is exactly the same as that shown in Figure 21.5, except that boiled (killed) yeast is used instead of living yeast.

Both sets of apparatus are left in a warm place for an hour or two. The mixture with living yeast will be seen to produce gas bubbles. The gas passes through the delivery tube and into the indicator in the second boiling tube.

If this tube contains limewater, it will turn cloudy (milky). If it contains hydrogen carbonate indicator, the indicator will change from orange to yellow. This shows that the gas is carbon dioxide. The time taken for the indicator to change colour is recorded and compared with the control (which will not change).

(If the bung is taken out of the first boiling tube and the liquid paraffin removed using a pipette, the tube will smell of alcohol.)

This method can be used to test predictions, such as:

■ the type of sugar (glucose, sucrose, maltose etc.) affects the rate of respiration of the yeast

■ the concentration of sugar affects the rate of respiration of the yeast

■ how temperature affects the rate of respiration of the yeast.

The rate can be found by timing how quickly the indicator changes colour, or from the rate of production of bubbles of carbon dioxide. You could plan experiments to test these hypotheses.

MAKING YOGHURT

Yoghurt is milk that has been fermented by certain species of bacteria, called **lactic acid bacteria**. The effect of the fermentation is to turn the liquid milk into a semi-solid food with a sour taste.

To make yoghurt, milk is first pasteurised at 85–95 °C for 15–30 minutes, to kill any natural bacteria that it contains, then homogenised to disperse the fat globules. The milk is then cooled to 40–45 °C and inoculated with a starter culture of two species of bacteria, called *Lactobacillus* and *Streptococcus*.

These bacteria produce lactic acid, as well as starting to digest the milk proteins. The culture is kept at this temperature for several hours while the pH falls to about 4.4 (these are the optimum conditions for the bacteria). The mixture coagulates (thickens) as the drop in pH causes the milk proteins to denature and turn into semi-solids.

When fermentation is finished, the yoghurt is stirred and cooled to 5 °C. Flavourings, colourants and fruit may then be added before it is packaged for sale.

The drop in pH (as the yoghurt forms) gradually reduces the reproduction of the lactic acid bacteria (although it doesn't kill them). It also helps to prevent the growth of other microorganisms, and so preserves the nutrients in the milk. The steps in yoghurt production are summarised in the flow chart (Figure 21.6).

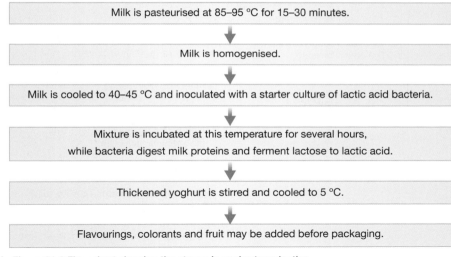

Milk is pasteurised at 85–95 °C for 15–30 minutes.

↓

Milk is homogenised.

↓

Milk is cooled to 40–45 °C and inoculated with a starter culture of lactic acid bacteria.

↓

Mixture is incubated at this temperature for several hours, while bacteria digest milk proteins and ferment lactose to lactic acid.

↓

Thickened yoghurt is stirred and cooled to 5 °C.

↓

Flavourings, colorants and fruit may be added before packaging.

▲ Figure 21.6 Flow chart showing the stages in yoghurt production.

INDUSTRIAL FERMENTERS

A **fermenter** is any vessel that is used to grow microorganisms used for fermentation. Even a baking tray containing a ball of dough could be defined as a fermenter!

Industrial fermenters are large tanks that can hold up to 200 000 dm³ of a liquid culture (Figure 21.7). They enable the environmental conditions such as temperature, oxygen and carbon dioxide concentrations, pH and nutrient supply to be carefully controlled so that the microorganisms will yield their product most efficiently. A simplified diagram of the inside of a fermenter is shown in Figure 21.8.

◀ Figure 21.7 An industrial fermenter holds hundreds of thousands of dm³ of a liquid culture.

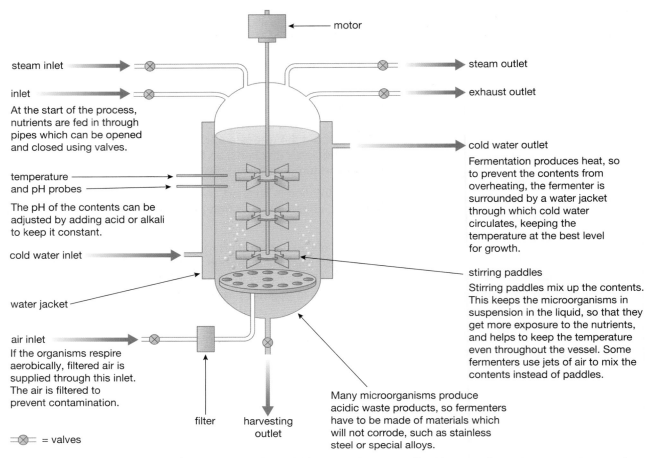

motor

steam inlet

inlet

At the start of the process, nutrients are fed in through pipes which can be opened and closed using valves.

temperature and pH probes

The pH of the contents can be adjusted by adding acid or alkali to keep it constant.

cold water inlet

water jacket

air inlet

If the organisms respire aerobically, filtered air is supplied through this inlet. The air is filtered to prevent contamination.

⊗ = valves

filter

harvesting outlet

steam outlet

exhaust outlet

cold water outlet

Fermentation produces heat, so to prevent the contents from overheating, the fermenter is surrounded by a water jacket through which cold water circulates, keeping the temperature at the best level for growth.

stirring paddles

Stirring paddles mix up the contents. This keeps the microorganisms in suspension in the liquid, so that they get more exposure to the nutrients, and helps to keep the temperature even throughout the vessel. Some fermenters use jets of air to mix the contents instead of paddles.

Many microorganisms produce acidic waste products, so fermenters have to be made of materials which will not corrode, such as stainless steel or special alloys.

▲ Figure 21.8 An industrial fermenter. Fermenters like this are used to make many products, such as the antibiotic penicillin.

Many microorganisms use an external food source from their growth medium to obtain energy. In doing this, they change substances in the medium. This is the modern meaning of fermentation. It is used by humans to make many important products. The use of microorganisms to make products useful to humans is called biotechnology.

When fermentation is completed, the products are collected through an outlet pipe. Before the fermenter is filled with new nutrients and culture, the inside of the tank and all the pipes must be cleaned and sterilised. This is usually done with very hot steam under high pressure.

If the inside of the fermenter and the new nutrients are not sterile, two problems are likely to develop. Firstly, any bacteria or fungi that manage to get in would compete with the organism in the culture, reducing the yield of product. Secondly, the product would become contaminated with waste products or cells of the 'foreign' organism. The methods used to prevent contamination by unwanted microorganisms such as filtering the air and sterilising the fermenter using steam, are known as 'aseptic precautions'.

CHAPTER QUESTIONS

More questions on growing useful organisms can be found at the end of Unit 6 on page 301.

SKILLS CRITICAL THINKING

1 Which of the following is *not* normally controlled in an industrial fermenter?

A Temperature   **B** Light intensity   **C** pH   **D** Oxygen

2 During the manufacture of yoghurt, the pH of the milk changes. What causes this change in pH?

A denaturing of the milk proteins

B pasteurisation of the milk

C the conversion of lactose into lactic acid

D yhe action of an enzyme in the milk

3 Yoghurt can be made at home by adding a small amount of old yoghurt to some fresh milk. What is present in the old yoghurt to cause this to happen?

A fat   **B** sugar   **C** protein   **D** bacteria

4 The diagram shows an industrial fermenter which is used to grow the mould *Penicillium* to make the antibiotic penicillin.

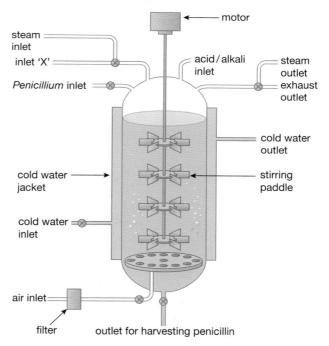

a Explain how the fermenter is sterilised before use.

SKILLS REASONING

b Why does air need to be pumped through the fermenter? Why is the air filtered?

SKILLS REASONING

c Explain how a steady temperature is maintained in the fermenter.

d What is supplied through the inlet marked 'X'?

e Suggest what would happen to the growth of *Penicillium* in the fermenter if the paddles stopped working.

SKILLS CRITICAL THINKING

5 Answer these questions about making yoghurt. Try at first to answer them without looking back to the section on page 285

a Why is the milk pasteurised at the start of the process?

b Why is the mixture of milk and bacteria incubated at 45 °C?

c What causes the milk to thicken?

d Why does fermentation eventually stop?

e Explain how making yoghurt is a way of preserving the nutrients from milk.

22 GENETIC MODIFICATION

In this chapter we will look at ways in which it is now possible to manipulate genes and produce genetically modified organisms – the science of 'genetic engineering'.

LEARNING OBJECTIVES

- Understand that the term 'transgenic' means the transfer of genetic material from one species to a different species

- Understand how restriction enzymes are used to cut DNA at specific sites and ligase enzymes are used to join pieces of DNA together

- Understand how plasmids and viruses can act as vectors, which take up pieces of DNA, and then insert this recombinant DNA into other cells

- Understand how large amounts of human insulin can be manufactured from genetically modified bacteria that are grown in a fermenter

- Understand how genetically modified plants can be used to improve food production

BIOLOGY ONLY

- Understand how cloned transgenic animals can be used to produce human proteins.

In Chapter 16 you saw that a gene is a section of a molecule of DNA that codes for the production of a protein. The coding strand of the DNA contains triplets of bases, each triplet coding for one amino acid. Different genes produce different proteins because each has a unique sequence of bases that codes for a unique sequence of amino acids (Figure 22.1).

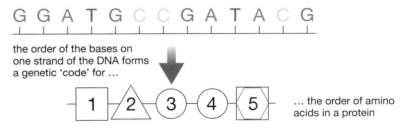

G G A T G C C G A T A C G

the order of the bases on one strand of the DNA forms a genetic 'code' for …

1   2   3   4   5   … the order of amino acids in a protein

▲ Figure 22.1 The role of DNA in protein synthesis

The protein that is produced could be:

- an enzyme that controls a particular reaction inside a cell or in the digestive system

- a structural protein like keratin in hair, collagen in skin or one of the many proteins found in the membranes of cells

- a protein hormone such as insulin

- a protein with a specific function such as haemoglobin or an antibody.

RECOMBINANT DNA

The production of **recombinant DNA** is the basis of genetic engineering. A section of DNA – a gene – is cut out of the DNA of one species and inserted into the DNA of another. This new DNA is called 'recombinant' DNA because the DNA from two different organisms has been 'recombined'. The organism that receives the gene from a different species is a **transgenic** organism.

KEY POINT

A transgenic organism is one that contains a gene that has been introduced from another species. For example, some bacteria have had human genes transferred to them that allow them to make human insulin. Some sheep secrete the protein alpha-1-antitrypsin (AAT) in their milk because they have the human gene that directs the manufacture of this substance. Because they contain 'foreign' genes, they are no longer quite the same organisms. They are transgenic.

The organism receiving the new gene now has an added capability. It will manufacture the protein that the new gene codes for. For example, a bacterium receiving the human gene that codes for insulin production will make human insulin. If these transgenic bacteria are cultured by the billion in a fermenter, they become a factory for making human insulin.

PRODUCING GENETICALLY MODIFIED (TRANSGENIC) BACTERIA

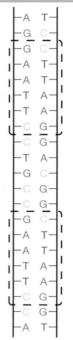

▲ Figure 22.2 Part of a DNA molecule containing the base sequence G-A-A-T-T-C. Notice that the sequence is present on both strands, but running in opposite directions.

The breakthrough in being able to transfer DNA from cell to cell came when it was found that bacteria have two sorts of DNA – the DNA found in their bacterial 'chromosome' and much smaller circular pieces of DNA called **plasmids** (see Unit 1, Figure 2.10).

Bacteria naturally 'swap' plasmids, and biologists found ways of transferring plasmids from one bacterium to another. The next stage was to find molecular 'scissors' and a molecular 'glue' that could cut out genes from one molecule of DNA and stick them back into another. Further research found the following enzymes that were able to do this.

- **Restriction endonucleases** (usually shortened to **restriction enzymes**) are enzymes that cut DNA molecules at specific points. Different restriction enzymes cut DNA at different places. They can be used to cut out specific genes from a molecule of DNA.

- **Ligases** (or DNA ligases) are enzymes that join the cut ends of DNA molecules.

Each restriction enzyme recognises a certain base sequence in a DNA strand. Wherever it encounters that sequence, it will cut the DNA molecule. Suppose a restriction enzyme recognises the base sequence G-A-A-T-T-C. It will only cut the DNA molecule if it can 'see' the base sequence on both strands. Figure 22.3 illustrates this.

Some restriction enzymes make a straight cut and the fragments of DNA they produce are said to have 'blunt ends' (Figure 22.3(a)). Other restriction enzymes make a staggered cut. These produce fragments of DNA with overlapping ends with complementary bases. These overlapping ends are called 'sticky ends' because fragments of DNA with exposed bases are more easily joined by ligase enzymes (Figure 22.3(b)).

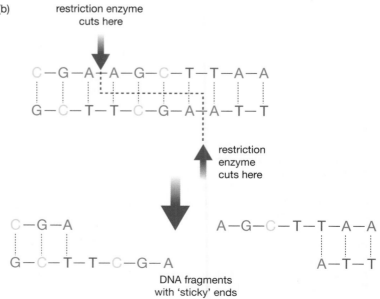

▲ Figure 22.3 How restriction enzymes cut DNA. (a) Forming blunt ends. (b) Forming sticky ends.

EXTENSION WORK

There is a lot more to producing recombinant DNA and transgenic bacteria than is shown here. You could carry out an some research to find out more about this topic.

Biologists now had a method of transferring a gene from any cell into a bacterium. They could insert the gene into a plasmid and then transfer the plasmid into a bacterium. The plasmid is called a **vector** because it is the means of transferring the gene. The main processes involved in producing a transgenic bacterium are shown in Figure 22.4.

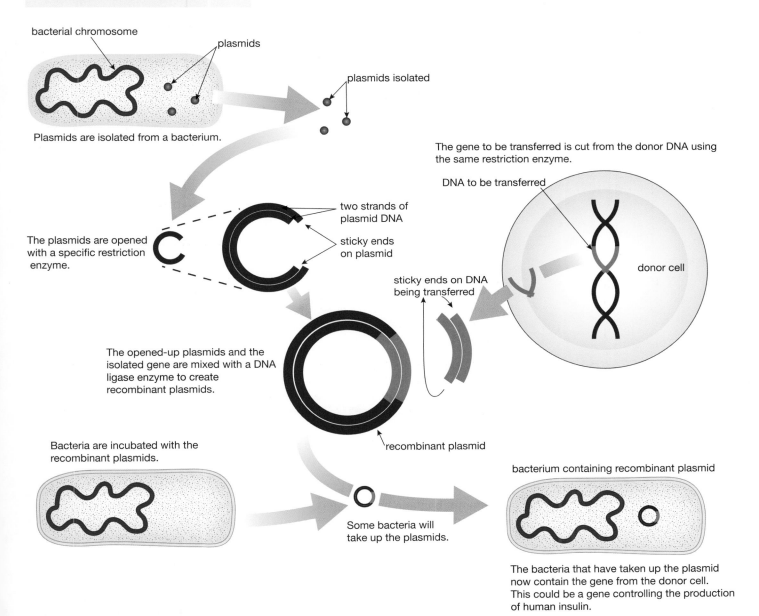

bacterial chromosome

plasmids

Plasmids are isolated from a bacterium.

plasmids isolated

The gene to be transferred is cut from the donor DNA using the same restriction enzyme.

DNA to be transferred

donor cell

The plasmids are opened with a specific restriction enzyme.

two strands of plasmid DNA

sticky ends on plasmid

sticky ends on DNA being transferred

The opened-up plasmids and the isolated gene are mixed with a DNA ligase enzyme to create recombinant plasmids.

recombinant plasmid

Bacteria are incubated with the recombinant plasmids.

bacterium containing recombinant plasmid

Some bacteria will take up the plasmids.

The bacteria that have taken up the plasmid now contain the gene from the donor cell. This could be a gene controlling the production of human insulin.

▲ Figure 22.4 Stages in producing a transgenic bacterium.

Another vector that has been used to introduce foreign DNA into bacterial cells is the **bacteriophage**. A bacteriophage, or 'phage', is a virus that attacks a bacterium. It does this by attaching to the cell wall of the bacterium and injecting its own DNA into the bacterial cell (Figure 22.5). This DNA becomes incorporated into the DNA of the host cell, and eventually causes the production of many virus particles.

If a foreign gene can be inserted into the DNA of the virus, the virus will inject it into the bacterium along with its own genes. Viruses were used as vectors in the early days of genetic modification, but most gene transfer is now carried out using plasmids.

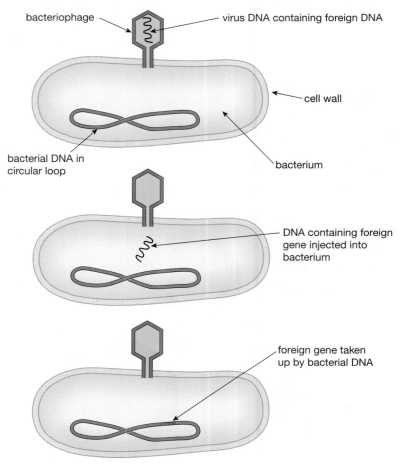

▲ Figure 22.5 A bacteriophage attacking a bacterial cell.

MAKING USE OF GENETICALLY MODIFIED BACTERIA

Different bacteria have been genetically modified to manufacture a range of products. Once they have been genetically modified, they are cultured in fermenters to produce large amounts of the product (see Chapter 21). Some examples are described below.

> **DID YOU KNOW?**
> More insulin is required every year because the number of diabetics increases worldwide each year and diabetics now have longer life spans.

1. **Human insulin** People suffering from diabetes need a reliable source of insulin. Before the use of genetic engineering to make human insulin, the only insulin available was extracted from the pancreases of other animals such as cattle or pigs. This is not quite the same as human insulin and does not give the same level of control of blood glucose levels.

2. **Enzymes for washing powders** Many stains on clothing are biological. Blood stains are largely proteins, grease marks are largely lipids. Enzymes can digest these large, insoluble molecules into smaller, soluble ones. These then dissolve in the water. Amylases digest starch, proteases digest proteins and lipases digest lipids. Bacteria have been genetically engineered to produce enzymes that work at higher temperatures, allowing even faster and more effective action.

3. **Enzymes in the food industry** One bacterial enzyme used in the food industry is glucose isomerase. This enzyme catalyses a reaction which converts glucose into a similar sugar called fructose. Fructose is much sweeter than glucose and so less is needed to sweeten foods. This has two advantages – it saves money (since less is used) and it means that the food contains less sugar and is healthier.

4. **Human growth hormone** The pituitary gland of some children does not produce enough of this hormone and they show a slow rate of growth. Injections of growth hormone from genetically modified bacteria restore normal growth patterns.

5. **Bovine somatotrophin** (BST) (a growth hormone in cattle) This hormone increases the milk yield of cows and increases the muscle (meat) production of bulls. Giving injections of BST to dairy cattle can increase the milk yield by up to 10 kg per day. To do this they need more food, but this increased cost is more than offset by the increased income from the increased milk yield (Table 22.1).

6. **Human vaccines** Bacteria have been genetically modified to produce the antigens of the hepatitis B virus. This is used in the vaccine against hepatitis B. The body makes antibodies against the antigens but there is no risk of contracting the actual disease from the vaccination.

Table 22.1 Effects of BST on milk yield.

	Feed / kg per day	Milk output / kg per day	Milk to feed ratio
without BST	34.1	27.9	0.82
with BST	37.8	37.3	0.99

Since the basic techniques of transferring genes were developed, many unicellular organisms have been genetically modified to produce useful products. Techniques for transferring genes into plants and animals have also been developed.

PRODUCING GENETICALLY MODIFIED PLANTS

The gene technology described so far can transfer DNA from one cell to another cell. In the case of bacteria, this is fine – a bacterium only has one cell. But plants have billions of cells and to genetically modify a plant, each cell must receive the new gene. So, any procedure for genetically modifying plants has two main stages:

■ introducing the new gene or genes into plant cells

■ producing whole plants from just a few cells.

Biologists initially had problems in inserting genes into plant cells. They then discovered a soil bacterium called *Agrobacterium*, which regularly inserts plasmids into plant cells. Now that a vector had been found, the rest became possible. Figure 22.6 outlines one procedure that uses *Agrobacterium* as a vector.

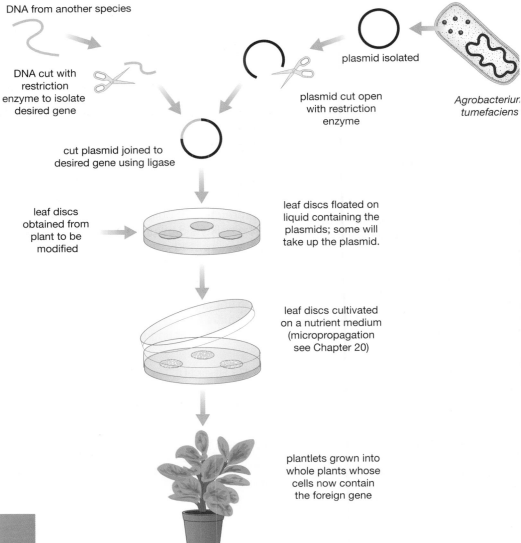

DNA from another species

DNA cut with restriction enzyme to isolate desired gene

plasmid isolated

plasmid cut open with restriction enzyme

Agrobacterium tumefaciens

cut plasmid joined to desired gene using ligase

leaf discs obtained from plant to be modified

leaf discs floated on liquid containing the plasmids; some will take up the plasmid.

leaf discs cultivated on a nutrient medium (micropropagation see Chapter 20)

plantlets grown into whole plants whose cells now contain the foreign gene

▲ Figure 22.6 Genetically modifying plants using *Agrobacterium*.

This technique cannot be used on all plants. *Agrobacterium* will not infect cereals and so another technique was needed for these. The 'gene gun' was invented. This is, quite literally, a gun that fires a golden bullet (Figure 22.7). Tiny pellets of gold are coated with DNA that contains the desired gene. These are then 'fired' directly into plant tissue. Research has shown that if young, delicate tissue is used, there is a good uptake of the DNA. The genetically modified tissue can then be grown into new plants using the same micropropagation techniques as those used in the *Agrobacterium* procedure. The gene gun has made it possible to genetically modify many cereal plants as well as tobacco, carrot, soybean, apple, oilseed rape, cotton and many others. There is, however, much debate about the rights and wrongs of GM crops. You may like to carry out some further research on this topic.

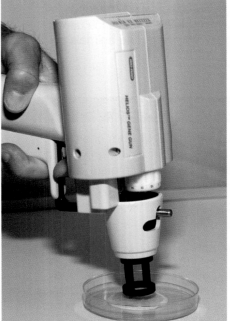

▲ Figure 22.7 A gene gun

MAKING USE OF GENETICALLY MODIFIED PLANTS

Large numbers of genetically modified plants are already available to plant growers and farmers.

Some fruit and vegetables have been engineered to have extended 'shelf lives' – they last longer before they start to go bad. Other crop plants have been modified to be resistant to herbicides (weedkillers). This allows farmers to spray herbicides at times when they will have maximum effect on the weeds, without affecting the crop plant.. There are concerns that this will encourage farmers to be less careful in their use of herbicides. In another example, genes from Arctic fish that code for an 'anti-freeze' in their blood have been transferred to some plants to make them frost resistant.

The gene gun allowed biologists to produce genetically modified rice called 'golden rice' (Figure 22.8). This rice has three genes added to its normal DNA content. Two of these come from daffodils and one from a bacterium. Together, these genes allow the rice to make beta-carotene – the chemical that gives carrots their colour. It also colours the rice, hence the name 'golden rice'. More importantly, the beta-carotene is converted to vitamin A when eaten. This could save the eyesight of millions of children in less economically developed countries, who go blind because they do not have enough vitamin A in their diet.

EXTENSION WORK

Golden rice sounds like a good idea, but there have been several problems with it. Some people believe that there are ethical and environmental reasons why golden rice should not be grown and that it is better to provide other, natural crops containing enough beta-carotene. You could carry out some research into the advantages and disadvantages of golden rice.

▲ Figure 22.8 Golden rice

KEY POINT

What if you could receive a 'vaccination' every time you ate a banana? Scientists are researching the possibility of transferring the genes that produce the antigens for certain diseases to bananas. If they succeed, then when you eat the banana, the antigens will stimulate an immune response. There will be no risk of you catching the disease – and no need to have needles stuck in you!

Genetically modified plants are also helping humans to resist infection. Biologists have succeeded in modifying tobacco plants and soybeans to produce antibodies against a range of infectious diseases. If these can be produced on a large scale, they could be given to people who are failing to produce their own antibodies. Other modified tobacco plants produce the hepatitis B antigens that could be used as the basis for a vaccine. There is always a risk with a vaccine containing viruses that they may somehow become infectious again. This could not happen with a vaccine containing only plant-produced antigens.

Besides the specific examples given, research into the genetic modification of plants provides, or hopes to provide, plants with:

- increased resistance to a range of pests and pathogens
- increased heat and drought tolerance
- increased salt tolerance
- a better balance of proteins, carbohydrates, lipids, vitamins and minerals.

In addition, some genetically modified oilseed rape plants will be used in large-scale production of biodegradable plastics and anti-coagulants.

One of the biggest achievements would be to modify crop plants like cereals and potatoes to allow nodules of nitrogen-fixing bacteria to form on their roots. At the moment only legumes (peas, beans and other plants with seeds in 'pods') can do this. Biologists know that the ability is genetically controlled. So far, they have not been able to transfer these genes to other plants. If they could, vast areas of infertile soil would be able to yield good crops of cereals without the need to use large quantities of fertilisers.

The bacteria in the root nodules would obtain nitrogen from the air in the soil and 'fix' it in a more usable form (usually ammonia). By doing this, they would make a supply of usable nitrogen available to the plants. The plants would convert this into plant protein and use the protein for growth. The cost of producing these crops would decrease dramatically.

PRODUCING GENETICALLY MODIFIED ANIMALS

Producing genetically modified animals presents some of the same problems as those connected with modifying plants. Animals, like plants, are multicellular. It is not enough simply to transfer a gene to a cell. That cell must then grow into a whole organism. The plasmid technology used to create genetically modified plants depends on the modified cells being grown into whole plants using micropropagation. No such micropropagation techniques exist for animals.

Scientists researching the production of genetically modified animals had to find other techniques. The most successful involves injecting DNA directly into a newly fertilised egg cell. This develops into an embryo, then an adult (Figure 22.9).

Research of this kind can produce beneficial results similar to those achieved by genetically modifying plants, such as:

- manufacture of human proteins, such as antibodies, blood clotting factors or alpha-1-antitrypsin (AAT) – see below

- increased production of a particular product, e.g. higher milk yield in cows, greater muscle mass in animals used for meat

- increased resistance to disease

- production of organs for transplantation (xenotransplantation).

DID YOU KNOW?

'Xenotransplantation' means transplanting organs from other animals into humans. Transgenic pigs have been produced with genes that code for the main human 'marker antigens'. The cells of the pig's organs therefore have these human antigens on their surface and the organs would be less likely to be rejected by a recipient. If this became possible on a large scale, it could help to overcome the shortage of donor organs for transplantation. However, many people have ethical concerns about using animal organs for transplanting into humans.

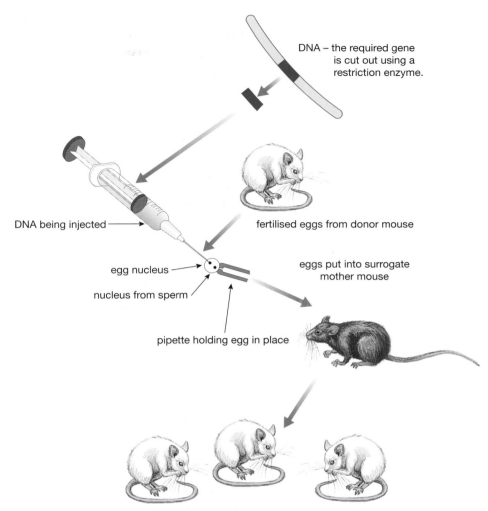

DNA – the required gene is cut out using a restriction enzyme.

DNA being injected

fertilised eggs from donor mouse

egg nucleus

nucleus from sperm

eggs put into surrogate mother mouse

pipette holding egg in place

baby mice are tested for the presence of the gene

▲ Figure 22.9 The procedures used in producing genetically modified animals.

BIOLOGY ONLY

CLONING TRANSGENIC ANIMALS

There is also the potential to **clone** genetically modified animals (see Chapter 20). This would make it possible to produce large numbers of animals, all genetically identical. The cloned animals could then be used for production of human proteins.

For example, the human protein alpha-1-antitrypsin (AAT) is involved in the immune response. When phagocytes fight an infection, they produce the enzyme trypsin. AAT is produced as a counter measure to stop trypsin attacking normal tissues.

KEY POINT

In the lungs, a deficiency of AAT causes a rare type of emphysema with symptoms similar to the emphysema caused by smoking – see Chapter 3.

Some people carry a mutation that means they do make enough AAT. The trypsin then damages normal tissues, mainly in the liver and lungs.

Transgenic sheep have been genetically modified to produce AAT in their milk. It may be possible to use this to treat people with AAT deficiency. If the sheep were to be cloned, this would increase the amount of AAT available.

END OF BIOLOGY ONLY

LOOKING AHEAD – THE POLYMERASE CHAIN REACTION

The polymerase chain reaction (PCR) is a laboratory process that amplifies DNA. This means that it rapidly produces a large number of copies of a piece of DNA, starting from as little as one molecule. The invention of PCR revolutionised gene technology and is used in virtually every application of the science.

Each cycle of PCR doubles the amount of DNA, producing an exponential increase in the numbers of copies (Figure 22.10). Thirty cycles produces over a billion copies of the original DNA.

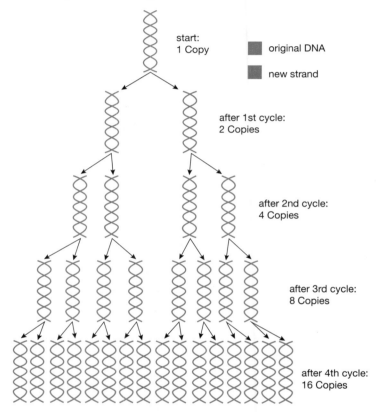

▲ Figure 22.10 PCR produces an exponential increase in the number of copies of a piece of DNA.

The procedure uses short lengths of single-stranded DNA called primers, which are precisely complementary to one end of each strand of the DNA you wish to amplify. The DNA is heated to separate the strands and the primers added to the ends of each strand. An enzyme called DNA polymerase builds up two new complimentary strands by adding nucleotides - one at a time - along each strand.

Nowadays, PCR is an automated process and can be performed in as little as 2–3 hours in a PCR machine. You just place small tubes containing the DNA sample, primers, free nucleotides, DNA polymerase and a buffer solution in the machine, switch on, and wait for the results.

CHAPTER QUESTIONS

More questions on gene technology can be found at the end of Unit 6 on page 301.

SKILLS   CRITICAL THINKING

1 Why is it relatively easy to genetically modify bacteria?

   A they reproduce slowly and accurately

   B they are able to take up plasmids

   C they only contain a single chromosome

   D they make restriction enzymes

2 The statements below show some stages in the production of human insulin from genetically modified bacteria.

   1. DNA for insulin inserted into plasmids

   2. bacteria cloned

   3. plasmids inserted into bacteria

   4. DNA for insulin cut out using restriction enzyme

   Which of the following shows the correct sequence of steps in the process?

   A 2 → 1 → 4 → 3

   B 4 → 2 → 3 → 1

   C 4 → 1 → 3 → 2

   D 2 → 3 → 4 → 1

3 Which of the following genetically modified plants have not been produced yet?

   A rice that is rich in beta-carotene to make vitamin A

   B plants resistant to weedkiller

   C plants containing an anti-freeze

   D cereals with nitrogen-fixing bacteria in root nodules

4 Which of the following enzymes is used to join together pieces of DNA?

   A ligase

   B DNA polymerase

   C protease

   D restriction enzyme

 ANALYSIS

5 The diagram shows the main stages in transferring the human insulin gene to a bacterium.

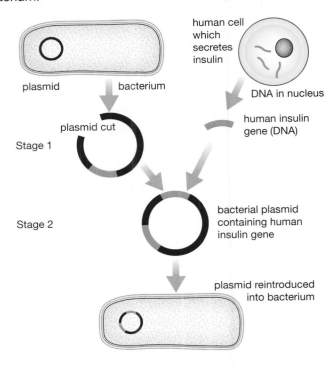

a Name the enzymes used at stages 1 and 2.

SKILLS CRITICAL THINKING

b What is the role of the plasmid in this procedure?

c How would the insulin-producing bacteria be used to produce significant amounts of insulin?

SKILLS REASONING

d Why is the insulin produced this way preferred to insulin extracted from other animals?

SKILLS INTRAPERSONAL

6 Carry out some research to find out more about the use of transgenic animals for organ transplantation. What are the arguments for and against its use? What are the ethical objections?

SKILLS CRITICAL THINKING

7 Producing genetically modified plants and animals is more complex than producing genetically modified bacteria.

a Describe two ways in which genes can be introduced into plant cells.

b How are these genetically modified cells used to produce whole organisms?

c What sort of animal cell is genetically modified and then used to produce the whole organism?

SKILLS INTRAPERSONAL

8 Write an essay about the importance of genetic engineering. In your essay you should make reference to:

a important potential benefits resulting from genetic engineering in animals and plants

b concerns about the risks resulting from genetic engineering in animals and plants.

UNIT QUESTIONS

SKILLS   CRITICAL THINKING   **1**

Some experimental sheep were genetically modified, using a human gene. The gene codes for a protein involved in blood clotting, called factor IX.

The gene for factor IX was cut from human DNA and injected into a sheep body cell. The nucleus of this body cell was transferred to an enucleated sheep ovum. The ovum was cultured in a dish and then implanted in the uterus of a surrogate sheep to complete its development.

Two lambs produced by this method grew to maturity. They produced milk from which Factor IX could be extracted.

a Name an enzyme that can be used to cut DNA.   **(1)**

b What is an *enucleated ovum*?   **(2)**

c The sheep produced by this method are called *transgenic*. What is meant by this term?   **(1)**

SKILLS   REASONING

d The transgenic sheep were reproduced by *cloning*. What are the advantages of producing the sheep in this way?   **(3)**

e Factor IX can be used to treat people with a blood clotting disorder called haemophilia. Why is important to have blood that clots?   **(2)**

SKILLS   CRITICAL THINKING

f Blood contains plasma, white blood cells, red blood cells and platelets. Which *two* components of the blood are involved with blood clotting?   **(2)**

Total 11 marks

SKILLS   CRITICAL THINKING   **2**

Some genes are transferred from one plant species to another. These genes are called 'jumping genes'. Environmentalists are concerned that genetically modified plants may transfer some of their genes to wild species. Explain their concern about genetically modified plants that:

a have genes that make them resistant to herbicides (weedkillers)   **(2)**

b have genes that make them resistant to pests   **(2)**

c have genes that increase the yield of the crop they can produce.   **(2)**

Total 6 marks

3 The diagram shows a fermenter of the type used to grow microorganisms used to produce human insulin. Insulin is used to treat people with diabetes.

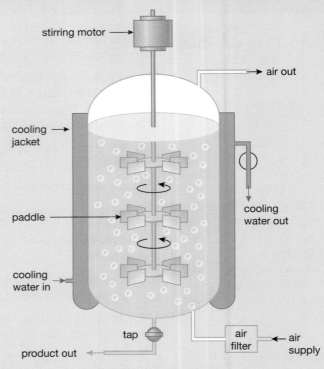

stirring motor

air out

cooling jacket

cooling water out

paddle

cooling water in

tap

air filter

air supply

product out

a Explain why air is pumped through the fermenter. (2)

b Why is it necessary to keep the temperature in the fermenter constant? (2)

c Suggest one other condition that must be kept constant inside the fermenter. (1)

d The fermenter is sterilised by steam. Suggest why this is better than using disinfectants. (2)

e Insulin can also be obtained from animal pancreases. Suggest two advantages of using microorganisms to produce the insulin needed by diabetics. (2)

Total 9 marks

4 a Explain the meaning of the following terms:
   i a plasmid
   ii a bacteriophage. (6)

b Explain the importance of the following enzymes in genetic engineering:
   i restriction enzymes
   ii ligase enzymes. (6)

Total 12 marks

APPENDIX A: A GUIDE TO EXAM QUESTIONS ON EXPERIMENTAL SKILLS

Copies of official specifications for all Edexcel qualifications may be found on the Edexcel website, www.edexcel.com. Past papers, marks schemes and examiners' reports are also available on the website.

WHY IS THIS APPENDIX IMPORTANT?

This appendix is designed to help you gain the 20% of marks allotted to the questions on experimental skills. These skills are tested in both Paper 1 and Paper 2, mixed up with theory questions. The practicals that you should know about are included in the biology content of the specification and fully described in this book. In the exams you will be asked questions based on these, although the apparatus shown may be slightly different. However, you will also be given questions about unfamiliar biological investigations. Don't be worried by these – you are not expected to have carried them out, just to be able to apply your knowledge and understanding to interpret experimental design or to analyse data.

Questions in the exams will generally cover the following areas:

- understanding safety precautions
- recognising apparatus and understanding how to use it
- recalling tests for certain substances
- manipulating data
- plotting graphs
- understanding experimental design
- planning experiments.

1. SAFETY PRECAUTIONS

As part of a question, you may be asked to comment on appropriate safety precautions. The most obvious precaution is to **wear eye protection**. Eye protection must be worn whenever chemicals or Bunsen burners are used. This will apply to many experiments or investigations, with a few exceptions such as fieldwork. Some other examples of safety precautions, and the reason they are taken, are:

Precaution	Reason
wash hands after handling biological material, such as plants, soil etc.	to avoid contamination
keep flammable liquids such as ethanol away from a naked flame	to avoid the ethanol catching fire
take care with fragile glassware such as pipettes, microscope cover slips, etc.	to avoid cutting yourself
do not touch electrical apparatus (e.g. a microscope with built-in lamp) with wet hands	to avoid getting an electric shock
use a water bath to heat a test tube of water, rather than heating it directly in a Bunsen flame	to avoid the heated liquid jumping out of the tube

2. RECOGNISING APPARATUS AND DEMONSTRATING AN UNDERSTANDING OF HOW TO USE THEM

One of the simplest types of question in the exams will require you to recognise common pieces of laboratory apparatus, such as a Bunsen burner, thermometer, measuring cylinder and stopwatch. You will also need to recognise particular 'biological' apparatus such as quadrat, microscope, bench lamp and so on, and how they are used. If you have been able to carry out most of the practical work in the specification, this kind of question is very straightforward, for example:

(a) What is the name of this piece of apparatus? **(1 mark)**

(b) Draw a line on the apparatus to show a volume of 30 cm^3 **(1 mark)**

Note: Use a pencil to draw any lines on diagrams, graphs etc. Then you can rub it out if you make a mistake. On this measuring cylinder a horizontal line should be drawn half way between the 20 and 40 cm^3 marks – it's that easy!

3. RECALLING TESTS FOR SUBSTANCES

There are really only five chemical tests in the specification. These are for:

1. starch – using iodine solution
2. glucose (or 'reducing sugar') – heating with Benedict's solution
3. protein – using the biuret test
4. fat (lipid) – using the emulsion test
5. carbon dioxide – using hydrogen carbonate indicator.

You should know the starch test (page 58) and how leaves can be tested for starch (page 136).

The tests for glucose, protein and fat are on page 58 The effects of carbon dioxide on hydrogen carbonate indicator are described on page 141.

Once you have learnt these, the questions are easy! For example:

John decided to test some milk for glucose.

(a) Describe the test he would do. **(2 marks)**

(b) What result would he see if glucose was present? **(1 mark)**

(c) Suggest how he might use the results to say how much glucose was present. **(1 mark)**

Note: You might get one mark for 'Use Benedict's solution', one for 'heating' and one for the colour produced (orange or similar). In (c) the depth of colour would show how much glucose was present.

4. MANIPULATING DATA

Some questions in an exam will involve manipulating data. This means you will be provided with some results from an experiment, and have to process it in various ways, such as:

- putting raw data from a notebook into an ordered table
- counting numbers of observations
- summing totals
- calculating an average
- calculating a missing value
- identifying anomalous results.

For example:

Kirsty monitored how her temperature changed during exercise. She took her temperature before she started the exercise, and every two minutes during the exercise. Here are her results:

Before exercise = 36.4 °C   At 12 minutes = 37.6 °C

37.3 °C after 10 mins, 2 mins = 36.8 °C.

After 4 mins = 37.1 °C, 6 min = 37.2 °C, 8 min = 36.9 °C

(a) Organise Kirsty's results into a table. **(4 marks)**

(b) Identify the time when an anomalous temperature reading was taken. **(1 mark)**

Marks for the answer are awarded like this:

(a)

Units in → header row *(1 mark)*

Readings → in order *(1 mark)*

Time / min	Temperature / °C
0	36.4
2	36.8
4	37.1
6	37.2
8	36.9
10	37.3
12	37.6

← Table with time and temperature headings *(1 mark)*

Two columns *(1 mark)*

(b) The anomalous temperature reading was taken at 8 minutes. (Note: 'anomalous' means that the result doesn't fit into the pattern of the other results. In this case the temperature reading at 8 minutes is lower than expected. You might have to circle an anomalous result in a table, or on a graph – see below.)

You may have to calculate totals, averages, or missing values from a table, for example in the table below, which shows the rate of germination when 25 seeds were placed in Petri dishes containing three different solutions:

Solution	Number of seeds germinated				Percentage germinated / %
	Dish 1	Dish 2	Dish 3	Total	
A	24	23	24	?	?
B	20	19	22	61	81.3
C	11	5	?	23	30.7

(a) Calculate the total number of seeds that germinated in dishes 1, 2 and 3 in solution A. **(1 mark)**

(b) Calculate the percentage of seeds that germinated in solution A. Show your working. **(2 marks)**

(c) Calculate the missing value for seeds that germinated in dish 3, solution C. **(1 mark)**

The answers are:

(a) Total = 24 + 23 + 24 = 71

(b) % = (total germinated / total number of seeds) × 100 = (71 / 75) × 100 = 94.7% (Use the same number of decimal places as in the given answers.)

(c) Missing value = 23 − (11 + 5) = 7

This is very straightforward arithmetic: just be careful!

5. PLOTTING GRAPHS

In biology, unlike in chemistry or physics, a line graph is usually constructed by joining the points with straight lines. It is rare to have data that produce a straight line, and 'best fit' lines are not usually appropriate either.

Take the following example:

Concentration of sucrose solution / mol per dm^3	Change in mass of potato tissue / %
0.0	+25.0
0.2	+16.4
0.4	+4.0
0.6	−5.0
0.8	−6.7
1.0	−12.2

The graph of these data looks like this:

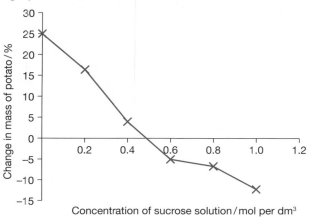

It is important to use a sharp pencil for drawing graphs. A pen is no good, and if you make a mistake with a pen you can't rub it out! Remember to label the axes, and give units.

You will probably have to describe the pattern shown by the graph, or identify a point on the graph where something happens. For instance in the graph above, the potato gains mass when it is placed in low concentrations of sucrose solution (below 0.5 mol per dm^3) and loses mass when in high concentrations (above 0.5 mol per dm^3). At 0.5 mol per dm^3 there is no gain or loss in mass. You'll probably have to answer questions about the biology behind the graph too: this one's all about osmosis.

Note that in many cases with 'biological' graphs the line doesn't pass through the 0, 0 coordinates.

You may also have to draw a bar chart from data you are given.

6. EXPERIMENTAL DESIGN

Expect questions on various aspects of experimental design, especially:

■ suggesting a suitable **Control** for an experiment

■ the meaning of **controlled variables** and how to control them

■ how to improve accuracy and precision

■ how to improve reliability.

(Note that here we are using a capital 'C' for Control, to distinguish it from 'controlled variables')

In a controlled experiment, only one factor is changed at a time. For example, imagine that you are carrying out an investigation into the effect of light intensity on the rate of photosynthesis of a piece of water plant, using this apparatus:

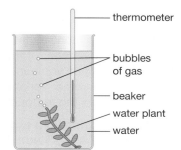

The factor that you want to change is the light intensity. You can do this by moving the lamp nearer to the beaker, or further away. But it is important that all other factors are kept constant. For example the temperature must be the same for all light intensities. You can check that it is by reading the thermometer. Moving the lamp closer might make the water in the beaker warmer – you would need to check that this doesn't happen. If the water temperature did increase, it might affect the rate of photosynthesis too. In that case, one of the important variables *hasn't* been controlled.

Other factors would need to be kept constant too, such as:

1. the size of the water plant (use the same piece throughout the experiment)

2. the wattage of the bulb in the lamp

3. the background lighting in the room (turn the lights off so the only light is from the lamp).

These are called **controlled variables**, because you, the experimenter, are controlling them (keeping them from changing) so that they can't affect the results.

Some biological experiments involve the use of a Control. This is an experimental set-up where the key factor is missing. For example, if you were investigating the effect of the enzyme amylase on starch, the experiment would involve mixing amylase with starch to see if the amylase breaks the starch down. It is possible (but unlikely!) that the starch might break down on its own, so a Control would be a tube with just starch, to check whether this happens. (A better control would be starch with *boiled* amylase, which is as near as possible to the experimental set-up but without the amylase being able to act as an enzyme.)

It is difficult to control variables in fieldwork investigations, such as when sampling with quadrats. For example, imagine you were comparing plants on and off a path using quadrats. Here you can control only certain variables, such as season and time of day when the sampling took place, and the size of quadrat used. But because this is not a laboratory investigation, you have no control over many variables, such as soil composition, water content, amount of trampling etc. All you can do is control the ones you can. However, you might be able to comment on other variables which it is impossible to control, such as the temperature or light intensity.

Hopefully, findings from experiments should be *accurate*, *precise* and *reliable*. Students often confuse these terms.

Accuracy means how close an experimental result is to its true value. Accuracy of the results depends mainly on the methods you use to obtain them. For example in the photosynthesis experiment above, the method of 'counting bubbles' is probably not a very accurate measure of photosynthesis, for more than one reason. Firstly, the bubbles might be different sizes; secondly, they might contain different mixtures of gases (not just oxygen). A more accurate method would be to collect the gas in a gas syringe, measure its volume, and then analyse it for its oxygen content. Clearly this is beyond what you could be expected to do for an experiment for International GCSE. Sometimes you have to accept that accuracy will be limited by the apparatus available. But you can always comment on how accuracy could be improved, and this might be an answer to part of an exam question.

Precision relates to the smallest division on the scale of the measuring instrument you are using. For example, a thermometer that reads to 0.1 °C is more precise than one that only reads to 1 °C. However, you have to be sensible about this when choosing suitable measuring instruments to use. If you are using a water bath consisting of a beaker of water heated by a Bunsen burner, a thermometer that reads to 0.1 °C is too precise. You wouldn't expect to be able to maintain the temperature any better than ±1 °C, so it is good enough to use a thermometer that reads to that degree of precision. Precision is also shown by the number of digits you give in a measurement, e.g. 9.998 g is more precise than 10 g.

Reliability is a measure of how similar the results are, if you carried out the same experiment several times. It tells you how confident you can be that your results are correct. For example, imagine you measured the

bubble rate in the photosynthesis experiment eight times with two different sets of apparatus and obtained these results:

	Apparatus 1	Apparatus 2
	Bubbles per minute	Bubbles per minute
1	26	14
2	25	19
3	27	52
4	22	44
5	28	12
6	25	10
7	26	37
8	26	21
Mean	25.6	26.1

Which mean value in the table is a more reliable measure of the bubble rate? You can tell that the mean for Apparatus 1 is more reliable, because the individual readings are more consistent. The only way you can tell if results are reliable is to carry out repeats.

You should make sure that you understand the difference between accuracy, precision and reliability.

7. PLANNING AN INVESTIGATION

You may have to give a short account of how you would carry out an investigation. This may be one based on an experiment in the book, or it might involve an unfamiliar situation. Take this example:

Describe an investigation you could carry out to find out the effect of changing the concentration of the enzyme amylase on the rate of starch digestion. **(6 marks)**

This is similar to the experiment described on page 58 in this book, but involves changing the concentration of amylase, rather than the temperature. In any question like this, you have to answer these points.

- What are you going to change?
- What are you going to keep constant (Control) during the experiment?
- What are you going to measure, and how?
- How are you going to check that the results are reliable?

A good answer would be:

'I will make up 5 different concentrations of amylase solution (e.g. 1%, 2%, 3%, 4% and 5%) using the same source of enzyme, such as fungal amylase. I will mix 10 cm³ of each concentration with 10 cm³ of 1% starch suspension, and time how long it takes for the amylase to digest the starch. To tell when the starch has been digested, I will remove a sample of the mixture at 30-second intervals and add one drop of iodine solution to it. The iodine solution will start off blue-black when the starch is present, but change colour to yellow when the starch has been digested, so I will time how long it takes for this to happen. I will use a water bath to keep the temperature of the reaction mixture at 40 °C. To ensure reliability, I will repeat the experiment three times at each concentration and compare the results.'

The marks (up to a maximum of 6) would be given for points like these:

- using two or more concentrations of amylase (the independent variable)
- using equal volumes of amylase (a controlled variable)
- using the same source of amylase, or explaining its source (a controlled variable)
- using the same concentration or volume of starch (a controlled variable)
- keeping the temperature constant (a controlled variable)
- use of iodine solution to show a colour change from blue-black to yellow
- measuring the time taken for the change (the dependent variable)
- repeating tests at each concentration (for reliability)
- using eye protection while carrying out the experiment.

Some students use the letters CORMS to help them check that they have included everything needed. CORMS stands for:

Change	What will you change? Give the values or range of the independent variable.
Organism	What will you control about the organism used, e.g. the same species of plant, the same batch of yeast, the same age of human subjects.
Repeats	State that you need to carry out repeats, for reliability.
Measure	What will you measure (the dependent variable)? How will you measure it?
Same	What factors will you keep the same (constant)? State at least two controlled variables.

The enzyme investigation above is a straightforward experiment that you could do yourself. However, you might get a question about an investigation that you couldn't do in a school laboratory, like this:

Genetically modified bacteria are grown in industrial fermenters to produce human insulin. Describe an investigation to find out if temperature affects the amount of insulin made by the bacteria. **(6 marks)**

Using the 'CORMS' headings, these features should be covered in your plan:

Change	Use fermenter at different temperatures or suggest a range of temperatures (e.g. 20 °C to 40 °C).
Organism	Use the same species of bacterium at each temperature and the same amount of bacteria.
Repeats	Carry out the experiment several times at each temperature to get reliable results.
Measure	Measure the insulin produced at each temperature. Measure the concentration, mass or volume of insulin produced.
Same	Keep any of these factors constant: pH, nutrients supplied to the bacteria, level of oxygen in the fermenter, sterile conditions.

Write an answer to this question using the CORMS table above to help you.

APPENDIX B: COMMAND WORDS

Command word	Definition
Add/Label	Requires the addition or labelling of a stimulus material given in the question, for example labelling a diagram or adding units to a table.
Calculate	Obtain a numerical answer, showing relevant working.
Comment on	Requires the synthesis of a number of variables from data/information to form a judgement.
Complete	Requires the completion of a table/diagram.
Deduce	Draw/reach conclusion(s) from the information provided.
Describe	To give an account of something. Statements in the response need to be developed, as they are often linked but do not need to include a justification or reason.
Determine	The answer must have an element that is quantitative from the stimulus provided, or must show how the answer can be reached quantitatively. To gain maximum marks, there must be a quantitative element to the answer.
Design	Plan or invent a procedure from existing principles/ideas.
Discuss	■ Identify the issue/situation/problem/argument that is being assessed within the question. ■ Explore all aspects of an issue/situation/problem/argument. ■ Investigate the issue/situation etc. by reasoning or argument.
Draw	Produce a diagram either using a ruler or freehand.
Estimate	Find an approximate value, number or quantity from a diagram/given data or through a calculation.
Evaluate	Review information (e.g. data, methods) then bring it together to form a conclusion, drawing on evidence including strengths, weaknesses, alternative actions, relevant data or information. Come to a supported judgement of a subject's quality and relate it to its context.

Command word	Definition
Explain	An explanation requires a justification/exemplification of a point. The answer must contain some element of reasoning/justification – this can include mathematical explanations.
Give/State/Name	All of these command words are really synonyms. They generally all require recall of one or more pieces of information.
Give a reason/reasons	When a statement has been made and the requirement is only to give the reason(s) why.
Identify	Usually requires some key information to be selected from a given stimulus/resource.
Justify	Give evidence to support (either the statement given in the question or an earlier answer).
Plot	Produce a graph by marking points accurately on a grid from data that is provided and then draw a line of best fit through these points. A suitable scale and appropriately labelled axes must be included if these are not provided in the question.
Predict	Give an expected result.
Show that	Verify the statement given in the question.
Sketch	Produce a freehand drawing. For a graph, this would need a line and labelled axes with important features indicated. The axes are not scaled.
State what is meant by	When the meaning of a term is expected but there are different ways for how these can be described.
Suggest	Use your knowledge to propose a solution to a problem in a novel context.
Verb proceeding a command word	
Analyse the data/graph to explain	Examine the data/graph in detail to provide an explanation.
Multiple choice questions	
What, Why	Direct command words used for multiple-choice questions.

GLOSSARY

abiotic factor physical or chemical factor affecting an ecosystem, e.g. light intensity or temperature

accommodation changes taking place in the eye which allow it to focus on objects at different distances

accuracy (of experimental results) closeness of an experimental result to its true value

acid rain rain with a pH less than 5.5, caused by pollutant gases such as sulfur dioxide and nitrogen oxides

active site area on the surface of an enzyme where the substrate attaches and products are formed

active transport movement of molecules or ions against a concentration gradient, using energy from respiration

adaptation feature of an organism that suits its structure to its function

adenosine triphosphate (ATP) chemical present in all cells which acts as an energy 'currency'. ATP is made by respiration and used up by any process that needs a supply of energy

ADH *see* antidiuretic hormone

adrenal glands pair of endocrine glands situated above the kidneys. Secrete adrenaline

adrenaline hormone secreted by the adrenal glands. Stimulates several organs in the 'fight or flight' response

aerobic respiration reaction that releases energy from food. Uses oxygen and produces carbon dioxide and water

agar jelly-like substance used as a culture medium for growing microorganisms

algae photosynthetic protoctists. Mostly unicellular, some multicellular forms (seaweeds)

algal bloom rapid increase in numbers of algal cells in an aquatic habitat. Often caused by eutrophication

alleles different forms of a gene

alveoli (singular = alveolus) microscopic air sacs in the lungs where gas exchange takes place

amino acid one of about 20 different molecules that form the building blocks of proteins

amnion membrane enclosing the embryo during pregnancy

amniotic fluid fluid secreted by the amnion that protects the embryo by acting as a shock absorber

amylase enzyme that digests starch into maltose

anaerobic respiration reaction that releases energy from food, without using oxygen. Produces lactate in mammals, carbon dioxide and ethanol in yeast

anther part of the stamen where pollen grains are produced

antibody protein produced by lymphocytes that binds with foreign antigens as part of the immune response

anticodon group of three bases on a tRNA molecule that are complementary to a codon on the mRNA

antidiuretic hormone (ADH) hormone released from the pituitary gland. Controls the water content of the blood by increasing reabsorption of water from the collecting ducts of the kidney into the blood

antigen chemical 'marker' on the surface of a cell that identifies the cell as 'self' or 'non-self'

anus outlet of the gut where faeces is expelled from the body

arteriole small artery

artery (plural = arteries) blood vessel with a thick muscular wall and a narrow lumen, carrying blood away from the heart

artificial insemination (AI) method of selective breeding, where semen is used to make an animal pregnant without sexual intercourse. e.g. using semen from prize bulls to inseminate cows

artificial selection *see* selective breeding

asexual reproduction reproduction that does not involve fusion of gametes. New organisms are produced by part of an organism separating from a single parent

assimilation manufacture of new substances in cells using the products of digestion

ATP *see* adenosine triphosphate

atria (singular = atrium) two upper chambers of the heart where blood enters the heart from the vena cava (right atrium) and pulmonary vein (left atrium)

auxin plant hormone involved in tropisms and other growth responses

axon long extension of a neurone that carries nerve impulses in a direction away from the cell body

bacteria (singular = bacterium) small single-celled organisms with no nucleus

bacteriophage virus that infects bacteria. Used as a vector in genetic engineering

balanced diet diet containing all the necessary food types and in the correct amounts and proportions to keep the body healthy

base (in DNA) one of four nitrogen-containing groups in the DNA molecule, called adenine, thymine, cytosine and guanine. Bases form complementary pairs linking the chains of the double helix

basement membrane (in Bowman's capsule) membrane in the wall of the Bowman's capsule that acts as a molecular filter during ultrafiltration in the kidney

beri-beri a cluster of symptoms caused primarily by thiamine (vitamin B1) deficiency

bicuspid valve valve in the heart between the left atrium and left ventricle. Prevents backflow of blood when the ventricle contracts

bile green liquid made by the liver and stored in the gall bladder. Causes lipids in the gut to form an emulsion, increasing their surface area for easier digestion by enzymes

bile duct tube carrying bile from the gall bladder to the duodenum

bioaccumulation build-up of pollutants such as insecticides in the fatty tissues of an organism

biodiversity the amount of variation shown by organisms in an ecosystem. Biodiversity is a measure of both numbers of species and abundance of each species

biological control use of another organism to control the numbers of a pest species

biomagnification increase in concentration of bioaccumulated substances along a food chain

biomass total mass of organisms, e.g. in an ecosystem

biotechnology use of microorganisms to make useful products

biotic factor biological factor affecting an ecosystem. e.g. food supply, predation

bladder muscular bag that stores urine before its removal from the body

blind spot area of the retina where the optic nerve leaves the eye. Contains no light-sensitive cells, so an image cannot be detected

Bowman's capsule structure consisting of a hollow cup of cells at the start of a kidney tubule. The site of ultrafiltration

bronchial tree branching network of air passages in the lungs

bronchioles small air passages leading from the bronchi to the alveoli

bronchitis lung disease caused by irritation of the bronchial tree and infection by bacteria, resulting in breathing difficulties

bronchi (singular = bronchus) tubes leading from the trachea to the lungs

capillary microscopic blood vessel that carries blood through organs and allows exchange of substances between the blood and the cells of the organ

capsule (of bacteria) slime layer covering some bacterial cells. Protects the bacterium and stops it drying out

carbohydrase enzyme that digests carbohydrates

carbohydrate organic compound composed of one or more sugar molecules

carbon monoxide toxic gas present in car exhaust fumes and cigarette smoke

carboxyhaemoglobin substance formed when carbon monoxide combines with haemoglobin, displacing oxygen from the haemoglobin

carcinogen something that causes cancer, e.g. a chemical or radiation

cardiac centre region in the medulla of the brain that controls heart rate

cardiac cycle sequence of events taking place in the heart during one heartbeat

cardiac muscle Specialised muscle making up the heart wall. Able to contract rhythmically without fatiguing

carnivore animal that feeds on other animals

cartilage tough tissue present in several places in the body, such as rings in the trachea and between the bones at a joint

catalyst chemical that increases the rate of a reaction but remains unchanged at the end of the reaction

cell basic structural unit of living organisms

cell membrane thin surface layer around the cytoplasm of a cell. Forms a partially permeable barrier between the cell contents and the outside of the cell

cell wall non-living layer outside the cell membrane of certain types of cell. Made of cellulose (plants and algae), chitin (fungi) or peptidoglycan (bacteria)

cellulose polysaccharide of glucose that forms plant cell walls

central nervous system (CNS) brain and spinal cord

cervix 'Neck' of the uterus

CHD *see* coronary heart disease

chitin substance that makes up the cell wall of fungi and the outside skeleton of insects

chlorophyll green pigment present in chloroplasts, which absorbs light energy during photosynthesis

chloroplast organelle found in some plant cells. The site of the reactions of photosynthesis

cholesterol lipid substance present in the blood and linked to coronary heart disease

choroid dark layer of tissue below the sclera of the eye. Contains blood vessels and pigment cells

chromatid one of two thread-like strands of a replicated chromosome. Each chromatid contains an exact copy of the double helix of DNA. Chromatids become visible at the start of mitosis and meiosis

chromosome thread-like structure found in the nucleus of a cell, made of DNA and protein. Contains the genetic information (genes)

cilia (singular = cilium) microscopic hair-like projections from the surface of some animal cells, such as those lining the trachea and bronchi. Beating of cilia moves mucus and trapped particles towards the mouth

ciliary muscle ring of muscle around the lens of the eye that alters the shape of the lens during accommodation

clone groups of cells, or organisms, that are genetically identical

CNS *see* central nervous system

codominance pattern of inheritance where neither allele of a gene is dominant over the other so that both alleles are expressed in the phenotype

codon triplet of bases on the mRNA molecule. Different triplets code for different amino acids in a protein

coleoptile protective sheath covering the first leaves of a cereal seedling. Used in tropism experiments

collecting duct last part of a kidney tubule, where water is reabsorbed before the final urine is produced

colon first part of the large intestine, where water is absorbed from the waste material in the gut

community all organisms of all species found in a particular area at a certain time

companion cell specialised cell lying next to a sieve tube in the phloem and controlling its activities

cone (cell) cell in the retina of the eye that is sensitive to different wavelengths of light and results in colour vision

Control part of an experiment which is set up to show that other variables are not having an effect on the outcome of the experiment

controlled variables variables in an experiment other than the independent variable, which are kept constant by the person carrying out the experiment so that they do not affect the results

consumer oganism that eats other organisms

cornea transparent 'window' at the front of the eye that allows light to enter the eye. Also (along with the lens) refracts the light as it enters the eye

coronary arteries small arteries supplying blood to the heart muscle

coronary heart disease (CHD) disease caused by a blockage of the coronary arteries due to a build-up of fatty material. It can cut off the blood supply to the heart and result in a heart attack

coronary veins small veins carrying blood away from the heart muscle

corpus luteum remains of an ovarian follicle after ovulation. Secretes progesterone

cortex (of kidney) outer part of the kidney, containing kidney tubules and blood vessels

cotyledons seed leaves. May act as food store in seed

cross-pollination transfer of pollen from an anther of one plant to a stigma of a different plant of the same species

cuticle thin layer of waxy material covering the epidermis cells of a plant

cutting method of producing new plants by taking a piece of a shoot and planting it in compost. An example of asexual reproduction

cytoplasm jelly-like material that makes up most of a cell

decomposition breakdown of the dead remains of other organisms, helping to recycle nutrients.

decomposer organism that feeds by breaking down the dead remains of other organisms. e.g. some bacteria and fungi

denaturing process where the structure of a protein is damaged by high temperatures (becomes denatured). If the protein is an enzyme, it will no longer catalyse its reaction

dendrites fine extensions of the dendrons of a neurone

dendron extension of the cytoplasm of a neurone that carries impulses towards the cell body

denitrifying bacteria type of bacteria in the nitrogen cycle that convert nitrates into nitrogen gas

deoxyribonucleic acid (DNA) chemical of which genes are made. Double helix composed of deoxyribose sugar, phosphates and four bases

dermis middle layer of the skin containing many sensory receptors

diabetes disease where the blood glucose concentration cannot be properly controlled. Caused by a lack of insulin

diaphragm muscular sheet separating the thorax from the abdomen. Involved in the mechanism that ventilates the lungs

dicot (dicotyledonous plant) plant with two seed leaves

dietary fibre indigestible plant material, mainly cellulose, in the diet. Helps to prevent constipation and bowel diseases

differentiation process taking place during the development of an embryo, where cells become specialised to carry out particular functions

diffusion movement of molecules or ions down a concentration gradient

digestion process by which food is broken down into simpler molecules that can be absorbed

diploid (number) number of chromosomes found in body cells. Diploid cells contain both chromosomes of each homologous pair.

disaccharide sugar made up of two monosaccharides. e.g. sucrose, which is a disaccharide of glucose and fructose

DNA *see* deoxyribonucleic acid

dominant (allele) allele of a gene that is expressed in the heterozygote

dorsal root part of a spinal nerve that emerges from the dorsal (back) side of the spinal cord

dorsal root ganglion swelling in the spinal nerve that contains the cell bodies of sensory neurones

double circulatory system blood circulatory system in mammals, where the blood passes through the heart to the lungs and returns to the heart before passing to the rest of the body

duodenum first part of the small intestine following the stomach

ecosystem community of living organisms together with their non-living environment

effector organ that brings about a response (a muscle or gland)

ejaculation release of semen during sexual intercourse

embryo multicellular structure formed by division of a zygote

emphysema lung disease where the walls of the alveoli break down and fuse together again, forming air spaces with a reduced surface area. It results in breathing difficulties

endocrine gland gland secreting a hormone into the blood stream

enzyme protein that acts as a biological catalyst

epidermis (in plants) outer layer of cells of a leaf or other non-woody parts of a plant

epidermis (in the skin) outer layer of skin, consisting of dead cells

erythrocyte red blood cell

eukaryotic cells that have a nucleus (the cells of all living organisms except bacteria)

eutrophication process where an aquatic habitat receives large amounts of minerals, either naturally or as a result of pollution by sewage or fertilisers

evolution change in form of organisms over the course of time. Process by which species develop from earlier forms during the history of the Earth.

excretion removal from the body of the waste products of metabolism

exocrine gland gland secreting a product through a duct

explant small piece of plant tissue used in micropropagation

faeces semi-solid indigestible waste that passes out of the gut via the anus

F_1 generation offspring formed from breeding the parent organisms

F_2 generation offspring formed from breeding individuals from the F_1 generation

fatty acid type of molecule that, with glycerol, is one of the building blocks of lipids

fermentation using the respiration of microorganisms to produce useful products

fermenter a vessel used to grow microorganisms

fertilisation fusion of male and female gametes to form a zygote

fetus unborn offspring of a mammal, in particular an unborn human embryo more than 2 months after fertilisation, when it shows recognisably human features

fibrin protein formed from fibrinogen during blood clotting

fibrinogen protein in blood plasma that forms insoluble fibres of fibrin during blood clotting

flaccid condition in a plant cell which has lost internal pressure, so that the cytoplasm no longer pushes against the cell wall

flagellum (plural = flagella) in animal cells such as sperm: tail-like structure that beats from side-to-side, producing movement. In some bacteria: structure with similar function but much smaller and quite different structure

follicle (in ovary) structure in the mammalian ovary that contains a single developing egg cell

follicle stimulating hormone (FSH) hormone made by the pituitary gland. Stimulates the maturation of eggs in the ovary and sperm production in the testes

food chain flow diagram showing the feeding relationships in an ecosystem

food web diagram showing the way in which several food chains are linked together in an ecosystem

fovea region at the centre of the retina of the eye where there is a high concentration of light-sensitive receptor cells

fructose monosaccharide sugar found in fruits

fruit structure containing a seed or several seeds. Formed from the ovary following fertilisation

FSH *see* follicle stimulating hormone

gall bladder organ that stores bile from the liver

gametes male and female sex cells, formed by meiosis

gene part of a chromosome, the basic unit of inheritance. A length of DNA that controls a characteristic of an organism by coding for the production of a specific protein

genetic engineering techniques used to transfer genes from the cells of a donor organism to those of a recipient

genome the entire DNA of an organism (the amount present in a diploid cell)

genotype alleles an organism has for a certain characteristic

geotropism growth movement of a plant in response to the directional stimulus of gravity

germination sequence of events taking place when the embryo in a seed begins to develop into a young plant

glomerular filtrate fluid that passes through the Bowman's capsule at the start of a kidney tubule

glomerulus ball of capillaries surrounded by the Bowman's capsule at the start of a kidney tubule

glucagon hormone released by the pancreas. Action of glucagon causes an increase in the concentration of glucose in the blood

glucose monosaccharide sugar, the main 'fuel' for respiration

glycerol molecule that, along with fatty acids, is a component of lipids

glycogen polysaccharide of glucose that acts as a storage carbohydrate in animals and fungi. Found in liver and muscles

grey matter tissue in the middle of the spinal cord and outer part of the brain. Consists mainly of nerve cell bodies

guard cells pair of specialised cells surrounding a stoma in the epidermis of a leaf. They change shape to open or close the stoma

habitat the place where an organism lives

haemoglobin chemical present in red blood cells that combines with oxygen and carries it around the body

hair erector muscle muscle attached to the base of each hair in the skin. The muscle contracts to pull the hair upright

haploid (number) number of chromosomes found in gametes. Haploid cells contain one chromosome from each homologous pair

hepatic portal vein blood vessel transporting the products of digestion from the ileum to the liver

herbivore animal that feeds on plants

heterozygous genotype with different alleles of a gene, e.g. Aa

histone protein associated with the DNA in a chromosome

homeostasis maintaining constant conditions in the body. Maintaining a constant internal environment

homeotherm animal that maintains a constant body temperature by physiological means (mammals and birds)

homologous pairs pair of chromosomes that carry genes controlling the same features at the same positions on each chromosome. The members of each homologous pair are the same size and shape

homozygous genotype with the same alleles of a gene, e.g. AA or aa

hormone in animals: chemical messenger that travels in the blood. Produced by endocrine glands
In plants: chemical messengers affecting growth

hydroponics crops grown with their roots in a solution of mineral ions rather than in soil

hyphae (singular = hypha) thread-like filaments of cells in fungi

hypodermis layer of skin below the dermis, containing fatty tissue

hypothalamus region at the base of the brain above the pituitary gland. Secretes hormones and monitors various 'drives' such as hunger and thirst

ileum last part of the small intestine, where the products of digestion are absorbed into the blood

immune response mechanism by which the body recognises and deals with an exposure to a pathogenic microorganism. Involves the production of memory cells that respond to a subsequent infection by dividing to give many antibody-producing cells

insulin hormone produced by the pancreas. Action of insulin results in a decrease in the concentration of glucose in the blood

intercostal muscles two sets of antagonistic muscles lying between the ribs. Contract and relax to move the ribs in order to ventilate the lungs

internal environment blood and tissue fluid

invertebrate animal without a vertebral column (backbone)

iris coloured part of the eye visible from the front. Muscles in the iris change the size of the pupil

kingdom major grouping used in classifying organisms, e.g. animals, plants, protoctists

kwashiorkor disease caused by starvation, resulting in body proteins being metabolised

lactate waste product of anaerobic respiration in muscle cells

lactic acid bacteria type of bacteria that produces lactic acid. Used in fermenting milk to make yoghurt and cheese

lacteal structure in the middle of a villus, containing lymph and forming part of the lymphatic system. The lacteal absorbs products of lipid digestion

lactose sugar found in milk. Disaccharide of glucose and galactose

leaching process whereby mineral ions (such as nitrates) are washed out of the soil by rain

ligase enzyme used to join pieces of DNA in genetic engineering

lignin woody material present in the cell walls of some plant cells, such as xylem vessels. Provides strength and makes the walls impermeable to water

limiting factor component of a process or reaction that is in 'short supply', so that it prevents the rate of reaction from increasing

lipase enzyme that digests lipids

lipid fats and oils. Most lipids are composed of fatty acids and glycerol

liver large organ in the abdomen that has many functions, including the storage of glycogen, manufacture of bile, and breakdown of amino acids

lock and key model model of enzyme action where the substrate is the 'key', fitting into the 'lock', which is the active site of the enzyme

loop of Henlé u-shaped part in the middle of a kidney tubule. Involved in concentrating the fluid in the tubule

lumen space in the middle of a tube such as an artery or a xylem vessel

luteinising hormone (LH) hormone produced by the pituitary gland, which stimulates the release of a mature egg from the ovary (ovulation). Also stimulates the ovary to make oestrogen.

lymphocyte type of white blood cell that produces antibodies

medulla (of brain) part of brain that controls basic body functions such as heart rate and breathing rate

medulla (of kidney) middle part of the kidney containing blood vessels, loops of Henlé and collecting ducts

meiosis Type of cell division that produces haploid cells (gametes)

memory cell cell formed from lymphocytes during the immune response. Remain in the blood for many years, producing long-lasting immunity to a disease

menstrual cycle monthly cycle of events preparing a woman's uterus for the possible implantation of a fertilised egg. Controlled by hormones from the pituitary gland

messenger RNA (mRNA) type of RNA that forms a copy of the template strand of the DNA during transcription

metabolism chemical reactions taking place inside cells

micropropagation growing plants from small pieces of plant tissue in test tubes under controlled conditions

microvilli minute projections from the surface membrane of some cells, such as those on the surface of the villi of the ileum, where they increase the surface area for the absorption of the products of digestion

minerals elements needed by the body and gained from food but are not present in carbohydrates, lipids or proteins

mitochondrion (plural = mitochondria) organelle that carries out aerobic respiration, releasing energy for the cell. Place where most of the cell's ATP is made

mitosis type of cell division that produces diploid body cells for growth and repair of tissues

monocot (monocotyledonous plant) plant with one seed leaf

monohybrid inheritance involving a single gene

monosaccharide 'Single' sugar such as glucose, which cannot be broken down to give a simpler sugar

motor neurone nerve cell that transmits impulses from the central nervous system to an effector organ

mRNA *see* messenger RNA

mucus Sticky liquid secreted by cells lining the trachea and bronchi to trap dust and bacteria

multicellular composed of many cells

mutagens factors that increase the rate of mutation, such as gamma radiation or certain chemicals present in cigarette smoke

mutation change in the structure of a gene or chromosome

mutualism relationship between two organisms where both organisms benefit from the relationship

mycelium network of fungal hyphae

myelin sheath covering made of a lipid material that surrounds an axon. Nerve cells that have a myelin sheath are described as myelinated

natural selection process where certain individuals in a population survive because they are better adapted to their environment. They are more likely to pass on their genes to their offspring. The mechanism of evolution

negative feedback process where a change in the body is detected and brings about events that return conditions to normal

nephron kidney tubule, the functional unit of a kidney

nerve impulse tiny electrical signal that passes down a nerve cell. Caused by movements of ions in and out of the axon

neuromuscular junction synapse of a nerve cell on a muscle

neurone nerve cell

neurotransmitter chemical released at the end of a neurone by the arrival of a nerve impulse. The neurotransmitter diffuses across a synapse, causing a new impulse in the following neurone

nicotine addictive drug present in tobacco and cigarette smoke

nitrifying bacteria bacteria in the nitrogen cycle that oxidise ammonia to nitrite and then nitrite to nitrate

nitrogen-fixing bacteria bacteria in the nitrogen cycle that convert nitrogen gas into ammonia. They may be free-living in soil, or live in root nodules of legumes

nucleus cell organelle that contains chromosomes. Controls the activities of the cell

oesophagus part of the alimentary canal between the mouth and the stomach

oestrogen female sex hormone secreted by the ovaries. Controls the development of the female sex characteristics and the repair of the uterine lining during the menstrual cycle

optic nerve nerve carrying impulses from the retina of the eye to the brain

organ structure in the body of an animal or plant that is a collection of different tissues working together to perform a function

organ system collection of different organs working together, e.g. the heart and blood vessels of the circulatory system

organelle part of the cell with a particular function, e.g. the nucleus

osmoregulation regulation of salt and water balance in the body

osmosis net diffusion of water molecules across a partially permeable membrane from a solution with a high water potential to a solution with a low water potential

ovary (plural = ovaries) in animals: female reproductive organ that produces ova (eggs).
In plants: female reproductive structure in the carpel of a flower, which contains ovules

oviduct tube leading from the ovary to the uterus. The oviduct is also known as the Fallopian tube.

ovulation release of an ovum from a follicle in the ovary

ovule structure within the ovary in plants. Cells in the ovules divide by meiosis to produce ova.

ovum (plural = ova) female gamete

oxygen debt the volume of oxygen that is needed to completely oxidise the lactate built up during a period of anaerobic respiration

oxyhaemoglobin haemoglobin bound to oxygen

palisade mesophyll layer of cells below the upper epidermis in a leaf. The main site of photosynthesis

pancreas gland discharging into the duodenum. Makes digestive enzymes and is also an endocrine organ, secreting the hormones insulin and glucagon

parasite animal or plant that lives in or on another organism (called the host) and gets nutrients from the host

partially permeable membrane membrane (e.g. the cell surface membrane) that is permeable to some molecules but not permeable to others

pathogen organism that causes disease, e.g. some bacteria

pedigree diagram showing a family tree for an inherited characteristic

pelvis (of kidney) funnel-like part of the kidney leading to the ureter

penicillin antibiotic obtained from the mould *Penicillium*

pepsin protease enzyme made in the stomach

period common name for loss of blood during the menstrual cycle

peristalsis waves of muscular contraction that push food along the gut

pesticide chemical used to kill pests

phagocyte cell capable of phagocytosis, e.g. some white blood cells

phagocytosis process by which cells engulf and digest material, e.g. white blood cells engulfing bacteria

phenotype how a gene is expressed. The 'appearance' of an organism resulting from its genotype

phloem plant transport tissue responsible for moving the products of photosynthesis from the leaves to other parts of the plant

photosynthesis process carried out in organisms containing chlorophyll. Light energy is used to drive reactions where carbon dioxide and water are used to make glucose and oxygen

phototropism growth movement of a plant in response to a directional light stimulus

pituitary gland gland at the base of the brain which secretes a number of hormones and substances that control the release of hormones from other endocrine glands

placenta organ in mammals which contains blood vessels of the embryo in close proximity to blood vessels of the mother. Allows exchange of gases, nutrients, waste products and other substances between the maternal blood and the embryo's blood

plant growth substances plant 'hormones' that affect various aspects of plant growth

plasmid small circular piece of DNA found in bacteria and used in genetic engineering

plasmolysed condition of a plant cell that has lost water by osmosis, resulting in the cell contents shrinking and the cell membrane and cytoplasm pulling away from the cell wall

platelets small fragments of cells in blood. Responsible for releasing chemicals involved in blood clotting

pleural cavity space between the pleural membranes

pleural fluid thin layer of liquid filling the pleural cavity

pleural membrane two layers of membrane forming a continuous envelope around the lungs

plumule embryonic shoot of a plant

pollen grain structure in plants that contains the male gamete

pollen tube tube that grows from a pollen grain and down through the style to allow the transfer of the male gamete to the ovule for fertilisation

pollination transfer of pollen from anther to stigma

pollution contamination of the environment by harmful substances that are produced by the activities of humans

polygene name given to a group of several genes working together to determine a characteristic, e.g. height in humans.

polygenic inheritance characteristics controlled by two or more genes working together

polysaccharide carbohydrate made of many sugar units, e.g. starch, which is a polysaccharide of glucose

population all the organisms of one species living in a particular habitat at a certain time

precision (of experimental results) smallest increment that can be usefully measured, i.e. the smallest division on the scale of any measuring instrument being used

predator animal that kills and eats other animals

primary consumer organism that feeds on producers

producer organism that makes its own food, e.g. green plants

progesterone female sex hormone made by the corpus luteum in the ovaries, and later by the placenta. Progesterone causes further thickening of the uterus lining during the menstrual cycle and maintenance of the lining during pregnancy. A drop in levels of progesterone stimulates menstruation

prokaryotic cells of bacteria, which are small and lack a nucleus or membrane-bound organelles

protease enzyme that digests proteins

protein organic substance made of chains of amino acids

protoctist a mixed group of eukaryotic organisms that are not plants, animals nor fungi. Mostly unicellular species

protozoa 'Animal-like' species of protoctists

puberty time when developmental changes take place in boys and girls that lead to sexual maturity

pulmonary circulation circulation of blood from the heart to the lungs via the pulmonary artery and back to the heart via the pulmonary vein

pupil hole in the centre of the iris that allows light to enter the eye

pyramid of biomass diagram in which blocks making up a pyramid represent the total mass of organisms at each trophic level

pyramid of numbers diagram in which blocks making up a pyramid represent the total numbers of organisms at each trophic level

quadrat square used in ecological investigations to sample animals or plants

quaternary consumer organism that feeds on tertiary consumers

radicle embryonic root of a plant

receptor cell or organ that detects a stimulus

recessive allele that is not expressed in the phenotype when a dominant allele of the gene is present (i.e. in the heterozygote)

recombinant DNA DNA made by genetic engineering, by combining DNA from two species of organisms

rectum last part of the large intestine, where faeces is stored

reflex action rapid, automatic, involuntary response to a stimulus

reflex arc nerve pathway of a reflex action

relay neurone short neurone that connects a sensory neurone with a motor neurone in the CNS

reliability (of experimental results) measure of how similar the results are, if the experiment is carried out several times

renal artery blood vessel that supplies blood to a kidney

renal vein blood vessel that takes blood away from a kidney

replication (of DNA) copying of DNA that takes place before cell division

respiration chemical reaction taking place in cells, where glucose is broken down to release energy

response reaction by an organism to a change in its surroundings

restriction endonuclease *see* restriction enzyme

restriction enzyme enzyme used in genetic engineering to cut out a section from a molecule of DNA

retina inner, light-sensitive layer at the back of the eye

ribonucleic acid (RNA) nucleic acid similar to DNA but made of a single strand, with ribose sugar and the base uracil instead of thymine. Involved in protein synthesis

ribosome tiny structure in the cytoplasm of cells, the site of protein synthesis

RNA *see* ribonucleic acid

rod (cell) light-sensitive cell in the retina which works in dim light, but cannot distinguish between different colours

root hairs specialised cells on the surface of a root that take up water and mineral ions from the soil

root nodules swellings on the roots of legumes, containing nitrogen-fixing bacteria

saliva digestive juice secreted into the mouth by the salivary glands. Contains the enzyme amylase

saprotrophic type of nutrition where dead organic material is digested outside the body of the organism by extracellular enzymes. Carried out by fungi and most bacteria

sclera tough outer coat of the eye

scurvy a disease resulting from a lack of vitamin C

secondary consumer organism that feeds on a primary consumer

secondary sexual characteristics changes taking place in the bodies of boys and girls at puberty

secretion release of a fluid or substances from a cell or tissue

seed structure that forms from the ovule following fertilisation. Contains the embryo plant and its food store

selective breeding process where humans cross-breed individual animals or plants that have been chosen because they show certain characteristics. Used to produce domestic animals and crop plants. Also known as artificial selection

selective reabsorption process taking place in a kidney tubule whereby different amounts of substances are absorbed from the filtrate into the blood

selectively permeable membrane a membrane that allows certain molecules or ions to pass through it by means of active or passive transport

self-pollination transfer of pollen from an anther to a stigma of the same flower or to another flower of the same plant

semen mixture of sperm from the testes and fluids from glands. Ejaculated during sexual intercourse

semilunar valves Valves present at the start of the aorta and the pulmonary artery. Prevents backflow of blood when the ventricles relax

sensory neurone nerve cell which carries impulses from a receptor into the CNS

sex chromosomes pair of chromosomes that determine sex in humans. XX in females, XY in males

sexual intercourse insertion of the penis into the vagina, followed by release of semen

sexual reproduction reproduction involving fusion of male and female gametes to form a zygote

shivering rapid, involuntary contractions of skeletal muscles which generates heat when a person is cold

sieve plate specialised end wall of sieve tube, with holes allowing connections between one cell and the next

sieve tube cell found in phloem, consisting of tube transporting the products of photosynthesis

single circulatory system blood circulatory system in fish, where the blood passes from the heart to the gills, then to the body before returning to the heart

sperm male gamete of an animal, with a tail for swimming and a head containing the nucleus

sphincter muscle ring of muscle in the wall of an organ, which holds back its contents. E.g. the sphincter muscle at the outlet of the bladder

spongy mesophyll layer of photosynthetic cells below the palisade layer in a leaf. Contains air spaces and is the main gas exchange surface

stamen male reproductive organ in plants, consisting of the anther and a stalk called the filament

starch polysaccharide of glucose that acts as a storage carbohydrate in plants

stem cell cell that can divide several times but remains undifferentiated. Present in the early embryo and in some adult tissues such as bone marrow

stigma part of the carpel of a flower which receives the pollen during pollination

stimulus change in the surroundings of an organism that produces a response

stomata (singular = stoma) pores in the epidermis of a leaf

style part of the carpel of a flower. Stalk connecting the stigma to the ovary, through which the pollen tube grows

substrate molecule upon which an enzyme acts

sucrose disaccharide made from glucose and fructose. The main sugar transported in the phloem of plants

suspensory ligaments fibres between the lens and ciliary body of the eye that hold the lens in position

sweat gland structure in the dermis of the skin that secretes sweat

synapse junction between two neurones

systemic circulation part of a double circulation that supplies blood to all parts of the body except the lungs

template strand (of DNA) strand of DNA that codes for the manufacture of proteins

tertiary consumer organism that feeds on secondary consumers

test cross cross of an organism showing the dominant phenotype with one showing the recessive phenotype. The F_1 from the cross shows whether the parent is homozygous dominant or heterozygous

testa seed coat

testis (plural = testes) male reproductive organ in animals that produces the male gametes

testosterone male sex hormone, made by the testes. Responsible for the development of the male secondary sex characteristics

thermoregulatory centre part of the brain that monitors core body temperature

thorax (in mammals) chest, which includes the ribcage enclosing the heart and lungs

tissue collection of similar cells working together to perform a function

tissue fluid watery solution of salts, glucose and other solutes. Surrounds all the cells of the body, forming a pathway for the transfer of nutrients between the blood and the cells

trachea 'Windpipe' leading from the nose and mouth to the bronchi

transcription process by which the information in the base sequence of a strand of the DNA is copied into a molecule of mRNA

transfer RNA (tRNA) type of RNA that carries amino acids to the ribosomes during translation of the mRNA

transgenic organism that has been engineered with a gene from another species

translation process by which the information in the base sequence of mRNA is used to produce the sequence of amino acids in a protein. Takes place at ribosomes

transpiration loss of water vapour from the leaves of plants

transpiration stream passage of water and minerals through the roots, stem and leaves of a plant

tricuspid valve valve in the heart between the right atrium and right ventricle. Prevents backflow of blood when the ventricle contracts

tRNA *see* transfer RNA

trophic levels different feeding levels in a food chain

tropical rainforests forests of tall, densely growing, broad-leaved evergreen trees in areas of high yearly rainfall. Rainforests form a belt around the Earth in countries near the Equator and have a very high biodiversity

tropism growth of a plant in response to a directional stimulus

tumour mass of cells produced by mutation and uncontrolled cell division

turgid description of a plant cell with a high internal pressure, so that the cytoplasm pushes against the cell wall

turgor condition of a plant when its cells are turgid

ultrafiltration filtration of the blood taking place in the Bowman's capsule of a kidney tubule, where the filter separates different-sized molecules under pressure

unicellular composed of a single cell

urea main nitrogenous excretory product of mammals

ureter tube carrying urine from the kidney to the bladder

urethra tube carrying urine from the bladder to the outside of the body

vaccination artificially supplying antigens to a person (e.g. as an injection) in order to stimulate an immune response and protect them against a disease-causing pathogen

vacuole membrane-bound space in a plant cell, filled with a solution of sugars and salts called cell sap

vascular bundle xylem and phloem grouped together in a stem or root

vasoconstriction narrowing of blood vessels in the skin. Decreases the blood flowing through the skin to reduce heat loss

vasodilation widening of blood vessels in the skin. Increases the blood flow through the skin to increase heat loss

vector (in genetic engineering) structure which can be used to transfer genes in genetic engineering, e.g. a plasmid

vein blood vessel with a thin muscular wall and a wide lumen, carrying blood towards the heart

ventilation movement of the air in and out of the lungs

ventral root part of a spinal nerve that emerges from the ventral (front) side of the spinal cord

ventricles two lower chambers of the heart, which pump blood out of the heart via the aorta (from the left ventricle) and pulmonary artery (from the right ventricle)

vertebrate animal with a vertebral column or 'backbone'

villi (singular = villus) tiny projections from the lining of the ileum that increase the surface area for the absorption of the products of digestion

viruses very small microorganisms that are not composed of cells. Virus particles consist of genetic material (DNA or RNA) surrounded by a protein coat

vitamins group of chemicals that are obtained from the diet and needed in small amounts to maintain health

water potential measure of the ability of water molecules to move in a solution. Pure water has the highest water potential

white matter tissue in the outer part of the spinal cord and the middle of the brain, consisting mainly of nerve cell axons

wilt changes taking place in a plant when its cells lose too much water and become flaccid. The leaves droop and collapse

xylem transport tissue carrying water and minerals up through a plant from the roots

zygote single cell resulting from fusion of a male and female gamete

INDEX

UNIT 1 ANSWERS

CHAPTER 1

1 ▶ B   2 ▶ A   3 ▶ A   4 ▶ D

5 ▶ a Diagram should show each part of a plant cell and its function, e.g. cell wall (maintains shape of cell), cell membrane (controls entry and exit of substances), cytoplasm (where metabolism/reactions take place), vacuole (stores dissolved substances), nucleus (controls activities of cell), chloroplasts (photosynthesis), mitochondria (respiration).

   b An animal cell lacks a cell wall, a large permanent vacuole and chloroplasts.

6 ▶ Description, in words or diagrams, should include the following points:
- enzymes are biological catalysts
- they speed up reactions in cells without being used up
- each enzyme catalyses a different reaction
- the production of enzymes is controlled by genes
- enzymes are made of protein
- the substrate attaches to the enzyme at the active site
- the substrate fits into the active site like a key in a lock
- this allows the products to be formed more easily
- intracellular enzymes catalyse reactions inside cells
- extracellular enzymes are secreted out of cells (e.g. digestive enzymes)
- they are affected by changes in pH and temperature.

7 ▶ a About 75 °C.

   b At 60 °C the molecules of enzyme and substrate have more kinetic energy and move around more quickly. There are more frequent collisions between enzyme and substrate molecules, so more reactions are likely to take place.

   c The microorganism lives at high temperatures, so it needs 'heat-resistant' enzymes with a high optimum temperature.

   d It is denatured.

8 ▶ Diffusion is the net movement of particles (molecules or ions) from a high to low concentration. It does not need energy from respiration. Active transport uses energy from respiration to transport particles against a concentration gradient.

9 ▶ The function of the motor neurone is to send nerve impulses to muscles and glands. It has a long axon, which conducts these impulses. It has a cell body with many extensions called dendrons and dendrites, which link with other neurones at synapses. At the other end of the neurone, the axon branches and forms connections with muscle fibres, called neuromuscular junctions.

The palisade cell's function is photosynthesis. Palisade cells are near the top surface of the leaf, where they are close to the sunlight. They have thin cell walls, so the light can easily reach the many chloroplasts that the cell contains.

10 ▶ a They carry out most of the reactions of respiration in the cell, providing it with energy.

   b Active transport. This uses the energy from the mitochondria.

   c Diffusion. The removal of glucose at A lowers the concentration inside the cell, so that the concentration at B is higher than inside the cell. Therefore glucose can diffuse down a concentration gradient.

   d Increases the surface area for greater absorption.

CHAPTER 2

1 ▶ D   2 ▶ A   3 ▶ B   4 ▶ C

5 ▶ a i Fungi   ii Protoctists
      iii Plants   iv Bacteria

   b Like most protoctists, *Euglena* is a microscopic, single-celled organism. It has features of both plant and animal cells: like plants, it contains chloroplasts; like animals, it can move.

6 ▶ a Diagram should show a core of DNA or RNA surrounded by a protein coat. (It may also have an outer envelope or membrane derived from the host cell.)

   b A virus can be considered either as living or as a chemical. It does not have any of the normal characteristics of living things, except that it is able to reproduce.

   c Viruses can reproduce only inside a host cell, by taking over the cell's genetic machinery to make more virus particles. So viruses are all parasites.

7 ▶ a An animal that does not have a vertebral column (backbone).

   b Fine, thread-like filaments forming the feeding network of cells of a fungus.

   c A type of nutrition used by most fungi and some bacteria, where the organism feeds on dead organic material by digesting it using extracellular enzymes.

END OF UNIT 1 QUESTIONS

1 ▶ a i nucleus, mitochondrion (both needed for 1)

      ii nucleus, chloroplast, mitochondrion (all needed for 1)

      iii nucleus, mitochondrion (both needed for 1).

   b The cells in a root have no chloroplasts because they don't receive any light and so can't carry out photosynthesis (1)

   c Nucleus controls the activities of the cell (1); chloroplast absorbs light energy for photosynthesis (1); mitochondrion carries out some reactions of respiration to release energy (1).

2 ▶ a The artery is an organ because it is made of several tissues (1); the capillary is made up of only one type of cell (1).

   b i Two from: Breaks down large insoluble molecules (1) into smaller soluble molecules (1) that can be absorbed (1)

ii (1 mark for organ, 1 mark for function).
Three from:
- mouth: chews / breaks down food into smaller pieces / produces saliva;
- oesophagus (gullet): move food from mouth to stomach;
- stomach: produces digestive enzymes;
- pancreas: produces digestive enzymes;
- liver: makes bile;
- ileum (small intestine) produces digestive enzymes / absorbs products of digestion;
- colon (large intestine): absorbs excess water;
- rectum: stores waste (faeces).

iii (1 mark for system, 2 marks for organs).
Two from:
- breathing system: trachea, lung, diaphragm;
- circulatory system: artery, vein, heart;
- musculoskeletal system: muscle, joint, (named) bone;
- nervous system: brain, spinal cord;
- reproductive system: testis, ovary, uterus, penis;
- excretory system: kidney, bladder.

3 ▶ a i 4 g (1). Mass at start was 100 g, decreased to 96 g due to oxygen lost (1).

ii Half this mass = 2 g (1). This loss in mass occurs by (approximately) 0.5 minutes / 30 seconds (1).

iii At the start there are a lot of enzyme and substrate molecules, so there are a lot of successful collisions (1). As the reaction proceeds, the number of substrate molecules decreases, so there are fewer successful collisions (1).

b i There would be no difference / 4 g formed (1); because the temperature affects only the reaction rate, not the end point (1).

ii The time would be shorter (1) because the rate of reaction is speeded up by the increase in temperature (1).

4 ▶ a 1 mark for each correct row in the table.

Feature	Active transport	Osmosis	Diffusion
particles must have kinetic energy	✗	✓	✓
requires energy from respiration	✓	✗	✗
particles move down a concentration gradient	✗	✓	✓

b i (As the temperature rises) ions gain kinetic energy (1), so they move faster (1).

ii Above this temperature the cell membranes are being denatured (1) so are more permeable to ions (1).

5 ▶ a i So that each of the two cells produced (1) will have the correct number of chromosomes / correct amount of DNA after the division (1).

ii The nucleus has divided into two (1).

b i They increase the surface area for absorption (1).

ii They (further) increase the surface area for absorption (1).

iii As the glucose moves out of the cell, the concentration inside the cell decreases (1) and increases the concentration gradient for diffusion of glucose into the cell (1).

6 ▶ a i $C_6H_{12}O_6 + 6O_2 \rightarrow 6CO_2 + 6H_2O$ (1 for each correct part).

ii It is the same (1), because there are six molecules of each / same number of molecules / same number of moles (1), 1 mole of any gas has the same volume (1).

iii Any sensible experimental error stated (1) with brief explanation (1).

iv No oxygen would be used up (1), so distance moved would be less / bead would not move (1).

7 ▶ (1 mark for each correct row)

Feature	Type of organism		
	Plant	Fungus	Virus
they are all parasites	✗	✗	✓
they are made up of a mycelium of hyphae	✗	✓	✗
they can only reproduce inside living cells	✗	✗	✓
they feed by extracellular digestion by enzymes	✗	✓	✗
they store carbohydrates as starch	✓	✗	✗

8 ▶ (One mark for each correct underlined term)
Plants have cell walls made of <u>cellulose.</u> They store carbohydrate as the insoluble compound called <u>starch</u> or sometimes as the sugar <u>sucrose.</u> Plants make these substances as a result of the process called <u>photosynthesis.</u> Animals, on the other hand, store carbohydrate as the compound <u>glycogen.</u> Both animals' and plants' cells have nuclei, but the cells of bacteria lack a true nucleus, having their DNA in a circular chromosome. They sometimes also contain small rings of DNA called <u>plasmids</u>, which are used in genetic engineering. Bacteria and fungi break down organic matter in the soil. They are known as <u>decomposers / saprotrophs.</u> Some bacteria are pathogens, which means that they <u>cause disease.</u>

9 ▶ a Germinating seeds produce heat (1) from respiration (1).

b To kill bacteria on the seeds (1)

c To allow oxygen into the flask (1)

d mass of seeds / number of seeds / age of seeds (1)

10 ▶ Any six for 6 marks, from:
- Use solution of ATP, compare with (control using) water (1)

- Same type of meat fibres / named type (1)
- Several replicates / number of replicates suggested, e.g. 10 (1)
- Measure length before treatment (1)
- Measure length after treatment / change in length / % change (1)
- Other controlled variables: temperature / volume of solutions / starting length (Max. 2)

UNIT 2 ANSWERS

CHAPTER 3

1 ▶ C   2 ▶ A   3 ▶ B   4 ▶ B

5 ▶

	Action during inhalation	Action during exhalation
external intercostal muscles	(contract)	relax
internal intercostal muscles	relax	contract
ribs	move up and out	(move down and in)
diaphragm	contracts and flattens	relaxes and becomes dome-shaped
volume of thorax	increases	decreases
pressure in thorax	decreases	increases
volume of air in lungs	increases	decreases

6 ▶ When we breathe in, the external intercostal muscles between our ribs contract, pulling the ribs up and out. The diaphragm muscles contract, flattening the diaphragm. This increases the volume in the chest cavity, lowering the pressure there, and causing air to enter from outside the body, through the nose or mouth. This is called ventilation. In the air sacs of the lungs, oxygen enters the blood. The blood then takes the oxygen around the body, where it is used by the cells. The blood returns to the lungs, where carbon dioxide leaves the blood and enters the air sacs. When we breathe out, the external intercostal muscles relax and the ribs move down and in. The diaphragm muscles relax, and the diaphragm returns to a dome shape. These changes decrease the volume of the chest cavity, increasing the pressure in the cavity, pushing the air out of the lungs.

7 ▶ a When the volume of the chest is increased by the movements of the ribs and diaphragm, the drop in pressure in the chest cavity draws air into the pleural cavity through the puncture in the chest wall, instead of through the mouth or nose into the lung.

  b Each lung is isolated from the other by being in a separate pleural cavity, so a pneumothorax on one side will not affect the opposite lung.

  c A tube is inserted through the chest wall into the pleural cavity on the side of the injured lung. This stops ventilation in that lung, while the other lung will be ventilated normally.

8 ▶ a The rings support the trachea so that it does not collapse during inhalation.

  The gap in the 'C' allows food to pass down the oesophagus, which runs next to the trachea, without catching on the rings.

  b The short distance allows easy diffusion of oxygen into the blood, and diffusion of carbon dioxide out of the blood.

  c The mucus traps bacteria and dirt particles. The cilia beat backwards and forwards to sweep these towards the mouth, preventing them entering the lungs.

  d Smoke contains carbon monoxide, which displaces oxygen from the haemoglobin of the red blood cells of the smoker.

  e The addictive drug in tobacco smoke is nicotine. Smokers who are trying to give up can use patches or gum to provide the nicotine they normally get from cigarettes, reducing the craving to smoke.

  f The large surface area is provided by the alveoli. It allows for efficient diffusion of oxygen into the large blood supply, and efficient removal of the waste product, carbon dioxide.

9 ▶ Bronchitis is a lung disease caused by irritation of the linings of the airways to the lungs, and may be made worse by bacteria infecting the bronchial system.

  Emphysema is a lung disease where the walls of the alveoli break down and then fuse together, reducing their surface area. (Both diseases may be caused by smoking.)

10 ▶ a Some points are:
   - non-smokers have a low death rate from lung cancer at all ages
   - the death rate from lung cancer among smokers increases with age
   - the death rate increases with the number of cigarettes smoked per day.
   - (Numbers should be used from the graph to illustrate any of these points.)

  b For 55-year-olds smoking 25 a day: about 4.5 per 1000 men (or 45 per 10 000 men).

  For 55-year-olds smoking 10 a day: about 1 per 1000 men.

  c Probably this investigation. The graph shows a direct relationship between number of cigarettes smoked and incidence of lung cancer, in one particular type of person (middle-aged male doctors): in other words, a more controlled group. In Table 3.2 the patients were matched for age, sex etc. but were from a more varied background. There could be other reasons for the correlation that had not been considered. However, they both show a strong link.

11 ▶ The leaflet should not be too complicated or have too much information so that it puts the reader off. It must have a clear message.

CHAPTER 4

1 ▶ D   **2 ▶** A   **3 ▶** D   **4 ▶** B

5 ▶ a Starch: take a sample of the water in a spotting tile and add a drop of iodine solution. The colour changes from orange to blue-black.

Glucose: take a sample of the water in a test tube and add blue Benedict's solution. Place the tube in a water bath and heat until it boils. A brick-red precipitate results.

b The starch molecules are too large to pass through the holes in the Visking tubing. Glucose molecules are smaller, so they can pass through.

c The blood.

d Large, insoluble food molecules are broken down into small, soluble ones.

6 ▶ a It is body temperature

b It had been broken down into smaller molecules called peptides (short chains of amino acids) forming the clear solution.

c The enzyme pepsin does not work in alkaline conditions, it is denatured.

d The experiment is looking at the effects of pepsin on the egg white. The Control is carried out without the enzyme; all other factors are the same. This shows that it is the enzyme that breaks down the protein. In other words, the egg white does not break down by itself.

e The enzyme works more slowly at a lower temperature. There are fewer collisions between enzyme and substrate molecules, because they have less kinetic energy.

f Hydrochloric acid kills bacteria in the food entering the stomach.

g By alkaline secretions in the bile and pancreatic juice.

7 ▶

Enzyme	Food on which it acts	Products
(amylase)	starch	maltose
(trypsin)	protein	peptides
lipase	fats	(fatty acids and glycerol)

8 ▶ Descriptions of any four of the following:
- length, which increases time and surface area for absorption
- folds in lining, which increase surface area
- villi covering lining, which increase surface area
- microvilli on lining cells, which increase surface area
- capillary networks in villi, where products are absorbed
- lacteals in villi, which absorb fats.

9 ▶ The account should include full descriptions of most of the following points:
- digestion of starch to maltose in the mouth, action of saliva in moistening food
- mechanical digestion by the teeth
- movement through the gut by peristalsis (diagram useful)
- digestion of protein by pepsin in the stomach and the role of hydrochloric acid
- emulsifying action of bile from the liver on fats

- pancreatic enzymes (amylase, trypsin, lipase) and their role in digestion of starch, protein and fats
- adaptations of the ileum for the absorption of digested food (see question 4)
- role of the colon in absorption of water.

10 ▶ a Energy = (20 × 18 × 4.2) = 1512 joules = 1.512 kilojoules.

b Energy per gram = 1.512 ÷ 0.22 = 6.872 kJ per g.

c There are several errors involved. Some major ones include:
- some of the energy from the burning pasta is used to heat the test tube, thermometer, etc
- much energy will be lost when heating up the air near the tube, or when transferring the pasta
- not all the energy in the pasta will be released when it burns
- some energy will be lost when evaporating the water from the tube
- measurement errors such as measurement of the volume of water and temperatures (although these are probably small compared with the other reasons).

d One way is to shield the tube inside (for example) a metal can, to reduce heat losses to the air (or use a calorimeter).

e Peanuts contain a large proportion of fat, which has a high energy content. Pasta is largely carbohydrate, which contains less energy per gram.

CHAPTER 5

1 ▶ B   **2 ▶** C   **3 ▶** A   **4 ▶** B

5 ▶ a Single: fish; double: human or other named mammal.

b i (Either) The blood passes once through the heart in a single system, and twice through the heart in a double system for every complete circulation of the body.

(Or) In a double system the blood flows from the heart through one circuit to the lungs, then back to the heart and out through another circuit to the rest of the body.

ii Double circulatory system pumps the blood twice per circulation so higher pressures can be maintained.

c Diffusion can take place because it has a large surface area compared with its volume and the distances for substances to move inside the cell are short.

6 ▶ a A red blood cell has a large surface area compared with its volume; contains haemoglobin; and has no nucleus, so more space is available for haemoglobin.

b i Oxygen dissolves in the liquid lining the alveoli and then diffuses down a concentration gradient through the walls of the alveoli and capillaries into the plasma and into the red blood cells.

ii Oxygen dissolves in the plasma and then diffuses down a concentration gradient through the walls of the capillaries into the muscle cells.

c Dissolved in plasma.

7 ▶ a Arteries have thick walls containing much muscle tissue and elastic fibres. These adaptations allow their walls to stretch and recoil under pressure.

b Veins have valves, thin walls with little muscle, and a large lumen; arteries have no valves (except at the start of the aorta and pulmonary artery), thick muscular walls with many elastic fibres, and a narrow lumen.

c Capillaries have thin walls / walls one cell thick, to allow exchange of materials. They have a very small diameter to fit between other cells of the body.

8 ▶ a A = left atrium, B = (atrioventricular) valves, C = left ventricle, D = aorta, E = right atrium.

b To ensure blood keeps flowing in one direction / prevent backflow of blood.

c i A; ii E

9 ▶ a i A (red blood cell), identified by its colour (red) and biconcave disc shape.

ii B (lymphocyte), identified by its colour (white) and large nucleus (to produce antibodies quickly).

iii C (phagocyte), identified by its colour (white), variable shape (shows it is flowing) and lobed nucleus.

b Platelets – blood clotting.

10 ▶ a C, heart rate is increasing so more blood can be pumped to muscles.

b E, brief jump in heart rate.

c A, lowest rate. B, increases from minimum to steady rate.

11 ▶ a i Low rate (75 beats/minute) because body is at rest, need for oxygen is low.

ii Rate increases because more blood carrying oxygen for respiration needs to be pumped to muscles.

iii Rate decreases as need for oxygen is reduced / lactate produced during exercise is removed (repaying oxygen debt).

b The shorter the recovery period, the fitter the person.

CHAPTER 6

1 ▶ D   2 ▶ B   3 ▶ C   4 ▶ D

5 ▶ a Changes that take place in the shape of the lens to allow the eye to focus upon objects at different distances away.

b The replacement artificial lens cannot change shape.

c The ciliary muscles contract and the suspensory ligaments slacken. The shape of the lens becomes more convex, refracting the light more.

6 ▶ a

Function	Letter
refracts light rays	G
converts light into nerve impulses	A
contains pigment to stop internal reflection	B
contracts to change the shape of the lens	E
takes nerve impulses to the brain	D

b i H

ii Contraction of circular muscles in the iris reduces the size of the pupil, letting less light into the eye. Contraction of radial muscles increases the size of the pupil, letting more light into the eye.

iii To protect the eye from damage by bright light, and to allow vision in different light intensities.

7 ▶ a i Sensory neurone

ii Relay neurone

iii Motor neurone

b The sensory neurone carries impulses from sensory receptors towards the central nervous system. The motor neurone carries impulses out from the CNS to effector organs (muscles and glands). The relay neurone links the other two types of neurone in the CNS.

c X: white matter, Y: grey matter, Z: dorsal root ganglion.

d Electrical impulses.

e The gap between one neurone and another is called a synapse. An impulse arrives at the end of an axon and causes the release of a chemical called a neurotransmitter into the synapse. The neurotransmitter diffuses across the synapse and attaches to the membrane of the next neurone. This starts an impulse in the second nerve cell.

8 ▶ a P: cell body, Q: dendrite, R: axon.

b Speed = distance/time
= 1.2 m / 0.016 s
= 75 m per s

c Mitochondrion

d i Insulation / prevents short circuits with other actions (Also speeds up conduction).

ii Person would not be able to control their muscle contractions / not be able to coordinate body movements / 'wrong' muscles would contract.

9 ▶ a A wide variety of answers are possible, such as:
- dust in the eye – secretion of tears
- smell of food – secretion of saliva
- touching a pin – withdrawal of hand
- attack by a predator – increased heart rate
- object thrown at head – ducking.

b Nature and role of receptor and effector correctly explained, e.g. for 'dust in the eye' above:

i The receptors consist of touch receptors in the eye. They respond by generating nerve impulses (which eventually stimulate the tear glands).

ii Tear glands are the effectors. They secrete tears, washing the irritant dust out of the eyes.

c Dust enters the eye and stimulates a touch receptor in the surface of the eye. The receptor sends nerve impulses along sensory neurones to the CNS (brain). In the CNS, impulses pass from sensory neurones to motor neurones via relay neurones. Impulses pass out from the CNS to the tear glands via motor neurones. These impulses stimulate the tear glands to secrete tears.

CHAPTER 7

1 ▶ B   **2** ▶ A   **3** ▶ B   **4** ▶ C

5 ▶ **a** 'Hormones' are chemical messenger substances, carried in the blood. 'Secreted' refers to the process where a cell makes a chemical that passes to the outside of the cell. 'Glands' are organs that secrete chemicals, and 'endocrine' glands secrete their products into the blood.

b A = insulin, B = adrenaline, C = testosterone, D = progesterone.

6 ▶ **a** Glucose has been absorbed into the blood following a meal (lunch!)

b The high concentration of glucose in the blood is detected by the pancreas, which secretes the hormone insulin into the blood. Insulin stimulates the uptake of blood glucose into the liver, where it is converted into an insoluble storage carbohydrate called glycogen.

c i Untreated diabetes leads to weakness and loss of weight, and eventually coma and death.

ii Coloured test strips to detect glucose in the urine, and direct measurement of blood glucose using a sensor.

iii Reducing the amount of carbohydrate in the diet, and injections of insulin.

CHAPTER 8

1 ▶ D   **2** ▶ A   **3** ▶ C   **4** ▶ C

5 ▶ **a** Maintaining constant conditions in the internal environment of the body.

b Removal of the waste products of metabolism from the body.

c Filtration of different sized molecules under pressure (as in the Bowman's capsule).

d Reabsorption of different amounts of different substances by the kidney tubule.

e An animal (mammal or bird) that generates internal (metabolic) heat to keep its temperature constant.

6 ▶ **a** X = glomerulus, Y = Bowman's capsule (or renal capsule), Z = loop of Henlé

b A = water, urea, protein, glucose, salt
B = water, urea, glucose, salt
C = water, urea, salt
D = water, urea, salt.

7 ▶ Description should include:
- increase in blood concentration
- receptors in hypothalamus of brain stimulated
- pituitary gland releases more ADH
- ADH travels in the blood to the kidney
- ADH causes collecting ducts of tubules to become more permeable to water
- more water reabsorbed into blood
- blood becomes more dilute, its concentration returns to normal

- negative feedback involves a change in the body that is detected and starts a process to return conditions to normal
- this is negative feedback because an increase in blood concentration is detected, action of ADH returns blood concentration to normal.

8 ▶ **a** Before the water was drunk, the volume of urine collected was about $80\,cm^3$. After drinking the water, the volume increased, reaching a peak of about $320\,cm^3$ after 60 min. After this, the volume decreased, until it reached the volume produced before drinking the water at about 180 min.

b At 60 minutes, the concentration of ADH in the blood was low. This made the collecting ducts of the kidney tubules less permeable to water, so less water was reabsorbed into the blood and more was excreted in the urine, forming a large volume of urine. By 120 minutes, the secretion of ADH had increased, causing the collecting ducts to become more permeable, so that more water was reabsorbed into the blood and less entered the urine.

c The volume would be less. More water would be lost in sweating, so less would be in the blood for production of urine.

d $150\,cm^3$ is produced in 30 minutes, which is $150 \div 30 = 5\,cm^3$ per minute.
- the filtration rate is $125\,cm^3$ per minute
- therefore $120\,cm^3$ is reabsorbed per minute
- so the percentage reabsorption is: $(120/125) \times 100 = 96\%$.

9 ▶

Changes taking place	Hot environment	Cold environment
(sweating)	increased sweat production so that evaporation of more sweat removes more heat from the skin	decreased sweat production so that evaporation of less sweat removes less heat from the skin
(blood flow through capillary loops)	vasodilation increases blood flow through surface capillaries so that more heat is radiated from the skin	(vasoconstriction decreases blood flow through surface capillaries so that less heat is radiated from the skin)
(hairs in skin)	hairs lie flat due to relaxed muscles, trapping less air next to the skin	hairs are pulled erect by muscles, trapping a layer of insulating air next to the skin
(shivering)	no shivering occurs	shivering occurs; respiration in muscles generates heat
(metabolism)	metabolism slows down, e.g. in organs such as the liver, reducing heat production.	metabolism speeds up, e.g. in organs such as the liver, generating heat.

10 ▶ a The average body temperature of birds is slightly higher than that of mammals. This is because they have a higher metabolic rate, needed for flight (note that the flightless birds have a lower body temperature).

b No. For example, the temperature of the camel and of the polar bear is the same, despite their different habitats.

c The fur traps air, providing insulation. The colour acts as camouflage (so they are not so easily seen by prey).

CHAPTER 9

1 ▶ C   **2 ▶** D   **3 ▶** D   **4 ▶** C

5 ▶ a A = placenta, B = umbilical cord, C = amnion, D = amniotic fluid, E = uterus (womb).

b The function of the placenta is the transfer of oxygen and nutrients from the mother's blood to the blood of the embryo / fetus, and removal of waste products such as carbon dioxide and urea from the fetus to the mother.

c Just before birth, contractions of the muscle of the uterus (E) causes the amnion to rupture, allowing the amniotic fluid (D) to escape. This is the 'breaking of the waters'.

d During birth, the cervix (F) becomes fully dilated, and strong contractions of the muscles of the uterus (E) pushes the baby out.

6 ▶ a Method B. the formation of a new individual (the bud) does not involve sex cells from sex organs (as shown in method A).

b In asexual reproduction, all the cells of the new individual are produced by mitosis from one cell in the parent. When cells divide by mitosis, all the new cells are genetically identical to the parent cell, and to each other.

c If *Hydra* is well adapted to its environment, and the environment is stable, asexual reproduction will produce offspring that are also well adapted. However, if the environment changes, they may not be well adapted and may die out. Sexual reproduction produces offspring that show variation, so some of the new *Hydra* may be better adapted to survive in the new conditions.

7 ▶ a i A   **ii** B   **iii** D   **iv** A

b i oestrogen

ii Approximately 29–30 days. This can be seen by counting the days from the start of the first menstruation (day 0) to the start of the next menstruation.

iii Fertilisation is most likely to have taken place about 15 days after the day when the last menstruation started. The last menstruation started on about day 57, so fertilisation probably took place on about day 72. (Note – this is very approximate!) After day 72 there is no menstruation, the uterus lining becomes thicker.

iv To prepare for implantation of the fertilised egg.

8 ▶ There is evidence for and against the involvement of pollutants in lowering of the sperm count, and indeed whether or not the count has become lower at all. A good account of the student's findings should be a balanced one, giving both sides of the argument. It should be illustrated with some graphs or tables of data.

9 ▶ a A = oestrogen, B = progesterone

b Corpus luteum

c To prepare for the implantation of a fertilised embryo

d 13

e Progesterone maintains the thickened uterus lining and prevents menstruation, as well as preventing further ovulation by inhibiting release of FSH and LH.

i Progesterone is secreted by the corpus luteum.

ii Progesterone is secreted by the placenta.

10 ▶

Name of hormone	Place where the hormone is made	Function(s) of the hormone
follicle stimulating hormone / FSH	pituitary (gland)	Stimulates growth of follicles in the ovary. Stimulates secretion of oestrogen by the ovary.
luteinising hormone / LH	pituitary (gland)	Stimulates ovulation.
oestrogen	ovary	Causes repair (thickening) of the lining of the uterus following menstruation.
progesterone	ovary (corpus luteum)	Completes the development of the uterus lining and maintains it ready for implantation of the egg. Inhibits the release of FSH and LH by the pituitary (and stops ovulation).

END OF UNIT 2 QUESTIONS

1 ▶ a (1 mark for each correct row)

Gas	Inhaled air / %	Exhaled air / %
nitrogen	(78)	(79)
oxygen	21	16
carbon dioxide	0.04	4
other gases (mainly argon)	(1)	(1)

b It increases in exhaled air (1) because carbon dioxide is produced in respiration (1).

c Excretion is getting rid of a waste product of metabolism (1); carbon dioxide is a waste product of respiration (1).

d i Short distance (1) allows rapid / efficient diffusion of oxygen and carbon dioxide (1).

ii Blood brings carbon dioxide and takes away oxygen (1) maintaining a diffusion gradient (1).

iii Increases the surface over which diffusion of oxygen and carbon dioxide can occur (2).

2 ▶ a i A = stomach (1) because it is an acidic pH (1).
B = small intestine (1) because it is an alkaline pH (1).
ii Protein (1).
b i Liver (1).
ii Proteins (1).
iii Proteins (from the urea) are a source of nutrients for the cattle (1).
iv The Bowman's capsule carries out ultrafiltration of the blood (1) allowing water and small solute molecules such as urea to pass through into the kidney tubule, but holding back blood cells and large molecules (1). The loop of Henlé is involved in concentrating the fluid in the tubule (1), so that urine with a high concentration of urea is produced at the end of the tubule (1).

3 ▶ a A = pulmonary vein, B = aorta, C = right atrium, D = left ventricle, E = renal vein (5).
b X (artery) has narrow lumen / muscular wall, Y (vein) has large lumen / little muscle (2).
c i Increases rate and volume of heartbeat (2).
ii Two from: increases breathing rate, diverts blood away from intestine to muscles, converts glycogen to glucose in the liver, dilates pupils, causes body hair to stand on end, increases mental awareness, increases rate of metabolism (2).
d Reflex action is automatic / involuntary (1), voluntary action is one a person chooses to carry out / is initiated by the brain (1).
e Lactate produced in muscles during exercise needs to be oxidised / removed / oxygen debt needs oxygen (1), oxygen is supplied by increased breathing rate and increased heartbeat (1).

4 ▶ a Labels: Cell membrane (1), <u>lobed</u> nucleus (1), cytoplasm (1)
b Two from: has a nucleus, irregular shape / not biconcave, no haemoglobin (2).
c Two from: ingest / engulf / surround (bacteria), digest / break them down, using enzymes (2).
d Three from: lymphocytes, make antibodies, specific to antigens, form memory cells (3).

5 ▶ a All chemical reactions taking place in cells can continue at a steady rate / metabolism doesn't slow down in cold conditions (1).
b i Arterioles: blood remains in core of body and doesn't lose heat (1). Sweat: no heat lost in evaporating the sweat (1). Shivering: increases heat production by respiration (1).
ii They have a lot of muscle fibres in their walls (1).
c i Antidiuretic hormone / ADH (1).
ii More water has been lost as sweat (1).
iii As concentration of water in blood decreases (1) ADH is released from the hypothalamus (1) and causes reabsorption of more water in kidney tubules (1).

6 ▶ a i B **ii** C **iii** B **iv** D **v** A (5).

b Pregnancy is most likely to result from sexual intercourse around the time of ovulation (1), i.e. in the middle of the menstrual cycle / around day 14 (1). If a couple avoid having sexual intercourse at this time, the woman is less likely to become pregnant (1).

7 ▶ a B (1). Cell division has reduced the chromosome number (1) from 46 to 23 / to the number present in gametes (1).
b The fertilised egg / zygote has 46 chromosomes (1). It divides by mitosis (1), so that all the cells of the body also have 46 chromosomes (1). In the sex organs, gametes are produced by meiosis (1), which halves the chromosome number to 23 (1). Fertilisation of an egg by a sperm restores the chromosome number to 46 (1).
c Any three for 3 marks, from:
- mitosis involves one division, meiosis involves two
- mitosis forms two cells, meiosis forms four
- mitosis forms cells with the same chromosome number as the parent cell / diploid, meiosis forms cells with half the chromosome number of the parent cell / haploid
- mitosis forms body cells, meiosis forms sex cells / gametes
- mitosis forms cells that are genetically identical, meiosis forms cells showing genetic variation.

8 ▶ Any six for 6 marks, from:
- rats given protein supplement / range of amounts of protein supplement, and rats given no supplement (Control)
- rats same age / same sex / same health / same variety
- several rats in each group (allow 6 or more per group)
- weigh before and after treatment / take other suitable measurement before and after treatment, such as circumference of leg muscles
- suggested time period for treatment (minimum one week)
- calculate (mean) % change in mass
- same diet (apart from supplement)
- same water / same amount of exercise / other suitable controlled factor.

UNIT 3 ANSWERS

CHAPTER 10

1 ▶ A   **2 ▶** D   **3 ▶** C   **4 ▶** A

5 ▶ a Iodine solution, turns from yellow-orange to blue-black.
b Only the green areas that are not covered would contain starch.
c Photosynthesis needs light and chlorophyll. These are only both available in green, uncovered areas.
d A storage carbohydrate. It is insoluble, so can be stored in cells and has no osmotic effects.

6 ▶

Part of leaf	Function	How the part is adapted for its function
palisade mesophyll layer	(main site of photosynthesis)	(cells contain many chloroplasts for photosynthesis)
spongy mesophyll layer	gas exchange surface: uptake of CO_2 and release of O_2 during photosynthesis, some photosynthesis	large surface area to volume ratio; air spaces between cells; many chloroplasts in cells for photosynthesis (but fewer than in palisade layer)
stomata	pores which exchange gases (CO_2, O_2 and water vapour) with the atmosphere	pores formed between two guard cells; guard cells can change shape to open and close pores
xylem	transport of water and minerals	cells consist of dead, hollow vessels, allows transport through the lumen of each vessel; lignified walls for strength, preventing cells collapsing under suction pressure
phloem	transport of products of photosynthesis	sieve tubes with sieve plates forming continuous tubes to transport solutes; cells living, so can exercise control over movement

7 ▶ **a** At 0200 hours (night) the grass respires, producing CO_2, but there is no photosynthesis. At 1200 hours (midday) photosynthesis in the grass exceeds respiration, so CO_2 is used up.

b At 0400 hours: light intensity. At 1400 hours: the concentration of CO_2 in the air.

8 ▶

Substance	Use
glucose	oxidised in respiration to give energy
sucrose	main sugar transported in the phloem
starch	storage carbohydrate
cellulose	makes up plant cell walls
protein	growth and repair of cells
lipid	energy store in some plants, e.g. nuts, seeds. Part of all cell membranes.

9 ▶ **a** The aeration tube supplies oxygen to allow the roots to respire. The foil stops light entering the tube, preventing the growth of algae.

b Phosphate.

10 ▶ **a**

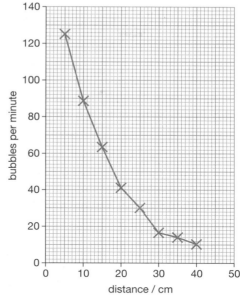

b About 52 bubbles per minute.

c • The gas is not pure oxygen, although it has a high oxygen content.
   • The bubbles may not be all the same size.
   • The water in the test tube may have increased in temperature as the lamp was brought nearer to the tube.

11 ▶ The account should include:
   • Description of photosynthesis as a chemical reaction where CO_2 and water are combined using light energy trapped by chlorophyll, forming glucose and oxygen.
   • Equation for the reaction.
   • Leaf adaptations: details of palisade mesophyll, spongy mesophyll, stomata and epidermis, xylem and phloem (diagram needed).
   • Photosynthesis supplies oxygen for respiration in animals and other organisms; it is needed at the start of food chains; how energy is harnessed by plants as the producers, and then passed to consumers (note: these topics are covered fully in Chapter 14).

CHAPTER 11

1 ▶ C   **2** ▶ B   **3** ▶ C   **4** ▶ A

5 ▶ **a** Loss in mass = (8.2 – 8.0) g = 0.2 g.
   Percentage change = (–0.2/8.2) × 100 = –2.4%.

   b Osmosis.   **c** Solution A.

   d Solution C.   **e** Solution B.

   f It is permeable to small molecules such as water, but not permeable to large molecules such as sucrose.

6 ▶ **a** Long, thin extension of the cell has a large surface area for the absorption of water and minerals.

   b Dead, lignified cells with hollow lumen, forming long tubes that carry water and minerals throughout the plant. The lignified walls are tough so that they don't collapse under pressure.

c 'Banana' shape with thicker cell wall on inside (around stoma) means that when the guard cells become turgid they change shape, bowing outwards, so opening the stoma for gas exchange.

7 ▶ a If a ball of soil is not left around the roots (e.g. if they are pulled out roughly), it will damage the root hair cells on the roots. This will mean the plant will not be able to absorb water so easily, causing it to wilt.

b If a cutting has too many leaves, it will lose too much water through transpiration and may wilt or die before it can establish new root growth.

c When stomata are in sunken pits in the leaf, a region of humid air is trapped in the pit. This reduces evaporation through the stomata, conserving water in the plant.

d Phloem contains products of photosynthesis, such as sugars, which provide food for the greenflies.

8 ▶ a A = epidermis, B = phloem, C = xylem.

b C. Xylem carries water up the stem. The dye is likely to be carried in this water.

9 ▶ a

Condition	Curve
1	(B)
2	A
3	D
4	C

b Humid air around the leaf reduces the diffusion gradient between the air spaces in the leaf and the atmosphere around the leaf. Moving air removes the water vapour that might remain near the stomata and slow down diffusion.

10 ▶ a Water forms a thin layer around the cells of the spongy mesophyll of the leaf, then evaporates from this layer and exits through the stomata. The water potential of the mesophyll cells falls, so more water passes from the xylem to the cells by osmosis. A gradient of water potential is set up, from the xylem to the cells.

b It would increase. A higher temperature would increase the rate of evaporation of water from the mesophyll.

c Many examples possible, for example:
- cacti have leaves reduced to spines
- leaves rolled into a tube with most stomata facing the inside of the tube
- sunken stomata in pits
- hairy leaves to trap layer of moist air round stomata.

11 ▶ a X = xylem, Y = phloem.

b Drawing should show a plant cell with root hair extension. Labels: cell wall, cytoplasm, vacuole, nucleus.

c Soil water contains few solutes, while there is a high concentration of solutes in the vacuole of the root hair cell. water therefore enters the cell by osmosis.

12 ▶ The description should include:
- uptake of water by osmosis from the soil through the root hairs

- the gradient of water potential across the root cortex, allowing water to move from cell to cell by osmosis
- passage of water into the xylem vessels in the root
- transport through the xylem to all parts of the plant
- evaporation of water vapour from the spongy mesophyll cells of the leaf, and loss through the stomata
- the water potential gradient in the mesophyll cells and water movement out of the xylem, the driving force for transpiration.

CHAPTER 12

1 ▶ B   **2 ▶** B   **3 ▶** D   **4 ▶** A

5 ▶ a i The direction of light and the direction of gravity.

ii The direction of gravity

b The stem grows towards the light, which allows more photosynthesis, and growth of the plant.

6 ▶

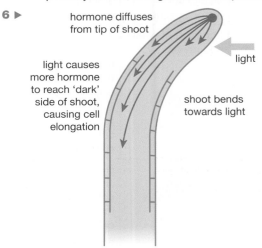

hormone diffuses from tip of shoot

light

light causes more hormone to reach 'dark' side of shoot, causing cell elongation

shoot bends towards light

7 ▶ a The coleoptile would not bend towards the light. The movement of auxin on the left (dark) side would be interrupted by the mica sheet.

b The coleoptile would grow (bend) towards the source of light. The greater amounts of auxin diffusing down the left side would be unaffected by the placement of the mica sheet. (It might even bend more than a control, with no sheet).

c The coleoptile would grow (bend) towards the source of light. The mica would not interrupt the movement of auxin away from the light.

8 ▶ a Decapitated coleoptiles would produce the least increase in length, because the tip is the source of auxin, which normally stimulates growth. No tip means that there is no auxin, so there will be reduced growth. The tip with the greatest growth is more difficult to predict. The coleoptiles with the tips covered would probably produce the most growth, since auxin is still made by the tip, but none is moved to the left side of the shoot, so there will be no bending, just upward growth.

b Decapitated coleoptiles – no bending, since no auxin produced.

Coleoptiles with tips covered – no bending, since light does not reach the region behind the tip and auxin remains evenly distributed either side of the shoot (you could argue that bending may still occur if the covers are not long enough down the coleoptiles to prevent this). Untreated coleoptiles – bend towards the right, because auxin is produced by the tip and diffuses away from the light on the left, stimulating growth on that side.

c Each coleoptile is a different starting length. Therefore to make for a fair comparison between treatments we need to find the increase in length in comparison to the starting length. We can do this by calculating a percentage increase.

CHAPTER 13

1 ▶ B   **2 ▶** A   **3 ▶** D   **4 ▶** D

5 ▶ a Stigma.
  b By the coloured petals, scent and nectar.
  c Pollen tube should be shown growing down through the rest of the style and entering the ovary.

6 ▶ a Independent: temperature
     Dependent: height of seedlings and % of seeds that germinated
  b (3.4 + 4.5 + 2.5 + 3.7 + 2.8 + 4.4 + 4.3 + 2.9 + 2.1 + 3.7) / 10 = 3.43 cm
  c Higher temperatures (20 °C or 30 °C) are needed for germination to take place. At a low temperature (4 °C) few seeds germinated or grew. Growth of seedlings was greater at 30 °C than at 20 °C.
  d Temperature affects the activity of enzymes and the rate of metabolic reactions. It increases the kinetic energy of molecules, so that there are more collisions between enzyme and substrate molecules, resulting in an increase in successful reactions. Germination depends on metabolic reactions, so temperature affects germination.
  e The light intensity is not controlled. Tube A is in the light, while B and C are in the dark. All three tubes should be in the light (or all three in the dark).

7 ▶ a This method of reproduction does not involve flowers / seeds / pollen and ovules, so is not sexual. It involves the tubers growing from body cells of the parent plant.
  b The tubers grow from body cells of the parent plant by mitosis, which produces cells that are genetically identical.
  c Growth may be affected by the environment of the plants, e.g. different soil minerals or different light intensity.
  d Sexual reproduction produces offspring that show genetic variation, allowing them to survive if the environment changes.

8 ▶ a A = stigma, B = ovary, C = anther, D = filament.
  b Any three of:
    • lack of large petals (no need to attract insects)
    • lack of brightly coloured petals (no need to attract insects)

    • exposed stamens (to catch the wind and blow pollen away)
    • exposed stigma (to catch windborne pollen)
    • stigma feathery (to catch pollen).
  c The pollen grain produces a pollen tube, which grows down through the tissue of the style and into the ovary. The pollen tube enters an opening in an ovule. The tip of the pollen tube breaks down and the pollen grain nucleus moves out of the pollen tube into the ovule, where it fertilises the nucleus of the egg cell (ovum).
  d Any four of:
    • large petals
    • brightly coloured petals
    • stamens enclosed within flower
    • stigma enclosed within flower
    • stigma sticky
    • nectaries present
    • large, sticky pollen grains.

9 ▶ a Method A. Fruits are produced by flowers via sexual reproduction, which introduces genetic variation.
  b Insect-pollinated. The flower has large, brightly coloured petals to attract insects.

10 ▶ a The banana plants reproduce asexually, so they are all genetically identical. Therefore all the plants are susceptible to the fungus, none is resistant to it.
  b If the plants reproduced sexually, this would introduce genetic variation. Some of the plants might then have resistance to the fungus, and would be able to survive.
  c Asexual reproduction is faster than sexual reproduction; so more banana plants can be produced more quickly. (Also, if the plants *are* resistant to a disease, they all will be, so won't be killed by it.)

END OF UNIT 3 QUESTIONS

1 ▶ a i Any four points from:
      As light intensity increases, the rate of photosynthesis increases (1). The rate of increase is faster at high CO_2 concentration than at low CO_2 concentration (1).
      (At both CO_2 concentrations) the rate of photosynthesis reaches a plateau / maximum / levels off (1). At low CO_2 concentration this happens below light intensity X (1) whereas at high CO_2 concentration it happens at / above light intensity X (1).
      The maximum rate of photosynthesis is higher at high CO_2 concentration than at low CO_2 concentration (1).
    ii Up to X the limiting factor is light (1), because increasing light intensity increases the rate of photosynthesis (1). Beyond X the limiting factor is CO_2 (1), as increasing light intensity has no effect on the rate of photosynthesis (1) whereas increased CO_2 increases the rate (1).
  b i Temperature, water availability.

ii Reactions are slow at low temperatures (1), because the molecules have little kinetic energy (1) and therefore there are fewer successful collisions between enzyme molecules and substrates (1). Water is a raw material for photosynthesis (1).

c The photosynthesis reaction uses / takes in light energy (1) and converts it into chemical energy stored in the glucose / starch produced (1).

2 ▶ a i To remove any water / sap on the outside of the cylinder (1).

ii To allow an average to be calculated / to check reliability of results (1).

iii So they all had the same surface area to volume ratio (1).

b i 3 mol per dm^3 sucrose solution has a lower water potential / lower concentration of water / higher concentration of solutes than potato cells (1), so water moves out of the cells and into the sucrose solution (1), resulting in a decrease in mass of the cylinder (1).

ii (Approximately) 0.75 mol per dm^3 (1), because there is no change in mass (1), as there is no net movement of water (1).

c Repeat experiment with more cylinders (1), use more concentrations of sucrose between 0 and 1 mol per dm^3 (such as 0.2 mol per dm^3, 0.4 mol per dm^3, etc.) (1).

3 ▶ a i A = xylem (1) because it carries water to the leaf (1).
B = phloem (1) since it is the other vascular tissue in the vein, but is not carrying water (1).

ii 1 = transpiration stream / under pressure / mass flow (1).
2 = osmosis (1).
3 = evaporation / diffusion (1).
4 = transpiration / evaporation (1).

b Any two adaptations and explanations from
- Palisade layer cells / spongy mesophyll cells contain many chloroplasts (1) which absorb light (1)
- Spongy mesophyll is a gas exchange surface (1) for exchange of CO_2 and O_2 (1)
- Stomata allow entry of CO_2 (1) a raw material for photosynthesis (1)
- Xylem supplies water (1), which is a raw material of photosynthesis (1)
- Phloem takes away (1) sugars / amino acids / products of photosynthesis (1)

c Carbon dioxide enters through the stomata (1) but stomata need to be closed to prevent loss of water (1).

4 ▶ a i (Positive) phototropism (1).

ii Any three from:
Auxin produced in tip of shoot (1) diffuses back down the shoot (1), auxin moves away from light source (1) causes growth on the dark side of the shoot (1).

iii The plant receives more light for photosynthesis (1).

b i Any two from:
Most curvature takes place at a wavelength of about 430 nm (1), light wavelengths above about 500–550 nm produce no curvature (1), there is a smaller increase in curvature with a peak at about 370 nm (1).

ii Any two for two marks from:
The tip / something in the tip only absorbs these wavelengths of light (1), cannot absorb other wavelengths (1), these wavelengths are present in sunlight (1).

c i Gravity (1).

ii Root grows towards gravity / positive geotropism (1), shoot (in some species) grows away from gravity / shows negative geotropism (1).

iii Shoots grow upwards towards light needed for photosynthesis (1), roots grow down towards source of water (1).

5 ▶ a i B (1). **ii** F (1). **iii** E (1).

b Any two for 2 marks:
- large petals
- brightly coloured petals
- stamens enclosed within flower
- stigma enclosed within flower.

c i H (1). **ii** G (1). **iii** C (1).

d i Pollination is the transfer of pollen from the anther to the stigma (1). Fertilisation is the fusion of the nucleus of the pollen grain with the nucleus of the ovum (1).

ii Self-pollination means transfer of pollen from the anther of a plant to the stigma of the same plant (1). Cross-pollination is when pollen is transferred to the stigma of another plant (1).

6 ▶ Any six points for 6 marks:
- pollen grains placed in sucrose solution / in range of concentrations of sucrose solutions, and pollen grains placed in water (Control)
- grains from same species / same plant / same flower
- stated number of grains in each treatment (minimum 10)
- (use microscope to) count the number of grains that germinate / grow pollen tube
- (after) suggested time period (minimum 1 hour)
- calculate % germination in each treatment
- same temperature / light intensity / other suitable controlled variable

UNIT 4 ANSWERS

CHAPTER 14

1 ▶ D **2 ▶** B **3 ▶** A **4 ▶** C

5 ▶ a Habitat: place where an organism lives; community: all the populations of living organisms in an ecosystem; environment: the non-biological components of an ecosystem; population: all the organisms of a particular species in an ecosystem.

b Plants = producers; animals = consumers; decomposers = breakdown of dead material.

6 ▶ a i Plankton.   **ii** Krill.

b Quaternary consumer / top carnivore.

c Very large amounts of photosynthesis / production by the plankton can support this number of trophic levels.

7 ▶ a Any two from:
- trees → moths → small birds → owls
- trees → moths → small birds → weasels
- trees → moths → small birds → shrews
- trees → moths → beetles → shrews

b Vole or small bird.

c Reduction in dead leaves means there will be fewer earthworms and beetles, so less food for shrews.

d In the pyramid of numbers there are only 200 trees, but each tree has a very large mass, and the pyramid of biomass shows the total mass of the trees.

8 ▶ a X = ammonia; Y = nitrate; Z = decomposer.

b Active transport.

c Bacteria that convert nitrogen gas into ammonia.

d In urine / faeces and in death.

9 ▶ a (125/3050) × 100 = 4.1%.

b As urine / faeces, and as heat from metabolic processes / respiration.

c Eaten by other herbivores, or ends up in dead matter / passes to decomposers.

10 ▶ a (For simplicity, crabs, shrimps and worms can be put together. Arrows should point in the direction of energy flow.)

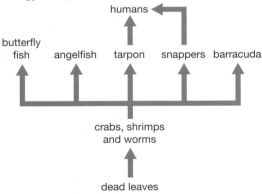

b Any suitable food chain with four organisms, such as:
- dead leaves → crabs → tarpon → humans
- dead leaves → shrimps → snappers → humans

c i Carbon dioxide.

ii Decomposers feed on the detritus; their respiration produces carbon dioxide as a waste product.

CHAPTER 15

1 ▶ B   **2 ▶** A   **3 ▶** A   **4 ▶** C

5 ▶ Because of the great increase in the human population, the need to produce food to sustain the population, and the industrial revolution and growth of technology.

6 ▶ a The concentration of carbon dioxide is increasing.

b The increase is due to increased burning of fossil fuels.

c In the summer there is more photosynthesis, which lowers the concentration of carbon dioxide. In the winter there is less photosynthesis, so carbon dioxide levels increase.

7 ▶ a Any two: carbon dioxide, methane, water vapour, CFCs

b Without a greenhouse effect, the temperature on the Earth's surface would be much colder than it is now, and life would not be able to exist. (One estimate is that the average temperature would be 30 °C lower.)

c Malaria is spread by mosquitoes, which are found in warmer regions of the world. If global warming occurs, mosquitoes will spread to more northerly parts of Europe.

8 ▶ a Rain washes fertiliser into the pond, causing the algae to grow.

b Rain washes the fertiliser down hill away from the pond.

c Algae are photosynthetic organisms (protoctists). An increased temperature increases their rate of photosynthesis, so they grow faster.

9 ▶ Sewage causes growth of bacteria in the water. The bacteria need oxygen for growth, using up the oxygen in the water, so that the fish suffocate / die.

10 ▶ a Pesticides kill pests (insects etc.) so less crop eaten; fertilisers supply minerals that increase the growth of crops.

b Use manure as fertiliser. After the crop has been harvested, dig in remains of plants, allowing them to decay and release nutrients. Use crop rotation including leguminous plants to produce nitrates. Use biological control methods to reduce pests.

END OF UNIT 4 QUESTIONS

1 ▶ a i Any of the following for 1 mark:
- plankton → sea butterfly → arrow worm → herring
- plankton → small crustaceans → large crustaceans → herring
- plankton → copepods → sand eel → herring

ii Primary consumer = sea butterfly / small crustaceans / copepods (1 mark for correct organism from food chain used).
Secondary consumer = arrow worm / large crustaceans / sand eel (1 mark for correct organism from food chain used).

iii Herring (1). It is a secondary consumer when it feeds on other small crustaceans, and a tertiary consumer when it feeds on sand eels or arrow worms (1).

b i Pyramid drawn correctly, with relative amounts of energy at each trophic level approximately correct (1).

ii (892/8869) × 100 = 10.1% (1 for correct values in calculation, 1 for answer).

iii (91/892) × 100 = 10.2% (1 for correct values in calculation, 1 for answer).

    iv (8869/0.1) × 100 = 8 869 000 kJ (1 for correct values in calculation, 1 for answer).

    v Two from: losses from respiration / in movement / as faeces / undigested food (2).

2 ▶

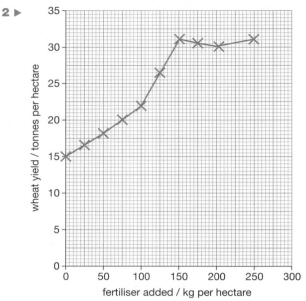

    a Axes correct way round, scales correct (1); axes labelled, with units (1); points plotted correctly (1); points joined with straight lines (1).

    b 150 kg per hectare (1). This amount gives maximum yield (1); any higher concentration would waste fertiliser / waste money (1) (since yield is not higher).

    c To make proteins (1).

    d Any of the following to a maximum of 5 marks:
- causes plants / algae to grow / form algal bloom
- reference to eutrophication
- plants /algae prevent light penetrating into the water
- submerged plants / algae underneath cannot photosynthesise so they die
- bacteria break down the dead plants / algae
- respiration of the bacteria uses up oxygen
- oxygen level of the water falls / water becomes anoxic
- aerobic animals / fish in the water die.

3 ▶ **a** **i** The insecticide becomes less effective / kills fewer insects over the three years (1). This is because some insects were resistant to the pesticide (1) so these reproduced / more resistant insects survived (1).

      ii Intermediate concentration (1), as almost as effective as the strongest c concentration (1) and will be cheaper / less polluting (1).

    b **i** When amounts of pesticide in body tissues build up over time (1).

      ii Named pesticide, e.g. DDT (1) accumulated in top carnivores / named example (e.g. osprey) (1) and caused death / other named problem (1).

      iii Could bioaccumulate in human tissues / cause illness / death (1).

4 ▶ **a** Plants carry out photosynthesis (1), which converts carbon dioxide into organic carbon compounds (1).

    b Combustion of fossil fuels, which increases carbon dioxide levels (1). Deforestation, which increases carbon dioxide levels (1).

    c **i** The bodies are broken down by respiration (1), which produces carbon dioxide (1).

      ii Insects chew bodies into smaller pieces (1), providing a larger surface area (1) for enzymes produced by decomposers (1).

      iii 4 marks for two sensible points from the curve, with reasons. e.g.
- curve 1 rises rapidly to a peak of CO_2 production by 7 days, whereas curve 2 shows little production during this time due to the slower action of decomposers on the intact bodies (2).
- curve 1 falls from the peak after 7 days due to material in the dead bodies being used up (1), while curve 2 shows little CO_2 production in this time (2).
- curve 2 starts to rise only at 9–12 days due to the slower action of decomposers on the intact bodies; CO_2 production in curve 1 has nearly fallen back to zero by 11 days (2).

5 ▶ **a** 2 marks for examples of competition, e.g. for same food source / nest sites, etc. (animals), light / minerals / water (plants) (2). Less well-adapted individuals die / best adapted survive (1) preventing population increasing / population numbers remain stable (1) (maximum 3 marks).

    b **i** 2 marks for two from:
- mineral ions / named ion, for healthy growth
- light for photosynthesis
- water for photosynthesis / turgidity / transport.

      ii To kill the weeds before they produce seeds (1) reducing need to use more herbicide later in season (1).

    c **i** Species A (1), because more beetles produced (1).

      ii The parasite kills species A (1) but does not affect numbers of species B (1). The first graph shows that species A is better at competing for resources than species B (1). The second graph shows that when species A is removed, species B can do better / increase in numbers (1).

6 ▶ **a** (88 600 − 886)/88 600 × 100 = 99% (1 mark for calculation, 1 mark for answer; allow 1 mark if answer given is 1%).

    b Sulfur dioxide and nitrogen oxides are acidic gases (1). They are blown long distances by winds (1) and dissolve in rain (so acidifying ground water) (1). (Deduct 1 mark if carbon monoxide given as acidic gas.)

    c Dissolved / suspended solids make water cloudy / dirty (1), preventing light reaching plants (1), so plants are unable to photosynthesise (1) and therefore die (1).

7 ▶ a Reduced growth / photosynthesis (1), affecting the appearance of the crop so not harvested / unfit for sale (1)

i

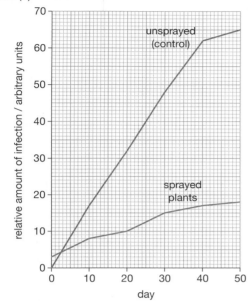

Axes correct way round, labelled (1), all points correctly plotted on both curves (2), key to each curve (1), points joined by straight lines (1)

ii Sprayed: 12 (1), unsprayed 40 (1).

iii The Control shows whether there is any infection without the fungicide (1). It is needed to be able to see how much effect the fungicide is having on the infection, i.e. as a comparison (1).

iv By day 30, the infection in the unsprayed (Control) plants was approximately the same as in Year 1 (1). However the infection in the sprayed plants had increased (1). This was probably because the plants had developed resistance to the fungicide (1).

UNIT 5 ANSWERS

CHAPTER 16

1 ▶ C   **2 ▶** A   **3 ▶** B   **4 ▶** D

5 ▶ a A = base / thymine; B = base / cytosine; C = deoxyribose / sugar; D = phosphate; E = nucleotide.

b Franklin used X-ray diffraction on DNA to find out about its structure. Watson & Crick used Franklin's data and other information to build a model of the structure of DNA.

c A always pairs with T, and C always pairs with G.

6 ▶ a i A gene is a length of DNA that codes for a protein.

ii Alleles are different forms of a gene.

b A chromosome is a structure in the nucleus of a cell composed of DNA (and proteins).

c i Both have 23 pairs of chromosomes in each cell.

ii Woman's skin cells contain XX sex chromosomes, man's contain XY.

7 ▶ a Flow diagram should have boxes showing the stages in order: (1) the two strands of the DNA separate; (2) each strand acts as a template for the formation of a new strand; (3) DNA polymerase assembles nucleotides alongside the two strands; (4) Two DNA molecules formed.

b i Caused by an addition, duplication or deletion of a base, resulting in all triplets of bases after the mutation being different and so different amino acids are coded for.

ii Caused by a change in one base in a triplet, by substitution or inversion, so that it codes for a different amino acid. Triplets after the mutation are not altered; so subsequent amino acids will not be affected.

8 ▶ a Five.

b AUG GAG CCA GUA GGG

c ATG GAG CCA GTA GGG

d The mRNA base sequence is converted into the amino acid sequence of a protein during a process called <u>translation</u>. The mRNA sequence consists of a triplet code. Each triplet of bases is called a <u>codon</u>. Reading of the mRNA base sequence begins at a <u>start codon</u> and ends at a <u>stop codon</u>. Molecules of tRNA carrying an amino acid bind to the mRNA at an organelle called the <u>ribosome</u>.

CHAPTER 17

1 ▶ D   **2 ▶** D   **3 ▶** C   **4 ▶** B

5 ▶ a Both types of division start with each chromosome copying itself / DNA replicating / DNA copying itself / chromatids forming.

Plus any two of:
- Mitosis produces two daughter cells, meiosis produces four daughter cells.
- Daughter cells from mitosis are genetically identical to each other and the parent cell; daughter cells from meiosis are genetically different from each other and the parent cell.
- Mitosis produces daughter cells with the same number of chromosomes as the parent cell / diploid to diploid; meiosis halves the chromosome number / diploid to haploid.

b Mitosis, they are formed by division of body cells to produce more body cells.

c Because the number of chromosomes per cell is reduced by half.

6 ▶ a They have been formed by mitosis, so are genetically identical.

b Meiosis is used to form pollen and egg cells, so fertilisation results in seeds that are genetically different from each other.

7 ▶ a Control.

b Plants from cuttings would be genetically identical, which is better in order to compare the effects of the treatment with nitrogen-fixing bacteria. Seeds would be genetically different, so their growth might depend on their genes, rather than the treatment.

c The nitrogen-fixing bacteria provide nitrates needed for growth. This is an environmental effect on growth, rather than a genetic one. Therefore the environment plays a big part in the growth of these plants.

8 ▶ a Meiosis, because sperm are gametes that are haploid / contain half the number of chromosomes of body cells.

b Mitosis, because body cells are dividing to produce more body cells with the normal chromosome number.

c Mitosis, because body cells are dividing to produce more body cells with the normal chromosome number.

d Meiosis, because pollen grains are gametes that are haploid / contain half the number of chromosomes of the plant's body cells.

e Mitosis, because the zygote must divide to produce more body cells with the normal chromosome number.

9 ▶ a Genetic – eye colour is inherited and not affected by the environment.

b Genetic – it depends on inheriting XX or XY chromosomes.

c Environmental – the pH of soil is a feature of the plant's environment.

d Both – genes determine whether a plant falls into the tall or dwarf categories, but environmental factors affect how well each plant grows.

e Both – genes affect the risk level, but environmental factors such as diet, smoking, etc. also have an effect.

10 ▶ a Chromosomes align themselves along the equator of the cell, attached to the spindle fibres.

b Spindle fibres shorten and pull chromatids towards opposite poles of the cell.

c Chromosomes reach the opposite poles of the cell. Nucleus starts to re-form.

CHAPTER 18

1 ▶ D　　**2 ▶ A**　　**3 ▶ D**　　**4 ▶ C**

5 ▶ a All tall.

b All tall.

c All tall.

d 3 tall : 1 short.

e 1 tall : 1 short (or 2 : 2).

f All short.

6 ▶ a i Homozygous.

ii Dominant gene hides the expression of the recessive gene when heterozygous; recessive gene expressed only in homozygous form.

b i B and b; **ii** all Bb.

c i Heterozygous.

ii

	B	b
B	BB	Bb
b	Bb	bb

Phenotypes = 3 black : 1 red.

7 ▶ a Gametes of parents = R and r

Genotypes of F1 = Rr

Genotypes of F1 parents = Rr and Rr

Gametes of F1 parents R, R and r, r

Genotypes of F2 =

	R	r
R	RR	Rr
r	Rr	rr

b A, B and C are red, D is yellow.

8 ▶ a Individual 8 has cystic fibrosis, but neither of his parents does, so they must be heterozygous and the allele must be recessive. If the allele was dominant, he would have to have inherited at least one dominant allele from one parent, so that parent would have cystic fibrosis too.

b 3 and 4 must be heterozygous for the gene, as they do not have the disease, but their son does. 11 must be homozygous for the gene, since she has the disease.

c i Probability that the next child is male is 1 in 2, or 0.5:

	X	Y
X	XX	XY
X	XX	XY

ii Let A = the normal allele of the gene and a = cystic fibrosis gene.

Individual 11's genotype = aa. Individual 10's genotype could be AA or Aa.

So there are two possible outcomes:

AA × aa

	A	A
a	Aa	Aa
a	Aa	Aa

Aa × aa

	A	a
a	Aa	aa
a	Aa	aa

Depending on whether 10 is AA or Aa, there could be no chance, or a 1 in 2 chance (0.5 probability) of their next child having cystic fibrosis. It could also be argued that if the genotype of 10 is unknown, the probability of the child having cystic fibrosis is 1 in 4, or 0.25.

9 ▶ a They must both be heterozygous. Let S = allele for short hair and s = allele for long hair.

	S	s
S	SS	Ss
s	Ss	ss

There is a 1 in 4 chance of producing a longhaired guinea pig (ss).

b Breed the shorthaired guinea pig with a homozygous longhaired guinea pig (ss). If it is heterozygous (Ss), both longhaired and shorthaired offspring will be produced (in a 1:1 ratio):

	S	s
s	Ss	ss
s	Ss	ss

If it is homozygous (SS), all offspring will have short hair:

	S	S
s	Ss	Ss
s	Ss	Ss

10 ▶ a A gene is a length of DNA, coding for the production of a protein. Alleles are different forms of a gene. The phenotype is the appearance of an organism, or the features that are produced by a gene. (The way that a gene is 'expressed'.)

b Let allele for red coat = R and allele for white coat = W (note that different letters are used, since this is a case of codominance).

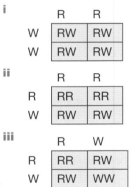

i

	R	R
W	RW	RW
W	RW	RW

ii

	R	R
R	RR	RR
W	RW	RW

iii

	R	W
R	RR	RW
W	RW	WW

c Ratios in (b) are:

i all roan;

ii 1 red: 1 roan;

iii 1 red : 2 roan : 1 white.

CHAPTER 19

1 ▶ D   **2 ▶** B   **3 ▶** A   **4 ▶** D

5 ▶ a It means that the organisms that are best adapted to their environment are more likely to survive and reproduce.

b Darwin and Wallace.

6 ▶ a An organism that causes disease.

b Fungi and bacteria.

c Random mutations produce some bacteria that are resistant to an antibiotic. If the antibiotic continues to be used, the resistant bacteria will survive and the non-resistant ones will be killed. The resistant bacteria have a selective advantage over the non-resistant bacteria; they quickly reproduce and cause disease.

7 ▶ a Rats with the resistant gene survived and reproduced, so now many more rats carry the gene. Rats without the gene did not survive to reproduce.

b It would decrease as it would not give an advantage; rats that don't have the gene will breed equally well. (In fact rats with the warfarin gene have a selective disadvantage when warfarin is not being used, although students will not know this.)

8 ▶ a They have a heavy beak, which is adapted to crush seeds.

b They have a long, narrow beak, which can be used to probe under the bark of trees for insects.

c Ancestors showed slight variations in their beaks. Where the variation enabled a bird to catch insects, or eat leaves and other food better than birds with other types of beak, the birds survived better and reproduced (survival of the fittest), passing on their genes for the adaptation. Eventually groups of birds became so different from members of other groups that they couldn't interbreed, and formed new species.

9 ▶ a As a result of (random) mutations.

b Selection pressure: a factor in the environment that affects the fitness of an organism. In this case the presence of toxic metals means that the non-tolerant plants will be killed and not reproduce to pass on their genes.

Selective advantage: varieties that survive in the presence of a selection pressure are said to have a selective advantage. In this example the plants that are tolerant to toxic metals have a selective advantage when compared with the non-tolerant plants.

Natural selection: the overall process that, when metals are present, results in fewer non-tolerant plants and an increase in the number of tolerant plants. If it continues, natural selection results in evolution.

c When there are no toxic metals, the metal-tolerant plants must have some sort of selective disadvantage over the non-tolerant ones. For example, they may need to use metabolic energy (ATP) to protect their cells against metals or get rid of metal ions. If there are no metal ions in the soil, this is a waste of resources.

CHAPTER 20

1 ▶ D   **2 ▶** C   **3 ▶** B   **4 ▶** A

5 ▶ a Both involve selection of which animals or plants survive to breed.

b In selective breeding the farmer / breeder does the selection. In natural selection it is the survival of the fittest in a habitat that leads to selection.

6 ▶ a 1) Plants have resistance to disease, so they are not killed by fungi, bacteria, etc.

2) Plants are better suited to climate, so can grow well in a particular location.

3) Plants have a better balance of nutrients; produce more nutritious food, or have a high vitamin content etc.

(Or any other correct reason.)

b Two from: quicker to produce large numbers of plants because only a few cells needed; plants can be produced at any time of year since grown inside; large numbers of plants can be stored easily until needed.

c All have same genes since produced by mitosis from cells of the same parent plant.

7 ▶ a Milk yield, and feed to milk conversion rate.

b Choose a cow with the best characteristics and give hormone / FSH injections to cause multiple ovulations. Collect ova and use IVF to fertilise with sperm collected from a bull with the best characteristics. Separate cells of embryos that develop and produce large numbers of embryos. Screen for sex (males) and implant into surrogate mother cows.

8 ▶ a Hybrid G was produced by selective breeding. Individual plants from pure lines of A and B were selected (for size of cobs) and crossed to produce hybrid E. Similarly, individual plants from pure lines of C and D were selected and crossed to produce hybrid F. Plants from hybrids E and F were then selected for their cob size, and crossed to produce hybrid G. (Crossing would be done by transfer of pollen from anthers to stigmas of plants.)

b Cob G is larger, it has more seeds and the cobs are more uniform size/shape.

c Any sensible suggestion, e.g. breed from each under identical environmental conditions, or sequence the genes to show differences.

9 ▶ The essay should include:
- examples of traditional selective breeding of crop plants or domestic animals
- advantages of this type of artificial selection, e.g. to crop yield, characteristics of animals
- cloning of plants and its advantages
- cloning animals and its uses
- causes for concern with cloned organisms (e.g. cloned plants all genetically identical, so susceptible to same pathogens; cloned animals like 'Dolly' may have genetic defects; ethical issues).

END OF UNIT 5 QUESTIONS

1 ▶ a Toxic copper ions (1), only copper-tolerant plants will grow and reproduce / non-tolerant plants will die (1).

b Predation by lions (1), only those wildebeest that are fast runners (or equivalent) will survive and reproduce / slow animals will be killed and not reproduce (1).

c Presence of pesticide (1), only those pests resistant to the pesticide will grow and reproduce / non-resistant pests will die (1).

2 ▶ a Tips of stems and side shoots removed (explants) (1); explants trimmed to 0.5–1 mm (1); put explants onto agar containing nutrients and hormones (1); when explants have grown transfer to compost in greenhouse (1).

b All have same genes since produced by mitosis from cells of the same parent plant.

c i Kinetin causes growth of shoots (1); auxin causes growth of callus and roots (1).

ii Use 2 mg per dm³ of auxin to cause growth of callus (1), then reduce to 0.02 mg per dm³ and add 1 mg per dm³ of kinetin until shoots have grown (1). Then use 2 mg per dm³ of auxin and 0.02 mg per dm³ of kinetin to grow roots (1).

d One advantage from: quicker to produce large numbers of plants because only a few cells needed; plants can be produced at any time of year since grown inside; large numbers of plants can be stored easily until needed. Disadvantage: all plants have same genes, so susceptible to same diseases / could all be affected at same time (2).

3 ▶ a Both 1 and 2 are tasters (1). If the gene was recessive, all their children would also be tasters, but 4 is a non-taster (1 mark for explanation or correct genetic diagram).

b Individual 3 is Tt (1), because if TT, she couldn't supply a 't' allele to have daughters who are non-tasters (1). Individual 7 is tt (1), because this is the only genotype that produces a taster (1).

c Individual 5 could be either TT or Tt (1), since her husband 6 is a non-taster (tt), and so she could donate a 'T' allele from either genotype to produce a son who is Tt (1 mark for explanation or correct genetic diagram).

d Individual 3 must have the genotype Tt (1). Individual 4 must be tt (1). So the cross produces a 1:1 ratio of tasters to non-tasters / probability is 0.5 that a child is a taster (1). (1 mark for correct genetic diagram):

	T	t
t	Tt	tt
t	Tt	tt

4 ▶ a D, C, B, E, F, A (all correct = 2 marks, 1 mark if one or more wrong).

b Mitosis (1), because there are only two cells produced / only one division / no reduction division / no pairing of homologous chromosomes (1).

c 46

d Any two of:
- mitosis produces two daughter cells, meiosis produces four daughter cells
- daughter cells from mitosis are genetically identical to each other and the parent cell; daughter cells from meiosis are genetically different from each other and the parent cell
- mitosis produces daughter cells with the same number of chromosomes as the parent cell / diploid to diploid; meiosis halves the chromosome number / diploid to haploid.

5 ▶ a From the nucleus of a mammary gland cell of sheep A (1).

b Nucleus of an egg is haploid / has half set of chromosomes; nucleus of an embryo is diploid / has full set of chromosomes (1).

c Sheep A.

d It does not involve fertilisation of an egg by a sperm (1); the embryo grows from a body cell nucleus (mammary gland cell nucleus) rather than from a zygote (1).

e Cloning (genetically modified) animals to produce human proteins (to treat diseases) (1). Cloning (genetically modified) animals to supply organs for transplants (1).

6 ▶ (One mark for each correct underlined term)

A gene is a section of a molecule known as <u>DNA / deoxyribonucleic acid</u>. The molecule is found within the <u>nucleus</u> of a cell, within thread-like structures called <u>chromosomes</u>. The strands of the molecule form a double helix joined by paired bases. The base adenine is always paired with its complementary base <u>thymine</u>, and the base cytosine is paired with <u>guanine</u>. During the process of transcription, the order of bases in one strand of the molecule is used to form <u>messenger RNA / mRNA</u> which carries the code for making proteins out to the cytoplasm.

7 ▶ 50 base pairs (1)

- 30 (G) bases (1) *(numbers of C and G must be the same)*
- 20 (T) bases (1) *(C+G = 60, rest = 40. T must be half the 40)*
- 20 (A) bases (1) *(numbers of T and A must be the same)*
- 100 deoxyribose sugar groups (1) *(the same as the number of bases)*

UNIT 6 ANSWERS

CHAPTER 21

1 ▶ A   **2 ▶** B   **3 ▶** C   **4 ▶** D

5 ▶ a Using (hot) steam under high pressure.

    b The air is needed to supply oxygen for aerobic respiration of the microorganisms. It is filtered to prevent contamination by unwanted microorganisms.

    c Microorganisms produce metabolic heat that could overheat the culture. The water jacket contains circulating cold water to cool the contents of the fermenter and maintain a constant temperature.

    d Nutrients.

    e Growth would be reduced. The paddles mix the contents, so that the *Penicillium* cells are exposed to more nutrients, achieving a faster rate of growth.

6 ▶ a glucose → ethanol + carbon dioxide

    b The fermentation air lock allows carbon dioxide to escape from the jar but prevents air from entering.

    c To raise the temperature of fermentation. Enzymes in the yeast will work more quickly if they are near their optimum temperature.

    d High concentrations of ethanol kill the yeast cells.

7 ▶ a To kill any natural bacteria in the milk.

    b It is the optimum temperature for growth and activity of the yoghurt bacteria.

    c Proteins in the milk coagulate due to the fall in pH.

    d The drop in pH reduces the growth of the lactic acid bacteria.

    e The low pH helps to prevent the growth of other spoiling microorganisms.

CHAPTER 22

1 ▶ B   **2 ▶** C   **3 ▶** D   **4 ▶** A

5 ▶ a 1 = restriction endonuclease / restriction enzyme; 2 = (DNA) ligase.

    b It is a vector, used to transfer the gene into the bacterium.

    c They are cultured in fermenters.

    d It is identical to human insulin and gives better control of blood glucose levels.

6 ▶ The account should discuss how far xenotransplantation has been developed and what advantages have been suggested for it. It should look at what the biological problems might be, and the ethical objections. It should be a balanced account.

7 ▶ a Use *Agrobacterium* to insert plasmids containing the required gene into plant cells or use a gene gun – firing a pellet of gold coated with DNA containing the required gene.

    b The plants are grown by micropropagation.

    c Egg cell.

8 ▶ Essay should describe a range of genetically engineered products, such as:

- products from bacteria: human insulin, enzymes, human growth hormone, etc
- genetically modified plants, such as 'golden rice' and crops resistant to herbicide
- genetically modified animals, e.g. sheep used to produce human proteins, xenotransplantation.

The benefits of each example should be discussed.

The risks from genetic engineering should also be discussed, such as:

- 'escape' of genes from crop plants into natural plant populations
- transfer of 'hidden' pathogens in xenotransplanted organs.

END OF UNIT 6 QUESTIONS

1 ▶ a Restriction endonuclease / restriction enzyme (1).

    b An egg cell / egg (1), with its nucleus removed (1).

    c An organism containing a gene / DNA / an allele from a different species (1).

    d Any three points for 3 marks:

- all sheep will be genetically identical / have same genes / have same DNA
- all sheep will produce Factor IX
- could be used to make more Factor IX
- faster reproduction of sheep
- only need to genetically modify the sheep once.

    e Prevents blood loss (1); prevents entry of pathogens / bacteria / microorganisms (1).

    f Plasma and platelets (2).

2 ▶ a It would not be possible to destroy these plants (1), and the genes could jump to other species so that they would also not be able to be destroyed (1).

b The plants could spread to other areas and would increase in numbers, as they were resistant to pests (1). The genes could jump to other species and they would also spread (1).

c The plants could spread to other areas and would compete with other species and take over a habitat (1). The genes could jump to other species and they would also compete with other species (1).

3 ▶ a To supply oxygen for aerobic respiration of the microorganisms (1).

b The temperature must be at the optimum for the enzymes in the microorganisms to work (1). If temperature is too low, reactions will be slow / if too high enzymes will be denatured / microorganisms killed (1).

c pH / supply of nutrients (1).

d Disinfectants are difficult to wash out of the fermenter (1) and might kill the microorganism being grown (1) (steam just leaves harmless water.)

e Two marks for any two from: human insulin works better than insulin from animal pancreases / there is no risk of transfer of pathogens using human insulin / using human insulin from microorganisms is acceptable to people who object to using animal tissues.

4 ▶ a i An organism that has had genes transferred to it (1) from another species (1).

ii A small ring of DNA (1) in (the cytoplasm of) a bacterium (1).

iii A virus (1) that infects bacterial cells (1).

b i Four points from the following, for maximum of 4 marks: restriction endonucleases are enzymes that cut DNA at specific points (1). They are used to cut out genes from the DNA (1) by recognising a certain base sequence (1). Different restriction enzymes cut DNA at different places (1). Use of the same restriction enzyme on a plasmid allows the DNA to be inserted into the plasmid (1).

ii Ligases are enzymes that join cut ends of DNA (1) allowing genes to be put into plasmids (1).

5 ▶ a Yeast / fungus (1).

b In beer making, the yeast respires to produce ethanol (1);

glucose → ethanol + carbon dioxide (1 mark per side of equation).

c Barley contains starch (1), which is broken down to maltose (1), which is used by the yeast for respiration (1).